This is an introduction to the field of many-body atomic physics suitable for researchers and graduate students. Drawing from three major subject areas, atomic structure, atomic photoionization, and electron-atom collisions, this book begins with an introduction to many-body diagrams, and continues with several chapters devoted to each subject area written by leading theorists in that field. Topics in atomic structure include the relativistic theory for highly charged atomic ions and calculations of parity nonconservation. Topics in atomic photoionization include single and double photoionization processes, and photoelectron angular distributions. Topics in electron atom collisions include the theory of electron impact ionization, perturbation series methods, target dependence of the triply differential cross section, Thomas processes, R-matrix theory, close coupling and distorted-wave theory. This volume has been prepared by leading atomic physicists as a tribute to Hugh Kelly, one of the pioneers of many-body theory.

Many-body atomic physics

Many-body atomic physics

Lectures on the application
of many-body theory to atomic physics

Edited by

J. J. Boyle
Harvard University

M. S. Pindzola
Auburn University

CAMBRIDGE
UNIVERSITY PRESS

CAMBRIDGE UNIVERSITY PRESS
Cambridge, New York, Melbourne, Madrid, Cape Town, Singapore, São Paulo

Cambridge University Press
The Edinburgh Building, Cambridge CB2 2RU, UK

Published in the United States of America by Cambridge University Press, New York

www.cambridge.org
Information on this title: www.cambridge.org/9780521470063

First published 1998
This digitally printed first paperback version 2005

A catalogue record for this publication is available from the British Library

ISBN-13 978-0-521-47006-3 hardback
ISBN-10 0-521-47006-4 hardback

ISBN-13 978-0-521-02199-9 paperback
ISBN-10 0-521-02199-5 paperback

Contents

Foreword		xiii
Contributors		xv
Preface		xvii
Acknowledgements		xxi

Part 1: ATOMIC STRUCTURE   1

1	**Development of atomic many-body theory**	3
	Ingvar Lindgren	
1.1	Background	3
	1.1.1 Nuclear MBPT	4
1.2	Development of atomic MBPT	5
	1.2.1 The degeneracy problem	8
	1.2.2 The all-order and coupled-cluster approach	10
	1.2.3 Relativistic MBPT and QED	13
1.3	Basic perturbation theory	14
	1.3.1 Brillouin-Wigner perturbation expansion	14
	1.3.2 Rayleigh-Schrödinger perturbation expansion	18
1.4	Diagrammatic representation of MBPT	20
	1.4.1 Second quantization and the graphical representation	21
	1.4.2 Wick's theorem	23
	1.4.3 Particle-hole representation	23
	1.4.4 Wick's theorem for products of normal-ordered operators	25
1.5	Closed-shell systems	26
	1.5.1 The linked-diagram expansion	27
	1.5.2 All-order approach	31
1.6	Open-shell systems	33
1.7	The coupled-cluster approach	37
2	**Relativistic many-body perturbation theory for highly charged ions**	39
	W. R. Johnson	
2.1	Introduction	39

2.2	Central-field Dirac equation	40
2.3	No-pair Hamiltonian	45
	2.3.1 Coulomb interaction	46
	2.3.2 Breit interaction	48
2.4	Perturbation theory	50
2.5	Angular reduction	52
2.6	Finite basis sets	54
2.7	Higher-order corrections	58

3	**Parity nonconservation in atoms**	**65**
	S. A. Blundell, W. R. Johnson, and J. Sapirstein	
3.1	Introduction	65
3.2	Overview of PNC effects in atoms	66
3.3	Relativistic MBPT in heavy neutral atoms	69
3.4	PNC calculations	77
	3.4.1 Mixed-parity HF	77
	3.4.2 Mixed-parity MBPT	80
	3.4.3 Sum-over-states approach for PNC amplitude	83
	3.4.4 Breit interaction	84
	3.4.5 Nuclear density	85
	3.4.6 Nuclear spin-dependent effects	86
	3.4.7 Electron-electron weak interaction	87
	3.4.8 Vector transition polarizability	87
3.5	Conclusions	88

Part 2: PHOTOIONIZATION OF ATOMS	**91**

4	**Many-body effects in single photoionization processes**	**93**
	J. J. Boyle and M. D. Kutzner	
4.1	Introduction	93
	4.1.1 The photoionization cross section	94
	4.1.2 The linked-cluster expansion	97
	4.1.3 An alternative expansion	102
4.2	The evaluation of first-order diagrams	103
	4.2.1 Ground- and final-state correlation diagrams	104
	4.2.2 The choice of the potential V	108
4.3	The evaluation of an infinite-order series of diagrams	114
	4.3.1 The coupled-equations technique	114
	4.3.2 The generalized resonance technique	120
4.4	Summary	123

5	**Photoionization dominated by double excitation in two-electron and divalent atoms**	**125**
	T.-N. Chang	
5.1	Introduction	125
5.2	B-spline-based configuration-interaction method	127

5.3   Configuration-interaction method for continuum spectrum   135
5.4   Applications   139

6   Direct double photoionization in atoms   150
Z. Liu
6.1   Introduction   150
6.2   Basic formalism   153
6.3   Correction to electron correlation interaction   160
6.4   Correction to dipole interaction   162
6.5   Future work   166

7   Photoelectron angular distributions   167
Steven T. Manson
7.1   Introduction   167
7.2   Theory   168
     7.2.1   General considerations   168
     7.2.2   Angular momentum transfer analysis   169
     7.2.3   The effects of approximations   172
     7.2.4   Application of the angular momentum transfer analysis   174
7.3   Illustrative examples   179
7.4   Final remarks   182

Part 3: ATOMIC SCATTERING
A. General considerations   183

8   The many-body approach to electron-atom collisions   185
M. Ya Amusia
8.1   Introduction   185
8.2   The diagrammatic method in the scattering problem   186
8.3   The Dyson equation and the optical model   191
8.4   Calculation of the elastic scattering cross section   193
8.5   Resonance scattering   199
8.6   Formation of negative ions and the scattering problem   201
8.7   Collisions with open shell and excited atoms   202
8.8   The scattering of muonic hydrogen by electrons   206
8.9   Inelastic electron-atom scattering   208
8.10   Concluding remarks   211

9   Theoretical aspects of electron impact ionization, (e, 2e), at high and intermediate energies   213
P. L. Altick
9.1   Introduction   213
9.2   The highest energies: the optical and impulsive limits   215
9.3   High energies: distorted waves and the second Born approximation   220
9.4   Conclusion   232

Part 3: ATOMIC SCATTERING
B. Low-order applications   235

**10 Perturbation series methods for calculating electron-
      atom differential cross sections**   237
D. H. Madison
10.1 Introduction   237
10.2 Theory   238
10.3 Practical considerations   240
      10.3.1 Bound state wavefunctions   240
      10.3.2 Distorting potentials   240
      10.3.3 Exchange distortion   242
      10.3.4 Relativistic effects   242
      10.3.5 Second order calculations   243
      10.3.6 Green's function   244
10.4 Results   244
      10.4.1 First order results   244
      10.4.2 Second order effects   245
      10.4.3 Convergence of perturbation series   248
      10.4.4 Comparison with close-coupling calculations   253
10.5 Conclusions   260

**11 Target dependence of the triply differential cross section
      for low energy (e, 2e) processes**   261
Cheng Pan and Anthony F. Starace
11.1 Introduction   261
11.2 Theory   265
      11.2.1 Partial wave expansion   265
      11.2.2 MBPT expansion for $\langle \Phi_f \mid \Delta H \mid \Psi_i \rangle$ 268
      11.2.3 Triply differential cross section for the $\theta_{12} = \pi$ case   272
11.3 Results and discussion   275
11.4 Concluding remarks   285

12 Overview of Thomas processes for fast mass transfer   287
J. H. McGuire, Jack C. Straton, and T. Ishihara
12.1 Introduction   287
12.2 The classical Thomas process   288
12.3 Quantum mechanics   289
12.4 Interpretations   293
12.5 Destructive interference of Thomas amplitudes   297
12.6 Summary   298
12.7 Appendix   299
      12.7.1 Overall momentum and energy conservation: the sharp ridge   299
      12.7.2 Conservation of intermediate energy: the broad ridge   300

Part 3: ATOMIC SCATTERING
C. All-order applications 303

13 ***R*-matrix theory: some recent applications** 305
Philip G. Burke
13.1 Introduction 305
13.2 Basic theory 306
13.3 Recent applications 311
   13.3.1 Electron scattering by ions of Fe 311
   13.3.2 Electron scattering at intermediate energies 314
   13.3.3 Multiphoton processes 320
13.4 Conclusions 324

14 **Electron scattering from atomic targets: application of**
   Dirac *R*-matrix theory 325
Wasantha Wijesundera, Ian Grant, and Patrick Norrington
14.1 Introduction 325
14.2 Theory 325
   14.2.1 Inner region 326
   14.2.2 Outer region 329
   14.2.3 Matching of inner and outer region solutions 331
14.3 Method of calculation and results 332
   14.3.1 Kr XXIX 332
   14.3.2 Hg 335
   14.3.3 Pb 343
   14.3.4 Cs 346
14.4 Conclusions 347

15 **Electron-ion collisions using close-coupling and distorted-**
   wave theory 349
D. C. Griffin and M. S. Pindzola
15.1 Introduction 349
15.2 Electron-impact excitation 350
   15.2.1 Theoretical methods 350
   15.2.2 Non-resonant excitation in K-like ions 355
   15.2.3 Electron-impact excitation including resonances for Na-like
      ions 357
15.3 Electron-impact ionization 359
   15.3.1 Theoretical methods 359
   15.3.2 Direct ionization of F^{2+} and W^+ 363
   15.3.3 Excitation-autoionization of Xe^{6+} and Fe^{15+} 364
   15.3.4 Dielectronic-capture resonances in C^{3+} 367
15.4 Summary 369

Appendix: Units and notation 371
References 376
Index 402

Hugh Kelly in his office at the University of Virginia Physics Department.
Photograph courtesy of *Inside UVA*.

Foreword

Hugh Padraic Kelly died on June 29, 1992 after a brave and lengthy struggle against cancer.

Hugh was a graduate of Harvard University, receiving an AB degree in 1953. He continued on to UCLA where he was awarded an M.Sc. degree in 1954. He served in the Marine Corps for three years before returning to graduate school at Berkeley. He worked there with Kenneth Watson, receiving his Ph.D. degree in 1963 and proceeding on to a postdoctoral fellowship with Keith Brueckner at the University of California, San Diego, where he began his seminal work on many-body theory. He was appointed to the faculty of the University of Virginia in 1965. He was a distinguished administrator, serving the University as Chairman of the Department of Physics, as Dean of the Faculty and as Provost.

Hugh was a very special person. He was from his first research paper the leader in the application of many-body perturbation theory in atomic and molecular physics using diagrammatic techniques. He was renowned internationally not only for his brilliant researches but also for his extraordinary personal qualities. He was modest, unassuming, always supportive of others. He had an abundance of creative ideas, which he freely shared. Hugh was the least competitive of people. He saw science as a joint enterprise in which he participated with his friends and students. For all that he was a major figure in scientific research and high level university administration, it is for his warmth and friendship and enthusiasm he will be remembered most.

Hugh was deeply involved in the education of graduate students. This book would have pleased him. Written by his friends and collaborators, it is intended to introduce the application of many-body theory to the new generation that comes after him.

Alexander Dalgarno
Harvard-Smithsonian Center for Astrophysics

Contributors

Philip L. Altick
Physics Department
University of Nevada
Reno, NV 89557, USA

Myron Ya Amusia
A. F. Ioffe Physical-Technical Institute
St. Petersburg, Russia

Steven A. Blundell
Centre d'Etudes Nucleaires Grenoble
Departement de Recherche
*   Fondamentale LIAA*
BP 85 X
38041 Grenoble CEDEX, France

James J. Boyle
Institute for Theoretical Atomic
*   and Molecular Physics*
Harvard-Smithsonian
*   Center for Astrophysics*
60 Garden Street
Cambridge, MA 02138, USA

Philip G. Burke
Department of Applied Mathematics
*   and Theoretical Physics*
The Queen's University of Belfast
Belfast BT7 1NN, United Kingdom

Tu-Nan Chang
Department of Physics
and Astronomy
University of Southern California
Los Angeles, CA 90089-0484, USA

Ian Grant
Mathematical Institute
Oxford University
24-29 St Giles'
Oxford, OX1 3LB, United Kingdom

Donald C. Griffin
Department of Physics
Rollins College
Winter Park, FL 32789, USA

Takeshi Ishihara
Institute of Applied Physics
University of Tsukuba,
Tsukuba,
Ibaraki 305, Japan

Walter R. Johnson
Department of Physics
University of Notre Dame
Notre Dame, IN 46556, USA

Contributors

Mickey D. Kutzner
Department of Physics
Andrews University
Berrien Springs, MI 49104, USA

Ingvar Lindgren
Department of Physics
Chalmers University of Technology
and the University of Gothenburg
S-412 96
Gothenburg, Sweden

Zuwei Liu
Department of Physics
University of Virginia
Charlottesville, VA 22901, USA

Don H. Madison
Laboratory of Atomic
and Molecular Physics
University of Missouri-Rolla
Rolla MO, 65401, USA

Steven T. Manson
Department of Physics and Astronomy
Georgia State University
Atlanta, GA 30303, USA

James H. McGuire
Department of Physics
Tulane University
New Orleans, LA 70118, USA

Patrick Norrington
Department of Applied Mathematics
and Theoretical Physics
The Queen's University of Belfast
Belfast BT7 1NN, United Kingdom

Cheng Pan
Department of Physics and Astronomy
University of Nebraska-Lincoln
Lincoln, NE 68588-0111, USA

Michael S. Pindzola
Department of Physics
Auburn University
Auburn, AL 36849, USA

Jonathan Sapirstein
Department of Physics
University of Notre Dame
Notre Dame, IN 46556, USA

Anthony F. Starace
Department of Physics and Astronomy
University of Nebraska-Lincoln
Lincoln, NE 68588-0111, USA

Jack C. Straton
Department of Physics
Tulane University
New Orleans, LA 70118, USA

Wasantha Wijesundera
Department of Applied Mathematics
and Theoretical Physics
The Queen's University of Belfast
Belfast BT7 1NN, United Kingdom

Preface

Hugh Padraic Kelly is generally recognized by the scientific community as a pioneer in the application of diagrammatic many-body perturbation theory to problems in atomic physics. His work began in the late 1950's and early 1960's with the study of correlation energies in light atoms. From that time, with numerous colleagues and students, he led the development of many-body methods for the treatment of a wide variety of problems in atomic structure, photoionization of atoms, and electron-scattering from atoms. On April 15-17, 1993, some of the leading atomic theorists in the world attended a workshop at the Institute for Theoretical Atomic and Molecular physics at the Harvard-Smithsonian Center for Astrophysics in order to review the application of many-body theory to atomic physics in memory of Hugh Kelly. The organizers of the workshop took advantage of this setting in order to ask the participants to contribute chapters for a textbook on theoretical many-body atomic physics – one that would promote the field to prospective students, as well as would pay tribute to the memory of the field's pioneer. This book is the result of that effort.

This book can be used as a supplement to seminar courses in atomic physics, many-body physics, or quantum chemistry. Each chapter is self-contained and there are cross references between the chapters. Where possible, consistent notation is used in the the various chapters and a key to this notation is included in the Appendix.

There are three main topics with a total of fifteen chapters. The first topic is Atomic Structure and includes chapters 1 through 3. In chapter 1, Ingvar Lindgren surveys the development of atomic many-body perturbation theory from its origin in nuclear physics showing the early role played by Hugh Kelly. Later developments, such as relativistic many-body perturbation theory, calculations performed within the framework of quantum electrodynamics, and coupled-cluster techniques are considered as well. Chapter 1 also includes a review of the mathematical fundamentals of many-body perturbation theory and the construction of many-body diagrams. In chapter 2, Walter Johnson provides a more detailed account of relativistic many-body perturbation theory, and of the use of finite element methods. He applies diagrammatic perturbation theory to the calculation of the energy levels of copper-like atomic ions in order to test our understanding of strong-field quantum electrodynamics. In chapter 3, Steven Blundell, Walter Johnson, and Jonathan Sapirstein review the application of relativistic many-body perturbation theory to the calculation of electroweak neutral current interactions in atoms. With further theoretical and experimental advances, parity non-conservation effects in atoms are likely to provide one of the most precise tests of the standard model of particle physics.

The second topic in the book is atomic photoionization and includes chapters 4 through 7. In chapter 4, James Boyle and Mickey Kutzner review the application of many-body perturbation theory to single photoionization processes in atoms. The emphasis in this chapter is on the evaluation of low-order many-body diagrams, as well as a short review of all-order techniques. Examples drawn from calculations performed on platinum, barium, xenon, and tungsten are provided. In chapter 5, Tu-Nan Chang presents a B-spline based configuration interaction method for the calculation of photoionization processes in two-electron and divalent atoms. This method has proven successful for the calculation of doubly-excited resonance profiles. In chapter 6, Zuwei Liu reviews the application of many-body perturbation theory to the calculation of direct double photoionization processes in atoms. In contrast to single photoionization, the process of double photoionization cannot occur without electron correlation. The focus of this chapter is on those cases where many-body perturbation theory has been applied successfully. In chapter 7, Steven Manson presents an overview of the theory of atomic photoelectron angular distributions. The separation of the angular momentum geometry and the dynamics is highlighted. Examples that illustrate the phenomenology and the physics to be learned from angular distributions are presented.

The third topic in the book is atomic scattering and includes chapters 8 through 15. This topic begins with a subheading of general considerations and includes chapters 8 and 9. In chapter 8, Myron Amusia reviews the application of many-body perturbation theory to electron–atom collisions. A variety of atomic processes are considered, including elastic and resonance scattering, the formation of negative ions, collisions with open-shell and excited atoms, and inelastic electron–atom scattering. In chapter 9, Philip Altick reviews the theoretical aspects of electron impact ionization at high and intermediate energies. The emphasis in this chapter is on the theoretical attempts to model the measured triply differential cross sections, which, minus the spin polarization, provide a complete description of the ionization process.

The next subheading in the topic of atomic scattering is low-order applications and includes chapters 10 through 12. In chapter 10, Don Madison applies Born series methods to the theory of electron impact excitation of atoms. Typical results of first and second order perturbation theory calculations for electron–alkali atom differential scattering are presented. In chapter 11, Cheng Pan and Anthony Starace review the application of many-body perturbation theory to the calculation of electron–atom triply differential cross sections at low energies. The near-threshold energy and angular dependence of two electrons escaping from a positive ion remains a difficult problem in atomic collision physics. In chapter 12, James McGuire, Jack Straton, and Takeshi Ishihara apply Born series methods to the theory of ion–atom charge and mass transfer collisions. The second Born term is found to be the largest Born term at high velocity and corresponds to the simplest allowed classical process — the Thomas process.

The final subheading in the topic of atomic scattering is all-order applications and includes chapters 13 through 15. In chapter 13, Philip Burke presents a summary of the basic R-matrix theory of atomic and molecular processes. Applications of the theory are made to low-energy electron scattering by iron ions, electron scattering by atoms and ions at intermediate energies, and multiphoton processes. In chapter 14, Wasantha Wijesundera, Ian Grant and Patrick Norrington provide a brief overview of the Dirac equation R-matrix theory for electron–atom collisions. Examples of calculations performed for the resonance structure and cross sections of electron-impact excitations of mercury, lead, and cesium are presented. In chapter 15, Donald Griffin and Michael Pindzola review the close-coupling and distorted-wave theories for the electron-impact excitation and ionization of atomic ions. It is demonstrated that independent-processes calculations have certain advantages for highly ionized species.

The contributors to this book gratefully acknowledge the support of the National Science Foundation of the USA and the Smithsonian Institution for their sponsorship of the workshop held on April 15-17, 1993 in Cambridge Massachusetts. As the organizers of the workshop, we would also like to extend our thanks to the contributors, who, in many cases, made considerable sacrifice in order to participate in the workshop and in this book. We would also like to thank Tom Gorczyca for editorial assistance in the atomic scattering section of the book. Finally, we would like to thank the Directors of the Institute for Theoretical Atomic and Molecular Physics at the Harvard-Smithsonian Center for Astrophysics for their support of this project. These include Alex Dalgarno, Eric Heller, Kate Kirby, and George Victor. Without their support, as well as the atmosphere provided by the Institute, this project would not have been possible.

As a group, we donate all royalties from the sale of this book to the Hugh P. Kelly Fellowship Fund at the University of Virginia, established in order to support education in theoretical physics.

James J. Boyle
Michael S. Pindzola

Acknowledgements

The work described in Chapter 5, Photoionization dominated by double-excitation in two-electron and divalent atoms, by T.-N. Chang, was supported by NSF under Grant No. PHY91-11420.

Chapter 8, The many-body approach to electron-atom collisions by M. Ya. Amusia, was written during the author's stay at the Université Paris-Sud. He is grateful to the Laboratoire de Photophysique Moléculaire and the Laboratoire de Spectroscopie Atomique et Ionique for their extended hospitality and to the Ministere de la Recherche et de la Technologie for financial support.

The research described in Chapter 10, The use of perturbation series methods for calculating electron-atom differential cross sections, by D. H. Madison, was supported by the NSF. The author would like to thank V. Bubelev for help in the preparation of the manuscript.

The work described in Chapter 11, Target dependence of the triply differential cross sections for low energy (e, 2e) processes, by Cheng Pan and Anthony F. Starace, was supported in part by NSF Grant No. PHY-9108002.

The work described in Chapter 12, Overview of Thomas processes for fast mass transfer, by J. H. McGuire, Jack C. Straton, and T. Ishihara, was supported in part by the Division of Chemical Sciences, Office of Basic Energy Science, Office of Energy Research, US Department of Energy.

Wasantha Wijesundera, Ian Grant, and Patrick Norrington, the authors of Chapter 14, Electron scattering from atomic targets: application of Dirac R-matrix theory, thank Farid Parpia for helpful discussions. All computation for the results presented in Chapter 14 was carried out on the Oxford University Computing Services VAX 8700/8800 cluster, the Rutherford Appleton Laboratory Computer Centre Cray X-MP/416 facility and the University of London Computer Centre Cray X-MP/28 facility. Wasantha Wijesundera was supported by an SERC research grant.

The research described in Chapter 15, Electron-ion collisions using close-coupling and distorted-wave theory, by D. C. Griffin and M. S. Pindzola, was supported in part by the Office of Fusion Energy of the US Department of Energy.

Part 1

ATOMIC STRUCTURE

Part 1

ATOMIC STRUCTURE

1
Development of atomic many-body theory

Ingvar Lindgren

Hugh Padraic Kelly was a pioneer in many-body perturbation theory (MBPT) and its application to atomic systems. He was the first to apply the new diagrammatic technique, developed mainly in field theory and nuclear physics, to problems in atomic physics. He introduced many new ideas, which have been widely used in different areas. In the first years, his work was concerned with the correlation energy of simple atoms and with the hyperfine interaction. In the last two decades he concentrated his efforts on the photoabsorption and photoionization problem, a field where he played a dominant role for a long time. In this chapter I shall give a brief review of the development of atomic MBPT and of the role played by Kelly. I shall also discuss later developments, including relativistic MBPT and QED calculations. I do not intend to give a full account of Kelly's work in different areas, since much of that will be covered in other chapters.

1.1 Background

Perturbation theory, which has been used for a long time in mathematical physics and astronomy, was introduced into quantum physics shortly after the advent of quantum mechanics in the 1920s. The most well-known of these schemes are the *Rayleigh-Schrödinger* and the *Brillouin-Wigner* expansions. In principle, these schemes can be used to any order, but they are, in actual atomic and molecular applications, almost prohibitive beyond fourth order, which may be insufficient to achieve the desired accuracy in many cases. During the last decades several new schemes have developed, which make it possible to apply perturbation theory to many-body systems in a more general manner. These schemes have their origin in quantum field theory, which was developed in the late 1940s and in the 1950s by Feynman [1, 2] Dyson [3, 4] and others.

1.1.1   Nuclear MBPT

The technique nowadays usually referred to as *many-body perturbation theory* (MBPT) was first introduced and applied in nuclear physics. The nucleus is a very complex many-body system with strongly interacting particles. In addition, there is no dominant central force, as there is in atomic systems. Nevertheless, it was found – probably to the great surprise of the scientific community – that the nucleus can be well described by a single-particle model in many cases. This is the *nuclear shell model,* proposed in the late 1940s by Goeppert Mayer [5] and Jensen *et al.* [6, 7] This model was analysed from a many-body point of view by Eden and Francis, [8] who introduced fictitious *"quasi-particles"* – distinct from the real nucleons – by means of a "transformed" or "effective" Hamiltonian, similar to that used in modern MBPT.

A particular problem in nuclear physics is the strong repulsive force at short distances between the nucleons, which has the consequence that ordinary perturbation theory cannot be applied. This problem was treated by Brueckner and collaborators in a series of papers in the 1950s. Brueckner and Levinson [9] introduced a *reaction operator,* closely related to the scattering matrix used in scattering theory by Lippman and Schwinger [10] and Watson. [11, 12] In this way an "effective" internucleon force was obtained, which contained the diagonal ("ladder") part of the true interaction to all orders. The remaining interaction could be handled in a perturbative way. This technique was further developed by Bethe [13] and Bethe and Goldstone, [14] and the first detailed numerical treatment was performed by Brueckner and Gammel. [15]

In a classic paper, Brueckner [16] applied the method of Brueckner and Levinson to nuclei with a large number of nucleons (*"nuclear matter"*). He found that the perturbative expansion of the energy contained terms which were *quadratic in the number of nucleons.* When the number of nucleons increases, such terms lead to an energy density, which increases without limit. Hence these terms are unphysical and must be cancelled by other contributions of the expansion. Brueckner called these terms *"reducible"* or *"unlinked"* clusters, since they could be expressed as products of energy contributions of lower order. Considering only two-particle interactions, it is easy to show that the unlinked contributions cancel in the fourth-order energy – the order where they first appear in a nontrivial manner – but Brueckner could, after lengthy calculations, show that they disappear also in the two next higher orders. He conjectured that they should cancel in all orders. This is the *linked-cluster theorem,* which was proved rigorously by Goldstone [17] shortly afterwards, using field-theoretical methods. Goldstone showed that the reducible energy clusters of Brueckner were in a diagrammatic description represented by diagrams which were *"unlinked,"* *i.e.*, which contained two or more disconnected pieces. The theorem is therefore also referred to as the *linked-diagram theorem*. It should be noted that the situation is similar in field theory, where the unlinked diagrams do not contribute to the energy but only to the phase of the wave function.

The first applications of the Brueckner theory were to infinite nuclear matter. It was indicated, however, in several of the papers mentioned above that the method should be applicable also to *finite systems*. The first detailed study of this kind was made by Bethe, [13] who showed that the Brueckner theory applied to finite nuclei could form a basis for the nuclear shell model. Of course, for infinite systems, which are translationally invariant, the standard basis functions are the plane-wave solutions of the Schrödinger equation. For finite systems the basis functions are instead solutions in some appropriate external field. To find the "best" field for this purpose is a major problem in the application of MBPT to any finite system. Bethe introduced the idea of starting from self-consistent fields of Hartree-Fock type, and this procedure was further developed by Brueckner *et al.* [18, 19] It is not the purpose to discuss here the development of the nuclear many-body theory in any detail. This development serves only as a historical background to the development in atomic physics, which took place later. For reviews on the development of the nuclear applications, the reader is referred to excellent reviews, such as those by Barrett and Kirson [20] or Ellis and Osnes. [21]

This introductory chapter will be organized in the following way. In the next section the development of atomic many-body perturbation theory will be reviewed, particularly in its non-relativistic formulation. (Relativistic MBPT and QED effects will be treated in the following chapters.) After this introductory review we shall devote the remaining part of the present chapter to an introduction to the diagrammatic formulation of MBPT, based on the Rayleigh-Schrödinger perturbation scheme and to the formalism of second quantization.

1.2   Development of atomic MBPT

The first application of MBPT to problems in atomic physics was made by Hugh Kelly in the early 1960s. [22, 23, 24, 25] Kelly applied the formalism that was developed by Goldstone. [17] The starting point is the many-electron Schrödinger equation

$$H\Psi = E\Psi , \qquad (1.1)$$

where H is the atomic Hamiltonian

$$H = \sum_{i=1}^{N} h_S(i) + \sum_{i<j=1}^{N} \frac{1}{r_{ij}},$$ (1.2)

N being the number of electrons, and h_S being the Schrödinger Hamiltonian for a single electron in the field of the nucleus

$$h_S = -\frac{1}{2}\nabla^2 - \frac{Z}{r}.$$ (1.3)

(Hartree atomic units are used throughout this chapter with $e = m = \hbar = 1$. See the Appendix.)

The many-electron Hamiltonian is separated in the standard way into an unperturbed Hamiltonian, H_0 and a perturbation, V,

$$H = H_0 + V.$$ (1.4)

H_0 is a sum of single-particle Hamiltonians

$$H_0 = \sum_{i=1}^{N} h_0(i); h_0 = -\frac{1}{2}\nabla^2 - \frac{Z}{r} + u(r),$$ (1.5)

and the perturbation becomes

$$V = \sum_{i<j=1}^{N} \frac{1}{r_{ij}} - \sum_{i=1}^{N} u(r_i),$$ (1.6)

where u is some appropriate potential.

The formalism of Goldstone is based on time-dependent perturbation theory, and after performing the time integrations the difference between the total energy E and the unperturbed energy E_0 can be expressed as a *linked-diagram expansion* (LDE)

$$E - E_0 = \Delta E = \sum_{n=0}^{\infty} \left\langle \Phi_0 \left| V \left(\frac{1}{E_0 - H_0} V \right)^n \right| \Phi_0 \right\rangle_{\text{linked}},$$ (1.7)

where Φ_0 is the unperturbed wave function of the state considered and an eigenfunction of the unperturbed Hamiltonian with the eigenvalue E_0

$$H_0 \Phi_0 = E_0 \Phi_0.$$ (1.8)

The summation in Eq. (1.7) is performed over *linked diagrams*, *i.e.*, diagrams with no disconnected pieces. The wave-function can similarly be expressed

$$\Psi = \sum_{n=0}^{\infty} \left(\frac{1}{E_0 - H_0} V \right)^n_{\text{linked}} \Phi_0.$$ (1.9)

Here also the summation is performed over linked diagrams, but with the term "linked" having a somewhat different definition. In section 1.4 below we shall discuss the LDE in more detail.

The Goldstone formalism is based on *time-ordered* diagrams rather than Feynman diagrams, which contain all possible time-orderings. The term "time-ordered" refers to the time-dependent perturbation formalism and implies that all possible time-orderings between the interactions are represented by separate diagrams.

Of course, for time-independent problems, this classification does not have any physical significance, but it is nevertheless a useful terminology for making a distinction between the two ways treating the perturbations. It can be shown by means of *Wick's theorem* [26] that for a disconnected diagram all possible time orderings between the interactions of the separate pieces appear. This leads to the *factorization theorem*, [27] which is useful in proving the linked-diagram theorem starting from Goldstone diagrams, as we shall demonstrate below.

An important ingredient in the MBPT for finite systems, which does not appear for infinite systems, is the *exclusion-principle violation* (EPV). As first stated by Goldstone and emphasized by Kelly, the unlinked diagrams mutually cancel only if the exclusion principle is abandoned in the intermediate states. This leads to additional linked diagrams. For practical applications EPV is an advantage, since the summations over the orbitals can then be performed independently in the different pieces of the diagram. (Among other things, this makes it possible to use the powerful angular momentum technique, developed by Yutsis *et al.* [28]) Kelly found that the EPV diagrams often dominate in higher orders, and he developed several techniques for summing the most important part of these diagrams to all orders.

Kelly's first application of atomic MBPT was on one of the most well-studied atoms of the periodic table, namely beryllium. [22] The ground state is a closed-shell system, $1s^2 2s^2$, and due to the near degeneracy between the $2s$ and the $2p$ states, the ground state is strongly correlated. Kelly used a single-electron basis set consisting of bound and unbound states, generated in the Hartree-Fock potential of the ground state. The unbound states were matched to plane-wave solutions at large radii. The calculations were carried out to third-order with some approximations. His result for the correlation energy, -0.0920 a.u., is remarkably good and represents about 97.5% of the correlation energy and compares well with later calculations (see Ref. [29] p. 417, or Refs. [30, 31]). Kelly also found that the higher-order excitations could be even better approximated by a series, which satisfied a simple differential equation. [23, 24] This was one form of what was later called the *independent-pair approximation*. Other related approximations were at about the same time discussed by

Brenig, [32] Nesbet, [33] (also p. 1 of Ref. [34]), Sinanoğlu [35] (also p. 237 of Ref. [34]), and later by Meyer [36] and Kutzelnigg. [37]

The formalism used by Kelly is applicable not only to closed-shell states but also to other systems with non-degenerate unperturbed states. Kelly applied this to evaluate the hyperfine structure of the oxygen atom [38, 39] (also p. 129 of Ref. [34]) and iron. [40] In these cases, Kelly evaluated all second- and third-order and some of the most important fourth-order contributions. The contact parameter obtained for oxygen was within 5% of the experimental value, and the other hyperfine parameters were even closer to the experimental ones.

1.2.1   The degeneracy problem

The original MBPT procedure introduced and developed by Brueckner, Goldstone and others was limited to systems with a *non-degenerate* unperturbed state, *i.e.*, essentially closed-shell systems. Early attempts to handle nuclei with open shells were made by Bloch and Horowitz, [41] but their treatment contained some unlinked contributions. A presentation of a fully linked treatment of degenerate (or quasi-degenerate) systems was first made by Brandow [42] and at about the same time by Sandars. [43] Brandow introduced the *folded diagrams* – by Sandars referred to as *backward diagrams* – which are characteristic of open-shell systems.

Brandow's treatment of the linked-diagram expansion is based on the Brillouin-Wigner (BW) expansion, where the *exact* energy of the state considered appears in the energy denominator. This leads to a fairly complicated double expansion. Sandars, on the other hand, based his treatment on the Rayleigh-Schrödinger (RS) expansion, which leads to a simpler structure. Sandars transformed the Schrödinger equation in the following way

$$(E_0 - H_0 - V)\boldsymbol{\Omega} \mid \Psi_0^a\rangle = -\sum_b \boldsymbol{\Omega} \mid \Psi_0^b\rangle\langle\Psi_0^b \mid V\boldsymbol{\Omega} \mid \Psi_0^a\rangle \,, \qquad (1.10)$$

a form which can be used to generate the RS expansion in a straightforward manner. Essentially the same equation was derived earlier in nuclear physics by Bloch [44]

$$(E_0 - H_0)\boldsymbol{\Omega}\mathbf{P} = V\boldsymbol{\Omega}\mathbf{P} - \boldsymbol{\Omega}\mathbf{P}V\boldsymbol{\Omega}\mathbf{P} \,, \qquad (1.11)$$

an equation nowadays known as the *Bloch equation*. $\boldsymbol{\Omega}$ is here the *wave operator*, or *Møller operator* [45], which transforms a group of unperturbed wave functions to the corresponding exact ones, $\Psi^a = \boldsymbol{\Omega}\Psi_0^a$. (This kind of operator was also used by Eden and Francis, quoted above [8].) $\mathbf{P}$ is the projection operator for the model space, which is spanned by the unperturbed states, Ψ_0^a. The Bloch equation is valid when the unperturbed states are degenerate (all with the same energy, E_0).

Sandars [43] proved the LDE for an open-shell system (with a single open shell), starting from Eq. (1.10). He also used the diagrammatic language to find expansions for various effective operators, taking advantage of the powerful graphical angular-momentum technique, introduced by Yutsis *et al.* [28] This technique is further exploited and developed in the book by Lindgren and Morrison. [29]

The wave operator can be used to define an *effective Hamiltonian*, [8] which operating on the unperturbed wave functions yields the corresponding exact energies, E^a,

$$H_{\text{eff}} = \mathbf{P} H \Omega \mathbf{P} \; ; \; H_{\text{eff}} \Psi_0^a = E^a \Psi_0^a \,. \tag{1.12}$$

Hence, the zeroth-order (unperturbed) states are eigenstates of the effective Hamiltonian. (In this formalism the effective Hamiltonian is *non-Hermitian*, which implies that the zeroth-order states are in general non-orthogonal. There exist several modified formalisms, which yield an effective Hamiltonian that is Hermitian. [46])

The Bloch equation (Eq. 1.11), or the equivalent equation used by Sandars (Eq. 1.10), is limited to the situation, where all unperturbed states are exactly degenerate, or, in other words, to atoms with a *single* open shell. In order to be able to treat systems with *several* open shells, it is necessary to generalize the procedure to cover also the case where the model space, which is spanned by the unperturbed states, is not completely degenerate. This was done independently by Lindgren [47] and Kvasnička [48] and led to a *"generalized Bloch equation"*

$$[\Omega, H_0] \mathbf{P} = V \Omega \mathbf{P} - \Omega \mathbf{P} V \Omega \mathbf{P} \,. \tag{1.13}$$

This generalized procedure is sometimes referred to as the method with a *multi-reference model space* or somewhat inadvertently as the method with a "quasi-degenerate" model space. It should be noted, however, that the splitting of the model space need not be small. Also large energy differences can be handled, which could appear, for instance, in dealing with inner-hole states. Small energy splittings of the model space, *i.e.*, true quasi-degeneracy, can be handled with the formalism for complete degeneracy by introducing a second perturbation, as demonstrated by Brandow and others.

The Bloch equation is a natural starting point for a formal treatment of MBPT. In the diagrammatic representation the first term, $V\Omega \mathbf{P}$, contains linked as well as unlinked diagrams. The unlinked diagrams are exactly cancelled by the unlinked part of the second term, $-\Omega \mathbf{P} V \Omega \mathbf{P}$, provided the *exclusion principle is abandoned* in the intermediate states. Only linked diagrams remain, including linked EPV diagrams, which is the *linked-*

diagram theorem, [47]

$$[\Omega, H_0]\, \mathbf{P} = (V\Omega - \Omega P V \Omega)_{\text{linked}}\mathbf{P}\,. \tag{1.14}$$

The linked part of the second term on the right hand side represents the folded diagrams. This part appears only for open-shell systems. For closed-shell systems, with a single state Φ_0 in the model space, it is easy to see the connection with the form of the LDE given above (see Eqs. 1.7 and 1.9).

1.2.2   The all-order and coupled-cluster approach

The main obstacle in making accurate atomic many-body calculations is usually the slow convergence of the perturbation expansion. Second-order results are often quite good, say within 10%-20% of the correlation, but in order to improve this result significantly it is usually necessary to go to fourth order or beyond. The number of diagrams increases drastically in each order, and a complete fourth-order calculation is monstrous. However, there are several ways of circumventing this problem.

It is well-known that pair-correlations dominate heavily for atomic systems, due to the two-body nature of the basic interaction. The most important higher-order effects appear between the individual pairs of electrons, so-called *intra*-pair correlations. This can be calculated to all orders relatively easily by means of the exact pair equation. In each pair equation, the remaining electrons are passive spectators, contributing only to the average potential. This is the *independent electron-pair approximation* (IEPA), introduced by Sinanoğlu ("exact pair theory") [35] and Nesbet ("atomic Bethe-Goldstone equation"). [33] This approach is, of course, exact for two-electron systems, but in order to improve the accuracy for many-electron systems, it is necessary to take into account – at least approximately – also the *inter*-pair correlation, *i.e.*, the interaction between different pairs of electrons.

Higher-order diagrams of a certain class can also be generated by an *iterative procedure*. This was first demonstrated for single excitations (core polarizations) [49] and first applied to the pair-correlation problem by Mårtensson. [50] By solving a set of coupled one- and two-electron equations, it is possible in this way to include the effects of single and double excitations to all orders of perturbation theory. [51] Single excitations are less important for closed-shell systems, due to the Brillouin theorem, but they are still significant when high accuracy is desired. [30, 51, 52] For open-shell systems, on the other hand, single-excitations are of vital importance and have to be included at every level of accuracy. The starting point for open-shell atomic MBPT calculations is usually the Hartree-Fock (HF) solution for a closed-shell state or some local potential, which includes the valence electrons in an averaged way. In neither case is the Brillouin theorem valid. Therefore, for open-shell systems single excitations appear in the second-order energy, while for closed-shell systems, using HF orbitals, they appear first in the fourth order.

It should be noted here that considering certain excitations (singles and doubles, say) in the linked-diagram expansion is *not* equivalent to considering such excitations using the original Schrödinger equation – or the Bloch equation, which is just a rewriting of the latter. This is due to the fact that, for instance, the cancellation of unlinked diagrams with double excitations in the last term of the Bloch equation, $-\Omega P V \Omega P$, requires *quadruple* excitations in the first term, $V\Omega P$. As a consequence, an all-order linked-diagram expansion with singles and doubles is not exact for the helium atom! Also some quadruple excitations of EPV type are needed. In the exact pair equation, the corresponding contribution is, of course, correctly included.

For many-electron systems, *true* quadruple excitations (not of EPV type) are quite important. They may represent several percent of the total correlation and are usually more important than triple excitations. The dominant quadruple excitations are of a special kind, which can be regarded as two independent double excitations. Such excitations can be included in a pair-correlation approach by expressing the wave operator in the form $\Omega = 1 + S + \frac{1}{2}S^2$, as first noted by Sinanoğlu. [35] A double excitation in S will then generate the dominating quadruple excitations in Ω due to the quadratic term. A generalization of these arguments leads to the "*exponential ansatz*" or exp(S) *formalism*, where the wave operator is expressed in exponential form

$$\Omega = e^S = 1 + S + \frac{1}{2}S^2 + \frac{1}{3!}S^3 + \cdots . \tag{1.15}$$

This approach is nowadays usually referred to as the *coupled-cluster approach* (CCA), S being the "cluster" operator.

The exponential ansatz has its roots in statistical mechanics and was introduced into nuclear physics in the late 1950's by Hubbard [53] and Coester and Kümmel. [54] (See also the review article by Kümmel *et al.* [55]) The procedure was introduced by Čížek [56] (also p. 35 of Ref. [34]) into quantum chemistry – a field where it has been widely used for many years. [57]

The CCA with singles and doubles combines the advantages of the standard LDE of being a good approximation for large systems and of the exact or independent pair approach of being exact for two-electron systems. For typical atoms it is possible to reach a 99% level of accuracy for the correlation energy with this approximation. [51, 57]

The advantage of CCA compared to standard MBPT is particularly important in molecular applications. For molecules it is of vital importance that the *size-consistency* and *separability* conditions [58, 59] are fulfilled, for instance, in treating dissociation processes. Each part of the molecule must then be treated on the same footing as the whole molecule, which is not the case in standard MBPT.

By means of the CCA it is easy to see the relation between different pair-correlation approaches. Omitting the non-linear coupled-cluster contributions entirely in the two-particle equation, leads to the all-order pair approach, described above. (This approach is sometimes called a "linear, coupled-cluster approach," which in itself is a contradiction.) In the independent-pair approach the coupled-cluster terms are approximated by the pair-correlation energy, and more sophisticated approximations lead to various CEPA schemes. [37] The disadvantage with the full CCA is that the complete coupled-cluster terms are quite time consuming to evaluate. One good and convenient approximation is to include only the diagrams which do contribute for two-electron systems, and to evaluate these completely (not only the EPV part as in Kelly's work). [50] These diagrams usually dominate and are easy to compute.

The coupled-cluster approach in its original form (1.15) is valid only for closed-shell systems. For general open-shell systems all parts of the cluster operators do not commute with each other, and this leads to unwanted contractions between such parts (connection between the corresponding diagram parts), when the exponential is evaluated. Several schemes have been suggested to remedy this situation. [60, 61, 62] A simple and convenient approach is to define the wave operator as a *normal-ordered exponential*, *i.e.*in terms of normal products between the cluster operators, as suggested independently by Lindgren [63] and Ey, [64]

$$\Omega = \left\{ e^S \right\} = 1 + \{S\} + \frac{1}{2}\left\{S^2\right\} + \frac{1}{3!}\left\{S^3\right\} + \cdots . \qquad (1.16)$$

In the normal product (here denoted by curly brackets) there is by definition no contraction between the operators. It can then be shown that the cluster operator satisfies an equation which is very analogous to the LDE equation (Eq. 1.14) [63]

$$[S, H_0] = (V\Omega - \Omega PV\Omega)_{\text{connected}} . \qquad (1.17)$$

The right hand side is entirely *connected*, which implies that each diagram on the right hand side contains only one connected piece.

The coupled-cluster approach is nowadays frequently used in atomic-molecular physics as well as in quantum chemistry, particularly in the non-relativistic formulation. For a review of the recent applications in these fields, the reader is referred to the proceedings of the Harvard symposium 1991, where numerous references to the original works can be found. [57]

1.2.3 Relativistic MBPT and QED

For heavy elements it is necessary to take relativistic effects into account. A rigorous relativistic treatment, however, has to be based on quantum-electrodynamics (QED), and we shall indicate only briefly here how such calculations can be performed. This will be discussed in more detail in the following chapters.

A simple approach to relativistic MBPT is to replace the Schrödinger single-electron Hamiltonian in the non-relativistic Hamiltonian (Eq. 1.2) by the corresponding *Dirac Hamiltonian*

$$H = \sum_{i=1}^{N} h_D(i) + \sum_{i<j=1}^{N} \frac{1}{r_{ij}} \; ; h_D = c\boldsymbol{\alpha} \cdot \mathbf{p} + \beta mc^2 - \frac{Z}{r} . \tag{1.18}$$

As is now well-known, this Hamiltonian suffers from the serious deficiency of not having a lower bound on the energy eigenvalues, due to the existence of negative energy single-electron states. This is the so-called *Brown-Ravenhall disease.* [65] A good approach for the relativistic problem is to limit the excitations to *positive* eigenstates, which formally can be expressed by introducing *projection operators* into the Hamiltonian [66]

$$H = \Lambda_{++} \left[\sum_{i=1}^{N} h_D(i) + \sum_{i<j=1}^{N} \frac{1}{r_{ij}} \right] \Lambda_{++} . \tag{1.19}$$

This Hamiltonian can then be treated in very much the same way as the non-relativistic Hamiltonian in a linked-diagram expansion or a coupled-cluster approach. Also the *Breit interaction* can be included in this Hamiltonian in order to take account of the magnetic interactions and retardation effects to lowest order

$$H = \Lambda_{++} \left[\sum_{i=1}^{N} h_D(i) + \sum_{i<j=1}^{N} \left(\frac{1}{r_{ij}} + B_{ij} \right) \right] \Lambda_{++} , \tag{1.20}$$

with

$$B_{ij} = \frac{-1}{2r_{ij}} \left[\boldsymbol{\alpha}_i \cdot \boldsymbol{\alpha}_j + \frac{(\boldsymbol{\alpha}_i \cdot \mathbf{r}_{ij})(\boldsymbol{\alpha}_j \cdot \mathbf{r}_{ij})}{r_{ij}^2} \right] . \tag{1.21}$$

In this approximation, often referred to as the *no-virtual-pair approximation* (NVPA), the effect of virtual pairs as well as radiative effects are omitted. This can be used in the MBPT as well as CCA formulations, essentially as in the non-relativistic case. During the last few years extensive calculations of this kind have been performed by the Notre Dame, [67, 68, 69, 70] Oxford, [71] and Göteborg groups [72] and others also. A relativistic coupled-cluster calculation for an open-shell system has recently been performed by Kaldor *et al.* [73]

Starting from QED, all perturbations can be expanded in powers of the fine-structure constant α and the nuclear charge, Z. It can then be shown that the NVPA in the form given here (with the Coulomb gauge) includes all effects to order $(Z\alpha)^2$ in atomic units. [74, 75] (In relativistic treatments it is customary to use so-called *relativistic units*, where the speed of light $c = 1$. The energy values in atomic units are transformed to relativistic units by multiplying by α^2).

The present experimental accuracy for heavy, highly charged ions is now so high that effects beyond the NVPA are clearly seen. Therefore, such ions form an excellent testing ground for advanced QED calculations, where radiative effects (Lamb shift) and effects of negative-energy states (virtual electron-positron-pair creation), in combination with many-body effects are taken into account. Calculations of this kind are now being performed by several groups. [76] Effects of this kind increase very rapidly with increasing Z. For helium-like systems, for instance, the effect of the Lamb shift on the ionization energy becomes comparable to the electron correlation already for $Z \approx 25$. For high Z the Lamb-shift effect is of the order α times the zeroth-order energy and of the same order as the *first-order* Coulomb and Breit contributions. At order α^2 several new effects appear, besides the second-order Coulomb and Breit contributions, such as the effect of virtual pairs and the "screening" of the Lamb shift (which is a "many-body effect"). [77] This implies that in order to evaluate all effects of order α^2 for heavy elements, it will be necessary to consider the combination of QED and many-body effects in more detail. Single-electron radiative effects can now be calculated with high accuracy for any nuclear charge, and the main emphasis is now to combine QED and MBPT in a proper and manageable manner.

After this mainly historical exposé of the development of atomic MBPT, we shall introduce – in a more tutorial way – some of the basic concepts, which are needed for an understanding of the different MBPT approaches and used in the following chapters. (For further details, the reader is referred to the book by Lindgren and Morrison [29] or to equivalent texts.)

1.3   Basic perturbation theory

1.3.1   Brillouin-Wigner perturbation expansion

We consider now the Schrödinger equation (1.1) with the Hamiltonian separated into an unperturbed Hamiltonian, H_0, and a perturbation, V, as in Eq. (1.4)

$$H = H_0 + V \, . \tag{1.22}$$

The unperturbed Hamiltonian, H_0, is assumed to be a sum of single-electron Hamiltonians (1.5),

$$H_0 = \sum_{i=1}^{N} h_0(i),$$ (1.23)

with a known spectrum of eigenstates

$$h_0 \phi_i = \varepsilon_i \phi_i.$$ (1.24)

The eigenfunctions of H_0 can be expressed as Slater determinants of these single-electron functions

$$H_0 \Phi^\beta = E_0^\beta \Phi^\beta \; ; \Phi^\beta = \mathbf{A} \{\phi_i(1), \phi_j(2), \cdots, \phi_v(N)\},$$ (1.25)

where $\mathbf{A}$ is an antisymmetrizing operator. The eigenvalues of H_0 are then equal to the sum of the eigenvalues of the single-electron states in the determinant

$$E_0^\beta = \sum_{i=1}^{N} \varepsilon_i.$$ (1.26)

For simplicity, we assume here that the unperturbed state, Ψ_0, corresponding to the exact state Ψ we consider, is nondegenerate. Ψ_0 is then identical to one of the basis functions

$$\Psi_0 = \Phi^\mu.$$ (1.27)

We shall also introduce Dirac notation and denote the state Φ^α by the *ket* $| \Phi^\alpha \rangle$ or simply $| \alpha \rangle$. The corresponding complex conjugate wave function is represented by the corresponding *bra* $\langle \Phi^\alpha |$ or $\langle \alpha |$. The basis set $\{\Phi^\beta\}$ is assumed to be complete and orthonormal,

$$\sum_\beta | \beta \rangle \langle \beta | = 1, \; \langle \alpha | \beta \rangle = \delta_{\alpha\beta}.$$ (1.28)

We also introduce *projection operators* for the unperturbed state (*"model space"*)

$$\mathbf{P} = | \alpha \rangle \langle \alpha |,$$ (1.29)

as well as for the *orthogonal space*

$$\mathbf{Q} = 1 - \mathbf{P} = \sum_{\beta \neq \alpha} | \beta \rangle \langle \beta |.$$ (1.30)

We assume here that the unperturbed state is normalized (*intermediate normalization*), which can be expressed

$$\langle \Psi_0 | \Psi \rangle = 1.$$ (1.31)

We can now generate a recursion formula for the exact wave function by first rewriting the Schrödinger equation in the form

$$(E - H_0)\Psi = V\Psi .$$ (1.32)

Operating from the left with the projection operator $\mathbf{Q}$ then gives

$$(E - H_0)\mathbf{Q}\Psi = \mathbf{Q}V\Psi ,$$

or

$$\mathbf{Q}\Psi = (E - H_0)^{-1}\mathbf{Q}V\Psi = \frac{\mathbf{Q}}{E - H_0}V\Psi ,$$ (1.33)

since $\mathbf{Q}$ commutes with H_0. The operator $\mathbf{R} = \frac{\mathbf{Q}}{E - H_0}$ is the (Brillouin-Wigner) *resolvent* for the Schrödinger equation. We introduce the order-by-order expansion

$$\Psi = \Psi^{(0)} + \Psi^{(1)} + \Psi^{(2)} + \cdots ,$$ (1.34)

where the zeroth-order term, $\Psi^{(0)}$, is identical with the unperturbed function, Ψ_0, and $\mathbf{Q}\Psi = \Psi^{(1)} + \Psi^{(2)} + \cdots$. Equation (1.33) then leads to the recursive formula

$$\Psi^{(n)} = \frac{\mathbf{Q}}{E - H_0}V\Psi^{(n-1)} \ (n \geq 1) ,$$ (1.35)

or

$$\Psi = \mathbf{P}\Psi + \mathbf{Q}\Psi = \left(1 + \frac{\mathbf{Q}}{E - H_0}V + \frac{\mathbf{Q}}{E - H_0}V\frac{\mathbf{Q}}{E - H_0}V + \cdots\right)\Psi_0 .$$ (1.36)

This is the *Brillouin-Wigner perturbation expansion*.

Formally, Eq. (1.36) can be written

$$\Psi = \mathbf{\Omega}_E\Psi_0 ,$$ (1.37)

where $\mathbf{\Omega}_E$ is an operator, called the *wave operator* (see Eq. (1.11)):

$$\mathbf{\Omega}_E = 1 + \frac{\mathbf{Q}}{E - H_0}V + \frac{\mathbf{Q}}{E - H_0}V\frac{\mathbf{Q}}{E - H_0}V + \cdots .$$ (1.38)

This operator plays a central role in the many-body perturbation theory.

By multiplying the Schrödinger equation (Eq. 1.1) from the left by Ψ_0 and integrating, using the intermediate-normalization condition of Eq. (1.31), we get the following expression for the energy

$$E = \langle\Psi_0 \mid H \mid \Psi\rangle = \langle\Psi_0 \mid H_0 \mid \Psi\rangle + \langle\Psi_0 \mid V \mid \Psi\rangle = E_0 + \langle\Psi_0 \mid V \mid \Psi\rangle .$$ (1.39)

With the expansion

$$E = E^{(0)} + E^{(1)} + E^{(2)} + \cdots ,$$ (1.40)

this gives

$$E^{(0)} = E_0^\alpha = E_0 = \langle \Psi_0 \mid H_0 \mid \Psi_0 \rangle \,, \qquad (1.41)$$

and

$$E^{(n)} = \langle \Psi_0 \mid V \mid \Psi^{(n-1)} \rangle \ (n \geq 1) \,, \qquad (1.42)$$

or

$$E = E_0 + \langle \Psi_0 \mid V + V \frac{Q}{E - H_0} V + V \frac{Q}{E - H_0} V \frac{Q}{E - H_0} V + \cdots \mid \Psi \rangle \,. \qquad (1.43)$$

Operating with the resolvent operator on the identity in Eq. (1.28) yields the *spectral resolution* of the resolvent:

$$\frac{Q}{E - H_0} = \frac{Q}{E - H_0} \sum_\beta \mid \beta \rangle \langle \beta \mid = \sum_{\beta \neq \alpha} \frac{\mid \beta \rangle \langle \beta \mid}{E - E_0^\beta} \,. \qquad (1.44)$$

We then get the following first few terms of the wave-function expansion

$$\Psi^{(1)} = \sum_{\beta \neq \alpha} \frac{\mid \beta \rangle \langle \beta \mid V \mid \alpha \rangle}{E - E_0^\beta} \,, \qquad (1.45)$$

and

$$\Psi^{(2)} = \sum_{\beta, \gamma \neq \alpha} \frac{\mid \beta \rangle \langle \beta \mid V \mid \gamma \rangle \langle \gamma \mid V \mid \alpha \rangle}{(E - E_0^\beta)(E - E_0^\gamma)} \,. \qquad (1.46)$$

Similarly, we get the first few terms of the energy expansion in Eq. (1.39)

$$E^{(1)} = \langle \alpha \mid V \mid \alpha \rangle \,, \qquad (1.47)$$

$$E^{(2)} = \sum_{\beta \neq \alpha} \frac{\langle \alpha \mid V \mid \beta \rangle \langle \beta \mid V \mid \alpha \rangle}{E - E_0^\beta} \,, \qquad (1.48)$$

and

$$E^{(3)} = \sum_{\beta, \gamma \neq \alpha} \frac{\langle \alpha \mid V \mid \beta \rangle \langle \beta \mid V \mid \gamma \rangle \langle \gamma \mid V \mid \alpha \rangle}{(E - E_0^\beta)(E - E_0^\gamma)} \,. \qquad (1.49)$$

The Brillouin-Wigner expansion has a very simple structure, but it has the disadvantage that each term depends on the *exact* energy (E) of the system. Since by assumption this is not known at the outset, the procedure leads to a *double* expansion. This is quite possible to handle in practice, though. One procedure is to first evaluate the first-order energy by means of the known zeroth-order wave function. Next the first-order wave function is evaluated, using the first-order energy, and this is used to evaluate the second-order energy. The first-order wave function and the second-order energy are then used in evaluating the second-order wave function, etc. For our further development, though, it will be more convenient to have a wave operator, which is *energy independent*. For that reason we shall consider an alternative form of perturbation expansion, known as the *Rayleigh-Schrödinger* expansion.

1.3.2   Rayleigh-Schrödinger perturbation expansion

We use the same Hamiltonian as before and the same kind of orthonormal basis functions as in Eq. (1.25)

$$H_0 \Phi^\beta = E_0^\beta \Phi^\beta \quad \text{or} \quad H_0 \mid \beta \rangle = E_0^\beta \mid \beta \rangle \,. \tag{1.50}$$

We shall now allow the unperturbed state to be *degenerate*. In order to make the procedure even more general, we shall consider a *group of states*,

$$H \Psi^a = E^a \Psi^a \quad (a = 1, 2, \cdots, d) \,, \tag{1.51}$$

corresponding to *one* or *several* eigenvalues of the unperturbed Hamiltonian. The corresponding eigenfunctions of H_0 form the *model space*. The projection operators for the model space and the corresponding orthogonal space are

$$\mathbf{P} = \sum_\alpha \mid \alpha \rangle \langle \alpha \mid \quad \text{and} \quad \mathbf{Q} = \sum_{\beta \neq \alpha} \mid \beta \rangle \langle \beta \mid \,. \tag{1.52}$$

and as before we have $\mathbf{P} + \mathbf{Q} = 1$.

The unperturbed wave functions (zeroth-order approximations) of the states considered lie in the model space, but in contrast to the non-degenerate case they may be represented by *linear combinations* of the basis functions. They are in the formalism presented here the *projections of the exact eigenfunctions on the model space*

$$\Psi_0^a = \mathbf{P} \Psi^a \,. \tag{1.53}$$

The *wave operator* $\mathbf{\Omega}$ has the opposite property and *transforms all the unperturbed states into the corresponding exact ones*

$$\Psi^a = \mathbf{\Omega} \Psi_0^a \,. \tag{1.54}$$

In order to find an equation for the wave operator, we first operate on the Schrödinger equation (Eq. 1.51) with $\mathbf{P}$ from the left

$$\mathbf{P} H \Psi^a = \mathbf{P} H \mathbf{\Omega} \Psi_0^a = E^a \Psi_0^a \,, \tag{1.55}$$

or

$$H_{\text{eff}} \Psi_0^a = E^a \Psi_0^a \quad \text{with} \quad H_{\text{eff}} = \mathbf{P} H \mathbf{\Omega} \mathbf{P} \,. \tag{1.56}$$

We then have an operator H_{eff}, which *operating on the unperturbed function generates the exact energy*. This is the *effective Hamiltonian* mentioned earlier in Eq. (1.12). By operating on Eq. (1.55) by $\mathbf{\Omega}$ from the left, we obtain

$$\boldsymbol{\Omega P H \Omega \Psi_0^a} = E^a \boldsymbol{\Omega \Psi_0^a} = E^a \Psi^a = H \Psi^a = H \boldsymbol{\Omega \Psi_0^a} . \tag{1.57}$$

Using the effective Hamiltonian, this equation can also be expressed

$$\boldsymbol{\Omega} H_{\text{eff}} = H \boldsymbol{\Omega} . \tag{1.58}$$

It should be noted that this equation is valid only when one operates to the right on the model space.

We now have an equation which does not depend explicitly on the energy. Using the partitioning from Eq. (1.22) and the fact that H_0 commutes with $\mathbf{P}$, we can rewrite this as

$$(\boldsymbol{\Omega} H_0 - H_0 \boldsymbol{\Omega}) \Psi_0^a = (V \boldsymbol{\Omega} - \boldsymbol{\Omega} P V \boldsymbol{\Omega}) \Psi_0^a ,$$

or as the operator relation

$$[\boldsymbol{\Omega}, H_0] \mathbf{P} = (V \boldsymbol{\Omega} - \boldsymbol{\Omega} P V \boldsymbol{\Omega}) \mathbf{P} \tag{1.59}$$

This is the *generalized Bloch equation* of Eq. (1.13). If the model space is completely degenerate with the energy E_0, this reduces to to the original Bloch equation (Equation 1.11)

$$(E_0 - H_0) \boldsymbol{\Omega P} = V \boldsymbol{\Omega P} - \boldsymbol{\Omega} P V \boldsymbol{\Omega P} . \tag{1.60}$$

Equation (1.60) can formally be solved by means of a resolvent (cf. Eq. 1.44)

$$\mathbf{R} = \frac{Q}{E_0 - H_0} = \sum_{\beta \neq \alpha} \frac{|\beta\rangle\langle\beta|}{E_0 - E_0^\beta} . \tag{1.61}$$

With the zeroth-order approximation $\boldsymbol{\Omega}^{(0)} = 1$, we get the successive approximations

$$\boldsymbol{\Omega}^{(1)} \mathbf{P} = \frac{Q}{E_0 - H_0} V \mathbf{P} , \tag{1.62}$$

$$\boldsymbol{\Omega}^{(2)} \mathbf{P} = \frac{Q}{E_0 - H_0} V \frac{Q}{E_0 - H_0} V \mathbf{P} - \left(\frac{Q}{E_0 - H_0}\right)^2 V \mathbf{P} V \mathbf{P} , \tag{1.63}$$

etc., and with the spectral resolution of the resolvent (Eq. 1.61)

$$\Psi_a^{(1)} = \boldsymbol{\Omega}^{(1)} \Psi_0^a = \sum_{\beta, \gamma \neq \alpha} \frac{|\beta\rangle\langle\beta|V|\Psi_0^a\rangle}{E_0 - E_0^\beta} , \tag{1.64}$$

$$\sum_a \Psi_a^{(2)} = \boldsymbol{\Omega}^{(2)} \Psi_0^a$$
$$= \sum_{\beta, \gamma \neq \alpha} \frac{|\beta\rangle\langle\beta|V|\gamma\rangle\langle\gamma|V|\Psi_0^a\rangle}{\left(E_0 - E_0^\beta\right)\left(E_0 - E_0^\gamma\right)}$$

$$-\sum_{\alpha,\beta} \frac{|\beta\rangle\langle\beta|V|\alpha\rangle\langle\alpha|V|\Psi_0^a\rangle}{\left(E_0 - E_0^\beta\right)^2}, \qquad (1.65)$$

etc.

Notice that the exact energy appearing in the Brillouin-Wigner expansion is here replaced by the unperturbed energy. However, now there are additional terms in the expansion. This might seem to be a complication, but we shall see that in the diagrammatic representation a simplification results.

Equations (1.60-1.65) are valid only for a degenerate model space. For the more general case, where the model space is not necessarily degenerate, we can express the successive approximations by means of the generalized Bloch equation (1.59):

$$\left[\Omega^{(1)}, H_0\right] P = VP - PVP = QVP, \qquad (1.66)$$

$$\left[\Omega^{(2)}, H_0\right] P = QV\Omega^{(1)}P - \Omega^{(1)}PVP, \qquad (1.67)$$

$$\left[\Omega^{(3)}, H_0\right] P = QV\Omega^{(2)}P - \Omega^{(2)}PVP - \Omega^{(1)}PV\Omega^{(1)}P, \qquad (1.68)$$

and so on.

The energy expansion is obtained by means of the effective Hamiltonian (Eq. 1.56)

$$E^a = \langle\Psi_0^a|H\Omega|\Psi_0^a\rangle = \langle\Psi_0^a|H_0|\Psi^a\rangle + \langle\Psi_0^a|V\Omega|\Psi_0^a\rangle, \qquad (1.69)$$

with the expansion

$$E_a^{(1)} = \langle\Psi_0^a|V|\Psi_0^a\rangle, \qquad (1.70)$$

$$E_a^{(2)} = \langle\Psi_0^a|V\Omega^{(1)}|\Psi_0^a\rangle, \qquad (1.71)$$

$$E_a^{(3)} = \langle\Psi_0^a|V\Omega^{(2)}|\Psi_0^a\rangle, \qquad (1.72)$$

and so on.

1.4   Diagrammatic representation of MBPT

We return now to the atomic Hamiltonian of Eqs. (1.2,1.3)

$$H = \sum_{i=1}^{N} h_S(i) + \sum_{i<j=1}^{N} \frac{1}{r_{ij}}; h_S = -\frac{1}{2}\nabla^2 - \frac{Z}{r}. \qquad (1.73)$$

The unperturbed Hamiltonian is assumed to be of the form (1.5)

$$H_0 = \sum_{i=1}^{N} h_0(i) \; ; h_0 = -\frac{1}{2}\nabla^2 - \frac{Z}{r} + u(r) \,, \tag{1.74}$$

where u is some appropriate potential, which gives the perturbation (1.6)

$$V = \sum_{i<j=1}^{N} \frac{1}{r_{ij}} - \sum_{i=1}^{N} u(r_i) \,. \tag{1.75}$$

We shall now introduce the diagrammatic representation of the perturbation expansion with this Hamiltonian, which is based on the formalism of second-quantization. (This follows closely the presentation in Ref. [29], chapter 11.)

1.4.1 Second quantization and the graphical representation

The basic concepts of second quantization are the operators which create $(a_i^\dagger)$ and absorb (destroy or annihilate) (a_i) single-electron states. A creation operator creates a single-electron state out of the vacuum, and the corresponding absorption operator destroys that state

$$a_i^\dagger \mid 0 \rangle = \mid \phi_i \rangle \; ; a_i \mid \phi_i \rangle = \mid 0 \rangle \,. \tag{1.76}$$

Many-electron states are formed by repeated application of the creation operators, and in order to satisfy the Pauli exclusion principle (antisymmetry) the operators must satisfy the *anticommutation relations*

$$\left\{ a_i^\dagger, a_j^\dagger \right\} = a_i^\dagger a_j^\dagger + a_j^\dagger a_i^\dagger = 0 \; ; \{a_i, a_j\} = 0 \text{ and } \left\{ a_i, a_j^\dagger \right\} = \delta_{ij} \,. \tag{1.77}$$

The last equation represents the contraction of the operators (when $i = j$).

Graphically, the particle creation and absorption are represented as shown in Fig. 1.1. The single-particle states are represented by full, vertical lines, directed upwards with creation (absorption) pointing out from (into) the dotted interaction line.

General one- and two-body operators can in second quantization be expressed

$$\mathsf{F} = \sum_{n=1}^{N} f_n = \sum_{i,j} a_i^\dagger a_j \langle i \mid f \mid j \rangle \,, \tag{1.78}$$

and

$$\mathsf{G} = \sum_{m>n=1}^{N} g_{mn} = \frac{1}{2} \sum_{i,j,l,k} a_i^\dagger a_j^\dagger a_l a_k \langle ij \mid g \mid kl \rangle \,. \tag{1.79}$$

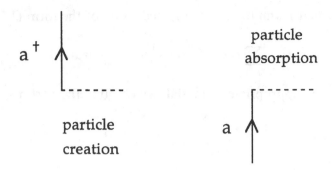

Fig. 1.1. Graphical representation of particle creation (outgoing orbital line) and absorption (incoming line).

f is here a general single-electron operator and g a two-electron operator,[*] and the matrix elements are defined

$$\langle i \mid f \mid j \rangle = \int d^3 x \phi_i^*(\mathbf{x}) f \phi_j(\mathbf{x}) ,$$

and

$$\langle ij \mid g \mid kl \rangle = \int \int d^3 x_1 d^3 x_2 \phi_i^*(\mathbf{x}_1) \phi_j^*(\mathbf{x}_2) g \phi_k(\mathbf{x}_1) \phi_l(\mathbf{x}_2) .$$

(Note that m and n refer here to the electrons and run from 1 to the number of electrons, N, while i, j, k, l, $\cdots$ run over all electron states.) The second-quantized forms of the unperturbed Hamiltonian (Eq. 1.74) and the perturbation (Eq. 1.75), will then be

$$H_0 = \sum_i a_i^\dagger a_i \varepsilon_i , \qquad (1.80)$$

and

$$V = \frac{1}{2} \sum_{i,j,l,k} a_i^\dagger a_j^\dagger a_l a_k \langle ij \mid r_{12}^{-1} \mid kl \rangle - \sum_{i,j} a_i^\dagger a_j \langle i \mid u \mid j \rangle . \qquad (1.81)$$

The factor of $1/2$ in the two-body term is due to the fact that all summations are independent. This implies that all terms (including exchange) will be *counted twice* in the sum. The operators appear here with *the absorption operators to the right of the creation operators,* which represents the *normal form* (with respect to the vacuum). Note also the ordering of the absorption operators. The individual terms can be represented graphically as shown in Fig. 1.2.

---

[*] In subsequent chapters, the variable "g" will explicitly refer to a two-particle coulomb interaction, although it is being used above to represent a general two-body operator.

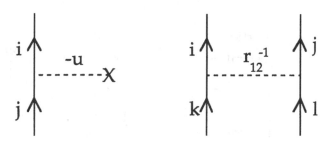

Fig. 1.2.   Graphical representation of the perturbation of Eq. (1.81).

1.4.2   Wick's theorem

A general second-quantized operator, A, which is not in normal form can be transferred to this form by moving the absorption operators to the far right, using the anticommutation relations from Eq. (1.77). This can be expressed more formally by means of *Wick's theorem* [26]

$$A = \{A\} + \left\{ \overset{\sqcap}{A} \right\} . \qquad (1.82)$$

Here, the first term on the right-hand side represents a transformation of the operator to normal order (without any contractions, but keeping the phase due to the parity of the permutation) and the second term represents all possible contractions within A (after normal-ordering the remaining operators).

1.4.3   Particle-hole representation

Instead of defining the states with respect to the empty vacuum, it is in atomic (and nuclear) physics usually more convenient to start from some close-lying closed-shell state, Φ, which we call the *reference state*. A general N-particle state can then be defined by means of creating *particles* and *holes* with respect to the reference state. A particle-hole (p-h) creation operator creates a particle outside the closed-shell core $(a^\dagger_{\text{part}}\Phi)$ *or* absorbs a particle in the core $(a_{\text{core}}\Phi)$, *i.e.* "creates a hole" in the core, and *vice versa* for a particle-hole absorption operator. (For instance, with the reference state being the neon ground-state configuration, $\Phi = 1s^2 2s^2 2p^6$, we can form the p-h state $1s^2 2s^2 2p^5 3s$ by absorbing a $2p$ electron and creating a $3s$ electron.) In the particle-hole formalism the normal order is defined as the order with *the particle-hole absorption operators* $(a_{\text{part}}, a^\dagger_{\text{core}})$ *to the right of the particle-hole creation* $(a^\dagger_{\text{part}}, a_{\text{core}})$ *operators.*

The graphical representation of particle-hole creation and absorption is shown in Fig. 1.3.

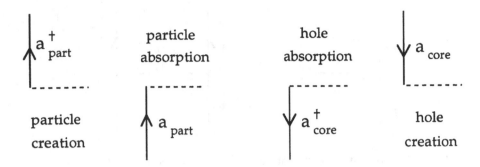

Fig. 1.3. Graphical representation of particle-hole creation and absorption. Particle operators are, as before, associated with orbital lines directed upwards and hole (core) operators with lines directed downwards. Note that particle-hole creation (absorption) operators are associated with lines above (below) the interaction line.

In order to transform a second-quantized operator of the standard form (Eqs. 1.80, 1.81) into the particle-hole normal order, we can use the anticommutation rules or Wick's theorem, given above, with the new definition of the normal order. After this procedure the perturbation of Eq. (1.81) can be expressed as

$$V = V_0 + V_1 + V_2 ,$$

$$V_0 = \sum_a^{\text{core}} \langle a \mid -u \mid a \rangle + \frac{1}{2} \sum_{a,b}^{\text{core}} \left(\langle ab \mid r_{12}^{-1} \mid ab \rangle - \langle ba \mid r_{12}^{-1} \mid ab \rangle \right) ,$$

$$V_1 = \sum_{i,j} \left\{ a_i^\dagger a_j \right\} \langle i \mid v \mid j \rangle ,$$

$$V_2 = \frac{1}{2} \sum_{i,j,k,l} \left\{ a_i^\dagger a_j^\dagger a_l a_k \right\} \langle ij \mid r_{12}^{-1} \mid kl \rangle . \qquad (1.83)$$

(For details, see Lindgren and Morrison, [29] chapter 11.) Notice here that a and b run over core states only, while i, j, $\cdots$ run over *all* states, core as well as non-core states. The curly brackets are used to denote the normal order in the p-h formulation. V_0 represents here a *zero-body operator* (pure number), V_1 a normal-ordered *one-body* and V_2 a normal-ordered *two-body* operator. Each *normal-ordered* operator can be represented by a diagram, and the graphical representation of the perturbation of Eq. (1.83) is given in Fig. 1.4. As before, there is a creation operator ($a^\dagger$) associated with each outgoing line and an absorption operator (a) with each incoming line. There is a matrix element associated with each interaction line. The *contractions* between operators correspond in the graphical representation to a *connection* between the orbital lines. There is also a summation over each such line. The circle containing a cross represents the "effective potential," v, defined by

$$\langle i \mid v \mid j \rangle = \langle i \mid -u \mid j \rangle + \sum_a^{core} \left(\langle ia \mid r_{12}^{-1} \mid ja \rangle - \langle ai \mid r_{12}^{-1} \mid ja \rangle \right) , \quad (1.84)$$

and is represented by the diagrams in Fig. 1.5. The second term in Eq. (1.84) is here identical to the Hartree-Fock potential, u_{HF}, of the reference state. Therefore, the effective potential can also be expressed as

$$\langle i \mid v \mid j \rangle = \langle i \mid u_{HF} - u \mid j \rangle . \quad (1.85)$$

This implies that *the effective potential vanishes, if the Hartree-Fock potential of the reference state is chosen to be in the unperturbed Hamiltonian.*

The wave operator can similarly be expressed as in second-quantized form

$$\Omega = 1 + \sum_{i,j} \left\{ a_i^\dagger a_j \right\} x_j^i + \frac{1}{2} \sum_{i,j,k,l} \left\{ a_i^\dagger a_j^\dagger a_l a_k \right\} x_{kl}^{ij} + \cdots . \quad (1.86)$$

The coefficients of the wave operator (the amplitudes of the excitations "x") are so far unknown, and the aim of our procedure is to find equations for these coefficients. This can be done completely algebraically, but we shall find it convenient to use the graphical representation. Before doing so, we shall develop the formalism somewhat further.

1.4.4 Wick's theorem for products of normal-ordered operators

If we have two operators A and B in normal form, we can normal order the operator product, AB, by means of Wick's theorem (Eq. 1.82). The p-h absorption operators of A have to be moved to the right of the p-h absorption operators of B, which can give rise to contractions. This can be expressed

$$AB = \{AB\} + \left\{ \overset{\sqcap}{AB} \right\} . \quad (1.87)$$

where the first term on the right-hand side represents the normal product without any contraction and the last term all possible contractions between the p-h absorption operators of A and the p-h creation operators of B. Graphically, this theorem is illustrated in Fig. 1.6. As before, contracted operators are represented by connected lines at the bottom of a figure. Since the p-h *creation (absorption)* operators are associated with the orbital lines *above (below)* the interaction line, we should connect lines at the bottom of diagram A with those at the top of diagram B (pointing in the same direction) in all possible ways (including no connection at all).

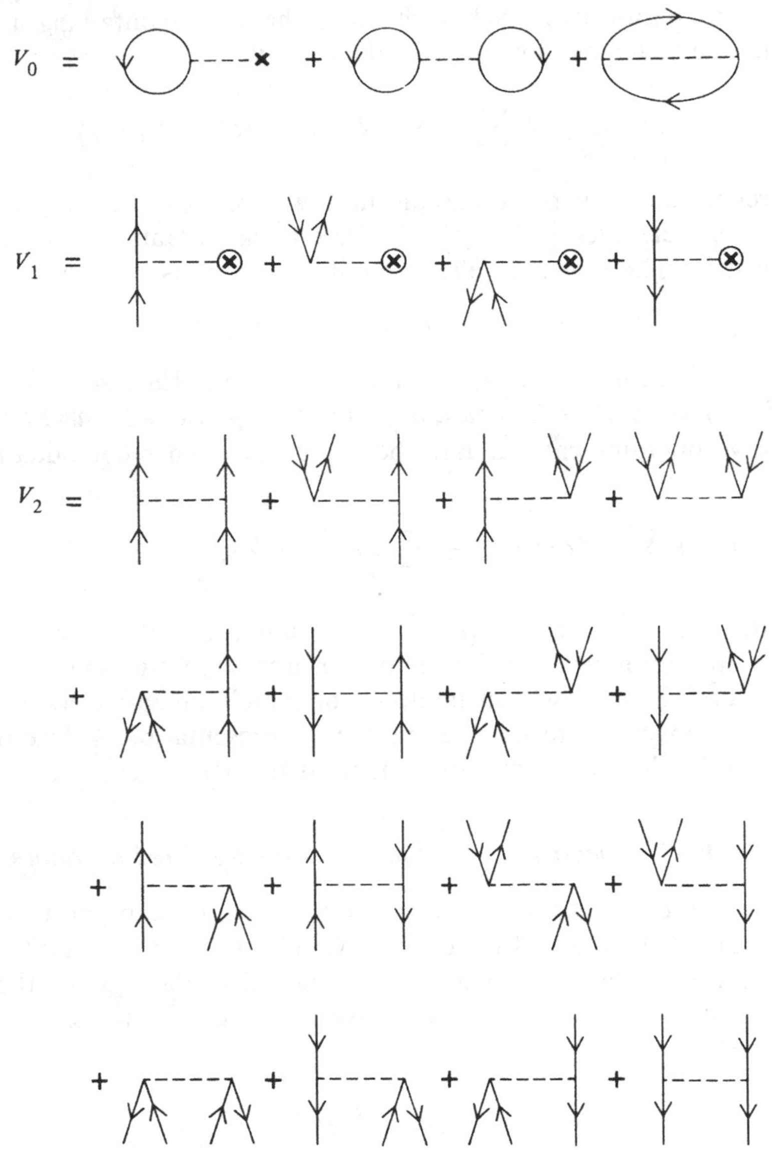

$$V_0 =$$

$$V_1 =$$

$$V_2 =$$

Fig. 1.4. The graphical representations of the perturbations itemized in Eq. (1.83). The total perturbation $V = V_0 + V_1 + V_2$ (Eq. 1.83).

1.5   Closed-shell systems

If we consider a closed-shell system, it is natural to choose the unperturbed state as the reference state (which defines the particles and holes). Since by definition no particles or holes are present in this state, p-h *absorption*

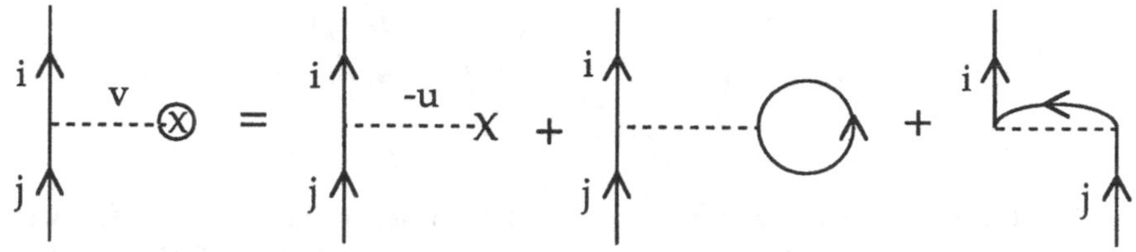

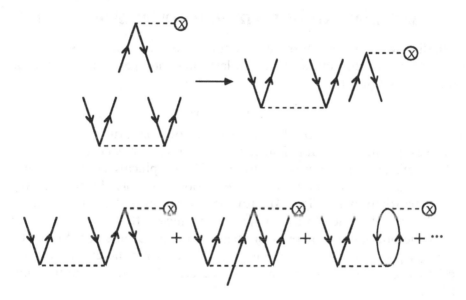

Fig. 1.5. Graphical representation of the effective potential of Eq. (1.84), when i and j are particle states and similarly for the other cases. This potential vanishes if the potential of the unperturbed Hamiltonian is chosen to be the Hartree-Fock potential of the reference state.

Fig. 1.6. Illustration of Wick's theorem for operator products in graphical form.

gives zero when operating on that state. Hence, there can be no free lines at the bottom of a diagram that operates on that state. Of the diagrams shown in Fig. 1.6, then, only the last one survives. (There are three more combinations of similar kind, not shown in the figure.)

1.5.1 The linked-diagram expansion

We now return to the Bloch equation (Eq. 1.59)

$$[\Omega, H_0]\, \mathbf{P} = (V\Omega - \Omega \mathbf{P} V\Omega)\, \mathbf{P} \tag{1.88}$$

and the successive iterations shown in Eqs. (1.66-1.68)

$$\left[\Omega^{(1)}, H_0\right] \mathbf{P} = V\mathbf{P} - \mathbf{P}V\mathbf{P} = \mathbf{Q}V\mathbf{P}, \tag{1.89}$$

$$\Omega^{(1)} = \qquad + \qquad$$

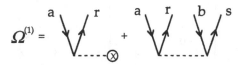

Fig. 1.7. Graphical representation of the first-order wave operator (or wave function) for a closed-shell system. The first diagram vanishes, if Hartree-Fock orbitals are used.

$$\left[\mathbf{\Omega}^{(2)}, H_0 \right] \mathbf{P} = \mathbf{Q} V \mathbf{\Omega}^{(1)} \mathbf{P} - \mathbf{\Omega}^{(1)} \mathbf{P} V \mathbf{P} , \qquad (1.90)$$

$$\left[\mathbf{\Omega}^{(3)}, H_0 \right] \mathbf{P} = \mathbf{Q} V \mathbf{\Omega}^{(2)} \mathbf{P} - \mathbf{\Omega}^{(2)} \mathbf{P} V \mathbf{P} - \mathbf{\Omega}^{(1)} \mathbf{P} V \mathbf{\Omega}^{(1)} \mathbf{P} . \qquad (1.91)$$

We shall now apply our graphical technique to these equations. It is easy to show that the commutator on the left hand side generally supplies an energy factor equal to

$$D = \sum \left(\varepsilon_{\text{out}} - \varepsilon_{\text{in}} \right) , \qquad (1.92)$$

where ε_{out} (ε_{in}) are the orbital energies of the outgoing (incoming) lines of the wave-operator diagram. This factor appears as an *energy denominator* in the corresponding expression for $\mathbf{\Omega}$. The graphical representation of the right hand side of the first-order equation is simply the same as that of the perturbation V (Fig. 1.4) with the restriction that the terms should lead from the $\mathbf{P}$ space to the $\mathbf{Q}$ space. This means that there are no free lines at the bottom and there must be at least one free line at the top. (The zero-body part, V_0, which represents a pure number is therefore excluded.) The first-order wave operator is then represented by the diagrams shown in Fig. 1.7.

The corresponding analytical expression is

$$\mathbf{\Omega}^{(1)} = \sum_{a,r} \left\{ a_r^\dagger a_a \right\} \frac{\langle r \mid v \mid a \rangle}{\varepsilon_a - \varepsilon_r} + \frac{1}{2} \sum_{abrs} \left\{ a_r^\dagger a_s^\dagger a_b a_a \right\} \frac{\langle rs \mid r_{12}^{-1} \mid ab \rangle}{\varepsilon_a + \varepsilon_b - \varepsilon_r - \varepsilon_s} , \qquad (1.93)$$

where we use a, b, $\cdots$ to denote core orbitals and r, s, $\cdots$ to denote *excited (virtual)* orbitals. (If Hartree-Fock orbitals are used, the first diagram in Fig. 1.7 vanishes. This is an illustration of Brillouin's theorem, which states that there are no first-order contributions due to single excitations from the Hartree-Fock wave function.) We have then found the *first-order* x coefficients of the wave operator in Eq. (1.86). It should be noted that the interpretation of the diagrams in Fig. 1.7 is different from those of the perturbation (Fig. 1.5), since there is an *energy denominator* associated with the wave-operator diagram, according to Eq. (1.92). The diagrams in Fig. 1.7 can also be used to represent the first-order wave function, simply by letting the wave operator operate on the unperturbed wave function.

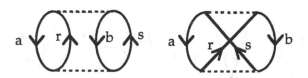

Fig. 1.8. Second-order energy diagrams for closed-shell systems.

The first-order wave operator can be used to evaluate the second-order energy according to Eq. (1.71)

$$E^{(2)} = \langle \Psi_0 \mid V\Omega^{(1)} \mid \Psi_0 \rangle \, . \tag{1.94}$$

This means that $\Omega^{(1)}$ is closed by the perturbation, *i.e.* closed diagrams (with no free lines) are formed, by Wick's theorem. In the Hartree-Fock case this gives rise to the two diagrams shown in Fig. 1.8.

The corresponding analytical expression is

$$E^{(2)} = \frac{1}{2} \sum_{a,b,r,s} \left[\frac{\langle ab \mid r_{12}^{-1} \mid rs \rangle \langle rs \mid r_{12}^{-1} \mid ab \rangle}{\varepsilon_a + \varepsilon_b - \varepsilon_r - \varepsilon_s} \right.$$
$$\left. - \frac{\langle ab \mid r_{12}^{-1} \mid sr \rangle \langle rs \mid r_{12}^{-1} \mid ab \rangle}{\varepsilon_a + \varepsilon_b - \varepsilon_r - \varepsilon_s} \right] \, . \tag{1.95}$$

The diagrams for the second-order wave operator can be obtained by means of the second-order Eq. (1.90) and Wick's theorem (Fig. 1.6). For simplicity, we assume that Hartree-Fock orbitals are used. The first term on the right hand side then leads to diagrams of the type shown in Fig. 1.9. Note here that one-body diagrams (with a single pair of free lines), such as diagram (c), can appear in the *second* order. Brillouin's theorem states that such effects are eliminated only in *first* order. The last two diagrams (d, e) are *disconnected*, *i.e.*, consist of two separate pieces, and we shall consider these diagrams further. The first of these (d) is simply equal to the zero-body part of V times the first-order wave operator, and this is exactly eliminated by the last term of Eq. (1.90). This is an illustration of the *linked-diagram theorem*, which states that diagrams with a disconnected, closed part, so-called *unlinked diagrams*, are eliminated from the expansion. The second disconnected diagram (e), on the other hand, which has no closed part, will remain in the expansion.

In a similar way third-order wave-operator diagrams can be constructed by means of Eq. (1.91). The second term on the right hand side $\Omega^{(2)}\mathbf{P}V\mathbf{P}$ will here cancel unlinked diagrams of the type just discussed, obtained when V_0 operates on $\Omega^{(2)}$. The $\Omega^{(2)}$ diagram, corresponding to the last diagram (e) in Fig. 1.9, will also give rise to unlinked diagrams of the type shown in Fig. 1.10(a) and (b). By taking into account all time-orderings

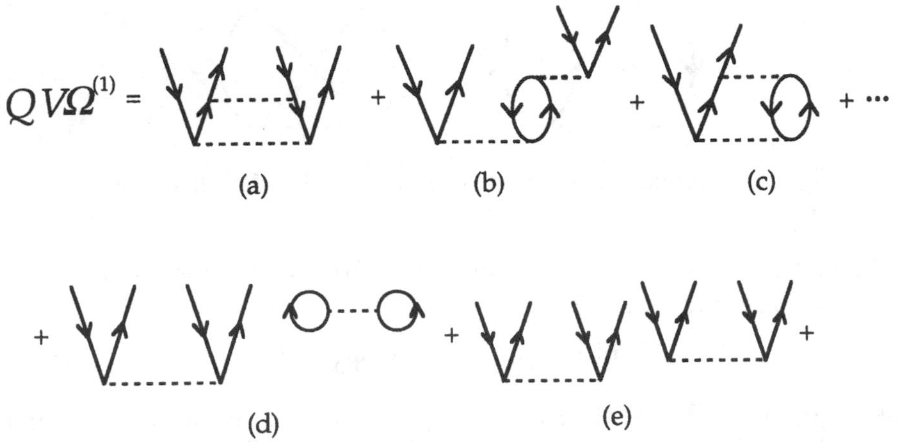

Fig. 1.9. Graphical representation of some second-order wave-operator diagrams. Diagram (d) is unlinked and eliminated from the expansion by the last term in Eq. (1.90).

and the factorization of the denominators, [27] it can be shown that these two diagrams can be expressed as a product of a first-order wave-operator diagram and a second-order energy diagram (Fig. 1.8), *provided that the exclusion principle is abandoned* in the intermediate state. The closed part represents here the second-order energy, and the corresponding diagram is eliminated by the last term in Eq. (1.91). Therefore, all unlinked diagrams are eliminated also in the third-order wave operator.

The third-order energy diagrams are obtained by closing the second-order wave-operator diagrams by the perturbation, in a similar way as in second order.

By continuing this procedure it can be shown that the unlinked diagrams, with a disconnected, closed part, cancel in all orders. This is the *linked diagram theorem* (LDT), which for a closed-shell system can be expressed by rewriting the Bloch as equation (Eq. 1.59) as

$$(E_0 - H_0)\,\mathbf{\Omega P} = (V\mathbf{\Omega})_{\text{linked}}\mathbf{P} . \qquad (1.96)$$

This can be written more explicitly in the way shown in section 1.2 (Eq. 1.9).

It should be stressed that the exclusion principle has to be abandoned in order to allow for the factorization, which is necessary for the elimination of the unlinked diagrams. This means that in our example contributions which violate this principle have to be included in the second-order wave operator. These do not contribute to the *total* wave operator, because of this violation, but it is important that *all* diagrams derived from this zero

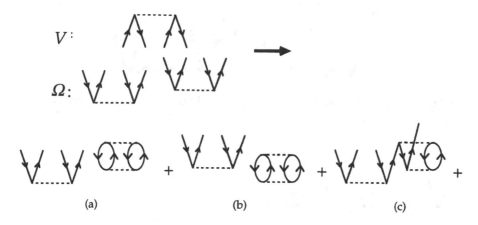

(a)   (b)   (c)

Fig. 1.10. When V operates on a disconnected wave-operator diagram, unlinked as well as linked diagrams are obtained. The unlinked diagrams (a,b) are eliminated by the last term on the right hand side of (1.91), provided the exclusion principle is abandoned. The linked diagrams (c, $\cdots$) remain, even if the exclusion principle is violated in the intermediate state (EPV diagrams).

contribution – *linked as well as unlinked* – are included. Since the unlinked diagrams are used to eliminate other unlinked diagrams, the linked part (Fig. 1.10(c), $\cdots$) must be included explicitly. These linked diagrams are the so-called *exclusion-principle-violating diagrams* (EPV), mentioned in section 1.2. Diagrams of this kind – with quadruple excitations in the intermediate state – contribute in LDE also for a two-electron system!

1.5.2 *All-order approach*

It is obvious that the order-by-order approach indicated above leads to a rapidly increasing number of diagrams for each increasing order of the expansion. In practice, this makes it almost prohibitive to use the method beyond the third-order wave function or the fourth-order energy. This difficulty can be circumvented, however, by starting from the Bloch equation directly – or in its linked-diagram form Eq. (1.96) – and thereby avoiding the perturbative expansion. Instead, we separate the wave operator into one-body and two-body parts according to Eq. (1.86) with the graphical representation given in Fig. 1.11.

The corresponding analytical expression is

$$\Omega = \sum_{a,r} a_r^\dagger a_a x_a^r + \frac{1}{2} \sum_{a,b,r,s} a_r^\dagger a_s^\dagger a_b a_a x_{ab}^{rs} + \cdots . \qquad (1.97)$$

As before, the indices a, b, $\cdots$ represent core orbitals and the indices r, s, $\cdots$ represent excited orbitals. Since the operator is already in normal

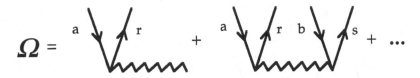

Fig. 1.11.   Graphical representation of the one-body and two-body parts of the wave operator in the closed-shell case.

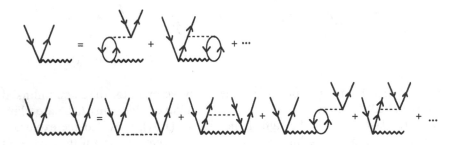

Fig. 1.12.   All-order diagrammatic representation of the one-body and two-body wave operator. Solving these coupled equations self-consistently, is equivalent to evaluating the corresponding effects to all orders of perturbation theory.

form, the signs for normal-ordering can be omitted.

Inserting the first-order wave operator (Eq. 1.97) into the Bloch equation (Eq. 1.96) and equating the one-body and two-body parts on the left and right hand sides, leads to an implicit relation between the wave-operator amplitudes, given graphically in Fig. 1.12. This can easily be transformed into a set of algebraic, linear equations for the coefficients of the wave operator, which can be solved by standard numerical methods. *This is equivalent to evaluating the corresponding effects to all orders of the perturbation expansion.*

The correlation energy, *i.e.*the energy beyond the first order, can according to Eqs. (1.69-1.72) be expressed

$$E_{\text{corr}}^a = E_a^{(2)} + E_a^{(3)} + \cdots = \langle \Psi_0^a \mid V \left(\Omega^{(1)} + \Omega^{(2)} + \cdots \right) \mid \Psi_0^a \rangle , \qquad (1.98)$$

which is in the all-order approach *exactly* represented by the two diagrams in Fig. 1.13 (assuming Hartree-Fock orbitals).

The corresponding analytical expression is

$$E_{\text{corr}}^a = \frac{1}{2} \sum_{a,b,r,s} \left[\langle ab \mid \frac{1}{r_{12}} \mid rs \rangle - \langle ab \mid \frac{1}{r_{12}} \mid sr \rangle \right] x_{ab}^{rs} . \qquad (1.99)$$

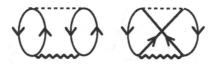

Fig. 1.13. Exact diagrammatic representation of the correlation energy for a closed-shell system in the Hartree-Fock representation.

1.6 Open-shell systems

The procedure described above for closed-shell systems can with some modifications be used also for general open-shell systems. In introducing the particle-hole formalism above, we defined two kinds of single-electron states, *core* or *hole* states, occupied in the reference state, and *particle* states not occupied in that state. The reference state is a closed-shell state, and when we consider a closed-shell system it is natural to choose the reference state to be identical to the unperturbed state.

For an open-shell system the situation is somewhat different. To be specific, we consider the *sodium atom* in its ground state, $1s^2 2s^2 2p^6 3s$, as an example. As the reference state, used to define the second-quantized operators, it is here natural to choose the closed-shell Na^+ ion, $1s^2 2s^2 2p^6$. This means that the states $1s$, $2s$ and $2p$ are defined as hole states (core orbitals) and all the remaining ones as particle states. We then have a particle state, $3s$, which is *occupied* in the unperturbed state we consider. In other words, we have a *third category* of single-particle states, which we shall denote *open-shell* or *valence states*. These are partly occupied in the model space. The remaining particle states, in our example $3p$, $4s$, $3d$, $4p$, $4s$, $\cdots$ are called *virtual states*, and they do not appear in the model space. All single-particle states of the same energy (same sub-shell) must belong to the same category.

The particle-hole formalism introduced above is still valid for open-shell systems, and the only modification we have to make is in its application to the new conditions. For closed-shell systems no particle-hole absorption operator could act on the model space, because no particles or holes are present in that space. This had the consequence that there could be no free lines at the bottom of operators, like the wave operator, operating on the model space. This is not true any longer, when we consider open-shell systems, due to the presence of valence states. We assume here that the valence states are *particle* states, although this restriction is by no means required by the formalism. With this assumption it is possible to absorb valence *particles* in the model space (but no holes), and consequently we must allow free valence lines (directed upwards) at the bottom of the wave operator. If we have only one valence particle, there can, of course,

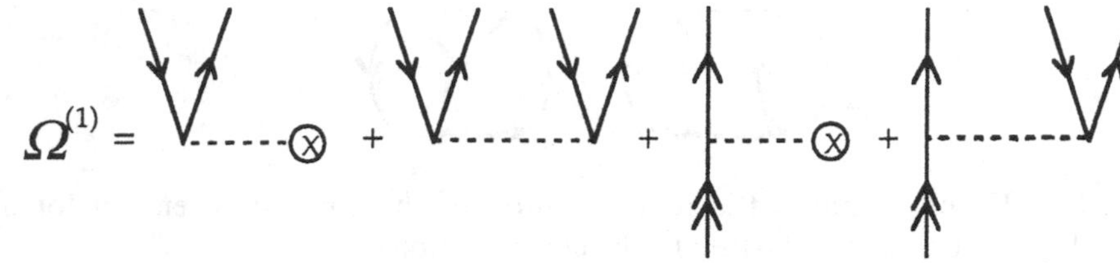

Fig. 1.14. First-order wave-operator diagrams for a system with a single valence electron.

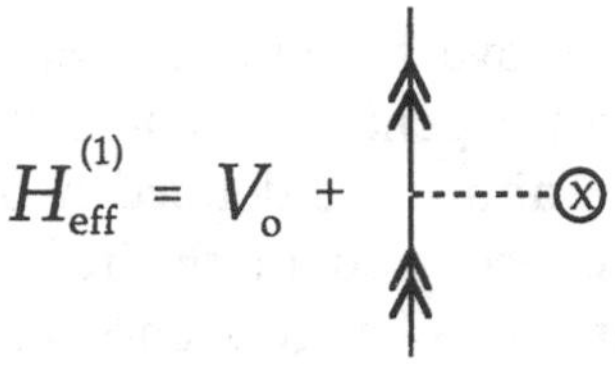

Fig. 1.15. First-order effective Hamiltonian for a system with a single valence electron.

be only one such line. We shall still assume that to be the case in the following. The first-order wave-operator diagrams are then identical to the perturbation diagrams that can operate on the model space (Fig. 1.4). We follow here the convention introduced by Sandars [43] of denoting valence states by *double* arrows. These are shown in Fig. 1.14. If the orbitals are generated in the Hartree-Fock potential of the reference state, there will, as before, be no effective potential.

The corresponding energy is obtained by constructing the diagrams of the effective Hamiltonian (Eq. 1.56)

$$H_{eff} = \mathbf{P}H\Omega\mathbf{P} = \mathbf{P}H_0\mathbf{P} + \mathbf{P}V\Omega\mathbf{P}, \qquad (1.100)$$

i.e., by *"closing"* the wave-operator diagrams by a final perturbation, V. It should be noted that "closing" means "going back to the model space," which in our case implies that there can also be an outgoing valence line. (This will create a valence particle, which is possible, without leaving the model space.)

The first-order effective Hamiltonian is the part of V (Fig. 1.4) that can operate within the model space. This is shown in Fig. 1.15. Here, V_0 represents the diagrams in Fig. 1.4 without free lines. If Hartree-Fock orbitals are used, the effective potential vanishes, and only the V_0 part remains in the first-order effective Hamiltonian (in our case with a single valence electron).

Note that the effective operator diagrams with free lines represent *operators*, not just pure numbers as in the closed-shell case. The first-

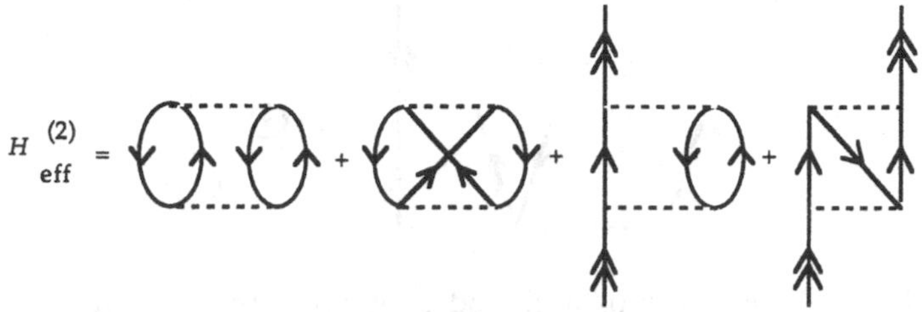

Fig. 1.16. Second-order diagrams of the effective Hamiltonian for a system with a single valence electron, assuming Hartree-Fock orbitals.

order effective Hamiltonian in Fig. 1.15 has the analytical expression

$$H_{\text{eff}}^{(1)} = V_0 + \sum_{pq} a_q^\dagger a_p \langle q \mid v \mid p \rangle , \qquad (1.101)$$

where p and q run over *valence* states.

The second-order diagrams of the effective Hamiltonian are obtained by closing the diagrams of the first-order wave operator in Fig. 1.14 by the perturbation, V. Assuming Hartree-Fock orbitals, the second-order diagrams are those shown in Fig. 1.16.

The analytical expression for the third diagram in Fig. 1.16 is given as an illustration:

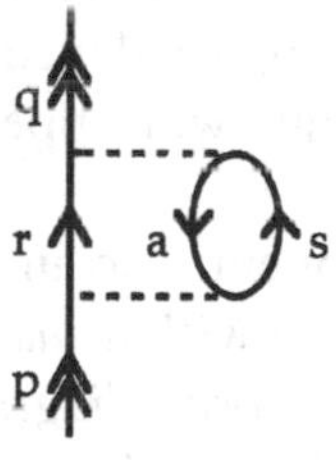

$$= a_q^\dagger a_p \frac{\langle qa \mid r_{12}^{-1} \mid rs \rangle \langle rs \mid r_{12}^{-1} \mid pa \rangle}{\varepsilon_p + \varepsilon_a - \varepsilon_r - \varepsilon_s}. \qquad (1.102)$$

The procedure can be continued to higher orders in very much the same way as in the closed-shell case. The second-order wave operator can be constructed by means of Eq. (1.90)

$$\left[\Omega^{(2)}, H_0 \right] \mathbf{P} = \mathbf{Q} V \Omega^{(1)} \mathbf{P} - \Omega^{(1)} \mathbf{P} V \mathbf{P} , \qquad (1.103)$$

first by operating with V on $\Omega^{(1)}$. A new type of unlinked diagram, shown in Fig. 1.17, then appears, in addition to those considered before. This is also cancelled by the corresponding part of $\Omega^{(1)} \mathbf{P} V \mathbf{P}$.

In addition, it should be observed that the *term* $\Omega^{(1)} \mathbf{P} V \mathbf{P}$ *is not completely unlinked* – as in the closed-shell case – *but contains also a linked part*. This is due to the fact that $\mathbf{P} V \mathbf{P}$ can contain free valence lines, and we can then get contractions between $\Omega^{(1)}$ and $\mathbf{P} V \mathbf{P}$, as illustrated in Fig. 1.18. These diagrams are for historical reasons usually drawn in a *"folded"*

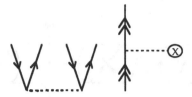

Fig. 1.17.   A new type of unlinked diagram in the open-shell case.

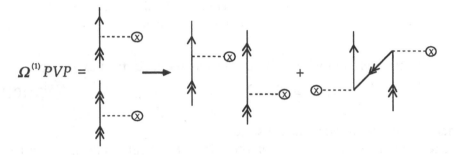

Fig. 1.18.   Folded diagrams will in the open-shell case appear in the second-order wave operator.

way, with the connecting valence line pointing backwards. These are the folded or backwards diagrams, mentioned in section 1.2. If Hartree-Fock orbitals are used, there will in the present case (single valence electron) be no folded diagram in second order. Such diagrams will appear first in third order.

It can be shown that all unlinked diagrams disappear also in the open-shell case, *if we still define unlinked diagrams as those with a disconnected, closed part* (see Fig. 1.17). *It should be noted that a closed diagram is now defined as a diagram which operates entirely within the model space, i.e., a diagram with no other free lines than valence lines.*

In the general open-shell case, the linked-diagram theorem can be formulated by means of the Bloch equation (Eq. 1.59) in analogy with Eq. (1.96), but we have now to keep also the second term on the right hand side, which gives rise to the folded diagrams

$$[\Omega, H_0]\, \mathbf{P} = (V\Omega - \Omega P V \Omega)_{\text{linked}}\, \mathbf{P} \qquad (1.104)$$

The formula (1.104) can be used to generate *all-order expressions* in the same way as in the closed-shell case. Still considering a single valence electron, the wave operator can be represented as shown in Fig. 1.19 (cf. Fig. 1.11). The all-order expressions can then be constructed in very much the same way as in the closed-shell case (Fig. 1.12). (For further details, see Lindgren and Morrison, [29] chapter 15.)

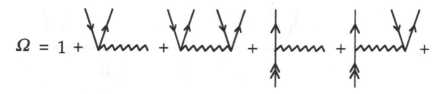

$$\Omega = 1 + \quad + \quad + \quad + \quad +$$

Fig. 1.19. All-order wave-operator diagrams for a system with a single valence electron.

1.7 The coupled-cluster approach

We shall conclude this survey of MBPT approaches by briefly describing also the so-called *coupled-cluster approach*. The same procedure as we used above can here be used with very small modifications.

We have seen that the wave-operator diagrams can contain several disconnected pieces, as long as none of them is closed. It can be shown that the wave operator can under very general conditions be written in the *exponential form* of Eq. (1.16) also for open-shell systems [63, 64]

$$\Omega = \left\{ e^S \right\} = 1 + \{S\} + \frac{1}{2}\left\{ S^2 \right\} + \frac{1}{3!}\left\{ S^3 \right\} + \cdots . \tag{1.105}$$

where the *"cluster operator,"* S, is completely connected. The curly brackets denote the normal order, as before. It can then be shown that the cluster operator satisfies an equation which is very analogous to the LDE equation (1.104) [63]

$$[S, H_0] = (V\Omega - \Omega P V \Omega)_{\text{connected}} \tag{1.106}$$

Therefore, the equations for S can be set up in very much the same way as those for Ω.

We can start by defining the S operator in second quantization in analogy with Eq. (1.86)

$$S = \sum_{i,j}\left\{ a_i^\dagger a_j \right\} s_j^i + \frac{1}{2}\sum_{i,j,k,l}\left\{ a_i^\dagger a_j^\dagger a_l a_k \right\} s_{kl}^{ij} + \cdots . \tag{1.107}$$

and the aim is to find equations for the s coefficients (excitation amplitudes). We introduce the graphical representation in Fig. 1.20, in analogy with the wave-operator representation in Fig. 1.19. We use here thick, solid lines to represent the cluster operator.

The equations for the various components of the cluster operator can now be obtained by means of the cluster equation (1.106). Formally, these equations look very much like the corresponding wave-operator equations. The only difference is that we still have Ω in the equation, which means that we shall use the expansion (1.105) to express it in terms of S. The

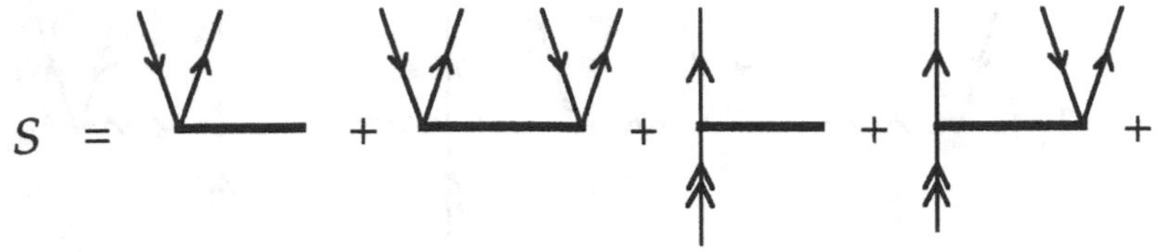

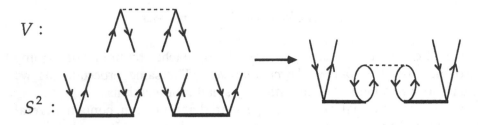

Fig. 1.20. Diagrammatic representation of the cluster operator for a system with a single valence electron.

Fig. 1.21. Construction of the special coupled-cluster diagrams from disconnected wave-operator diagrams.

first term of the expansion gives rise to diagrams which are completely analogous to those of Ω. However, when V operates on the second term in the expansion, $\frac{1}{2}\{S^2\}$, a new type of diagram will appear, as illustrated in Fig. 1.21. These are the special *coupled-cluster* diagrams. They are similar to the EPV diagrams given above (Fig. 1.10), but the important difference is that in the coupled-cluster diagrams exclusion-principle-allowed as well as exclusion-principle-violating contriutions appear.

In our example in Fig. 1.21 we have constructed a diagram contributing to the two-body part of S (two pairs of free lines). The corresponding effect is *not* included in the all-order two-body or pair procedure, the reason being that there is an intermediate four-particle excitation. In the traditional many-body approach, this would require inclusion of *four-body* effects in the expansion of the wave operator. This kind of four-body effect, which in the coupled-cluster approach can handled within the pair approximation, is normally the most important one. In addition, the CCA in the pair approximation is – in contrast to the standard LDE – *exact* for He-like systems. Therefore, the CCA combines the advantage of the LDE of being accurate for many-electron systems and that of the exact pair approach of being exact for two-electron systems. This demonstrates the power of the coupled-cluster approach.

2

Relativistic many-body perturbation theory for highly charged ions

W. R. Johnson

2.1 Introduction

In the thirty years since Kelly's seminal papers [22, 24, 25] appeared in print, many-body perturbation theory [17] has become one of the most important and widely used methods for determining energy levels and properties of atoms. [29] The theory has developed in two directions during recent years. Firstly, it has been extended to include infinite classes of terms in the perturbation expansion using coupled-cluster [54] and other all-order methods. These extensions were discussed in the first chapter of this book. Secondly, the theory has been modified to include relativistic corrections [78, 79] in an *ab initio* way starting from the Dirac equation. Kelly contributed to both of these extensions; indeed, two of his recent papers [80, 81] were devoted to relativistic all-order methods.

Relativistic many-body theory is based on the *no-pair* Hamiltonian, an approximate Hamiltonian derived from QED by ignoring the effects of virtual electron-positron pairs. [65, 66, 82, 83] The no-pair Hamiltonian accounts for the electron-electron Coulomb interaction and the Breit interaction, but does not include QED effects such as the electron self-energy and vacuum-polarization. Relativistic atomic structure calculations based strictly on QED are difficult if not intractable; they have been carried out only for the simplest many-electron atom, helium. [84] The dominant QED corrections to energies of many-electron atoms have been evaluated in a few cases; [85, 86] however, a systematic account of QED corrections to the no-pair Hamiltonian has not as yet been given for atoms other than helium.

Multiconfiguration Dirac-Fock (MCDF) methods [87, 88] are also extensively used to determine energy levels of relativistic atomic systems. These methods are complementary to those of relativistic MBPT. The MCDF methods account for the dominant correlations between electrons in open valence shells in an efficient way. The correlation between valence electrons and the atomic core are, however, difficult to treat using MCDF methods. By contrast, MBPT accounts in a simple and systematic way for the correlations within a closed-shell atom or between a single valence electron and a closed-shell ionic core, but are difficult and awkward to apply for atoms with complex shell structure. We concentrate here on closed-shell atoms and atoms with one valence electron; systems for which MBPT is particularly well suited.

In this chapter, we describe the no-pair Hamiltonian, develop the rules for higher-order perturbation theory calculations, explain how these rules are implemented in practical situations, and present results of applications of the rules to determine energy levels of low-lying states of some highly-ionized systems. We illustrate the theory with specific calculations for copper-like ions taken from Ref. [70]. Similar calculations have also been carried out for lithium-like [68] and for sodium-like [69] ions. The presentation is intended to be introductory; it is directed toward readers with a knowledge of elementary quantum mechanics including the Dirac equation.

2.2   Central-field Dirac equation

In our discussion of the relativistic many-body problem, we employ second quantization techniques. Many-electron atomic states in second quantization are constructed from single-electron states; the energies ε_i and the orbitals $\phi_i(\mathbf{r})$ of single-electron (and single-positron) states satisfy the central-field Dirac equation

$$h(\mathbf{r})\,\phi_i(\mathbf{r}) = \varepsilon_i\,\phi_i(\mathbf{r})\,. \qquad (2.1)$$

Since all of our subsequent calculations are based on Eq. (2.1), we start our discussion with a review of this equation.

The Dirac Hamiltonian $h(\mathbf{r})$ in Eq. (2.1) is given by*

$$h(\mathbf{r}) = c\boldsymbol{\alpha}\cdot\mathbf{p} + (\boldsymbol{\beta}-1)mc^2 - \frac{e^2 Z}{r} + U(r)\,, \qquad (2.2)$$

where the first two terms represent the electron's kinetic energy, the third term its interaction with the nucleus, and the fourth term accounts for its interaction with the remaining electrons in an approximate way. We discuss the choice of $U(r)$ in more detail later in this subsection. The electron's rest energy mc^2 has been subtracted from the Dirac Hamiltonian to make comparisons with nonrelativistic calculations easier. The 4×4

---

* Cf. Eq. (1.18).

Dirac matrices in Eq. (2.2), α and β are given by

$$\alpha = \begin{bmatrix} 0 & \sigma \\ \sigma & 0 \end{bmatrix}, \quad \beta = \begin{bmatrix} 1 & 0 \\ 0 & -1 \end{bmatrix}, \tag{2.3}$$

where the 2×2 matrix σ is the Pauli spin matrix.

For inner electrons of heavy atoms, it is important to account for the finite size of the nucleus. This is done by replacing the nuclear potential $-e^2 Z / r$ in Eq. (2.2) by the potential of a distributed charge $V_{\text{nuc}}(r)$. For a nucleus with atomic number $Z > 9$ and nucleon number A, the nuclear charge distribution can be described approximately [89] by a uniform ball of charge with radius

$$R_{\text{nuc}} = (1.079 \, A^{1/3} + 0.736 \pm 0.065)\text{fm}, \tag{2.4}$$

leading to an electron-nucleus potential energy function

$$V_{\text{nuc}}(r) = \begin{cases} -e^2 Z \, (3 - (r/R_{\text{nuc}})^2)/2R_{\text{nuc}}, & r < R_{\text{nuc}}, \\ -e^2 Z / r, & r > R_{\text{nuc}}. \end{cases} \tag{2.5}$$

A more precise evaluation of nuclear-size effects can be made using potentials constructed from empirical nuclear charge distributions obtained from electron-nucleus scattering experiments or from X-ray measurements in muonic atoms. [90]

It is not difficult to show that the total angular momentum vector, $\mathbf{J} = \mathbf{L} + \mathbf{S}$, where $\mathbf{L}$ is the orbital angular momentum and $\mathbf{S}$ is the 4×4 spin angular momentum matrix,

$$\mathbf{S} = \frac{\hbar}{2} \begin{bmatrix} \sigma & 0 \\ 0 & \sigma \end{bmatrix}, \tag{2.6}$$

commutes with the single-particle Hamiltonian $h(\mathbf{r})$. We may, therefore, classify the eigenstates of h according to the eigenvalues of energy, $\mathbf{J}^2$ and of J_z. Eigenstates of $\mathbf{J}^2$ and J_z are constructed using the two-component representation of $\mathbf{S}$. They are

$$\Omega_{j\ell m}(\hat{\mathbf{r}}) = \sum_{\mu} \langle \ell, m - \mu, \tfrac{1}{2}, \mu | j, m \rangle Y_{\ell m - \mu}(\hat{\mathbf{r}}) \chi_{\mu}. \tag{2.7}$$

In Eq. (2.7), $Y_{\ell m_\ell}(\hat{\mathbf{r}})$ is a spherical harmonic and χ_μ is a two-component eigenfunction of σ_z. The quantities $\langle \ell, m_\ell, \tfrac{1}{2}, \mu | j, m \rangle$ are Clebsch-Gordon coefficients. The two-component angular momentum eigenfunctions $\Omega_{j\ell m}(\hat{\mathbf{r}})$ are referred to as spherical spinors.

There are two possible values of ℓ in Eq. (2.7) for each j: $\ell = j + \tfrac{1}{2}$ and $\ell = j - \tfrac{1}{2}$. The corresponding spherical spinors have opposite parity. We introduce the operator

$$K = -2\mathbf{L} \cdot \mathbf{S} - 1, \tag{2.8}$$

and note that

$$K \, \Omega_{j\ell m} = \kappa \, \Omega_{j\ell m}, \tag{2.9}$$

where $\kappa = -(j+\frac{1}{2})$ for $j = \ell + \frac{1}{2}$, and $\kappa = j+\frac{1}{2}$ for $j = \ell - \frac{1}{2}$. The operator K has integer eigenvalues κ; the absolute value of κ determines j and the sign of κ determines ℓ (or the parity). We may, therefore, introduce the more compact notation

$$\Omega_{\kappa m}(\hat{r}) \equiv \Omega_{j\ell m}(\hat{r}). \tag{2.10}$$

The spherical spinors satisfy the orthogonality relations

$$\int \Omega_{\kappa m}^{\dagger}(\hat{r})\Omega_{\kappa' m'}(\hat{r})d\Omega = \delta_{\kappa\kappa'}\delta_{mm'}. \tag{2.11}$$

We seek a solution to the Dirac equation (2.1) in the form

$$\phi_{n\kappa m}(\mathbf{r}) = \frac{1}{r} \begin{pmatrix} iP_{n\kappa}(r) & \Omega_{\kappa m}(\hat{r}) \\ Q_{n\kappa}(r) & \Omega_{-\kappa m}(\hat{r}) \end{pmatrix}, \tag{2.12}$$

and we find that the radial functions $P_{n\kappa}(r)$ and $Q_{n\kappa}(r)$ satisfy a pair of coupled first-order radial differential equations:

$$\begin{pmatrix} V & \hbar c \left(\dfrac{d}{dr} - \dfrac{\kappa}{r} \right) \\ -\hbar c \left(\dfrac{d}{dr} + \dfrac{\kappa}{r} \right) & V - 2mc^2 \end{pmatrix} \begin{pmatrix} P_{n\kappa}(r) \\ Q_{n\kappa}(r) \end{pmatrix} = \varepsilon_{n\kappa} \begin{pmatrix} P_{n\kappa}(r) \\ Q_{n\kappa}(r) \end{pmatrix} \tag{2.13}$$

where $V(r) = V_{\text{nuc}}(r) + U(r)$. With the aid of this equation, the radial functions for a given value of κ, but different values of n, can be shown to satisfy the orthogonality relations:

$$\int_0^\infty [\, P_{n\kappa}(r)P_{n'\kappa}(r) + Q_{n\kappa}(r)Q_{n'\kappa}(r) \,]dr = \delta_{nn'} . \tag{2.14}$$

The normalization condition in Eq. (2.14) for $n' = n$ is chosen so that the orbital $\phi_{n\kappa m}(\mathbf{r})$, itself, is properly normalized. The radial eigenfunctions and the associated eigenvalues $\varepsilon_{n\kappa}$ can be determined to high accuracy by solving Eq. (2.13) numerically using standard finite-difference methods, once $U(r)$ is specified.

One important choice of $U(r)$ in many-body calculations is the Hartree-Fock (HF) potential $V_{\text{HF}}(r)$. For a closed-shell atom or ion, the HF potential is a non-local potential defined by its action on an arbitrary orbital $\phi_a(\mathbf{r})$ through the equation

$$V_{\text{HF}}\phi_a(\mathbf{r}) = \sum_b \int d^3\mathbf{r}' \frac{e^2}{|\mathbf{r} - \mathbf{r}'|} \left[\phi_b^\dagger(\mathbf{r}')\phi_b(\mathbf{r}')\phi_a(\mathbf{r}) - \phi_b^\dagger(\mathbf{r}')\phi_a(\mathbf{r}')\phi_b(\mathbf{r}) \right] . \tag{2.15}$$

In this equation, the sum extends over the quantum numbers $b = (n_b, \kappa_b, m_b)$ of all occupied orbitals in the closed core. The Hartree-Fock potential for a closed-shell atom or ion is spherically symmetrical; therefore, with a suitable angular reduction, it can be expressed as an operator function of r acting on the two-component radial wave function of the orbital with quantum numbers n_a, κ_a:

$$R_a(r) = \begin{pmatrix} P_{n_a\kappa_a}(r) \\ Q_{n_a\kappa_a}(r) \end{pmatrix}. \tag{2.16}$$

To carry out the angular reduction of Eq. (2.15), we must expand $1/|\mathbf{r} - \mathbf{r}'|$ in spherical coordinates. With the aid of such an expansion, one can evaluate the sum over magnetic substates m_b to obtain

$$\sum_b \int d^3\mathbf{r}' \frac{e^2}{|\mathbf{r} - \mathbf{r}'|} \phi_b^\dagger(\mathbf{r}')\phi_b(\mathbf{r}') = \sum_{n_b\kappa_b} [j_b] \, v_0(b,b,r), \tag{2.17}$$

where $[j_b] = 2j_b + 1$ is the occupation number of subshell b. The function $v_0(b,b,r)$ is a special case of the Hartree screening functions $v_k(a,b,r)$, which are defined as

$$v_k(a,b,r) = e^2 \int_0^\infty \frac{r_<^k}{r_>^{k+1}} \left[P_a(r')P_b(r') + Q_a(r')Q_b(r') \right] dr', \tag{2.18}$$

where $r_>$ is the larger of r and r', and $r_<$ is the smaller of the two. From Eq. (2.17), it follows that the first (direct) term in the HF potential can be written

$$V_{\mathrm{HF}}^{\mathrm{dir}} R_a(r) = \sum_{n_b\kappa_b} [j_b] \, v_0(b,b,r) R_a(r). \tag{2.19}$$

Similarly, the second (exchange) term in Eq. (2.15) is found to be

$$V_{\mathrm{HF}}^{\mathrm{exc}} R_a(r) = \sum_{n_b\kappa_b} [j_b] \sum_L \Lambda_{\kappa_b L \kappa_a} v_L(b,a,r) \, R_b(r), \tag{2.20}$$

where the angular factor $\Lambda_{\kappa_b L \kappa_a}$ is

$$\Lambda_{\kappa_b L \kappa_a} = \frac{\langle \kappa_b \| C_L \| \kappa_a \rangle^2}{[j_a][j_b]}. \tag{2.21}$$

The quantity $\langle \kappa_b \| C_L \| \kappa_a \rangle$ is the reduced matrix element of the normalized spherical harmonic

$$C_{LM}(\hat{\mathbf{r}}) = \sqrt{\frac{4\pi}{2L+1}} \, Y_{LM}(\hat{\mathbf{r}}). \tag{2.22}$$

We use the Wigner-Eckart theorem to express the m-dependence of the angular matrix element in terms of a reduced matrix element as

$$\langle \kappa_b, m_b | C_{LM} | \kappa_i, m_i \rangle = \int d\Omega\, \Omega_{\kappa_b m_b}^\dagger(\hat{r})\, C_{LM}(\hat{r})\, \Omega_{\kappa_i m_i}(\hat{r})$$

$$= (-1)^{j_b - m_b} \begin{pmatrix} j_b & L & j_i \\ -m_b & M & m_i \end{pmatrix}$$

$$\times \langle \kappa_b \| C_L \| \kappa_i \rangle, \tag{2.23}$$

where the large parentheses designate a Wigner $3j$ symbol. The reduced matrix element is given by

$$\langle \kappa_b \| C_L \| \kappa_a \rangle = \sqrt{[j_a][j_b]}\,(-1)^{j_b + \frac{1}{2}} \begin{pmatrix} j_b & j_a & L \\ -\frac{1}{2} & \frac{1}{2} & 0 \end{pmatrix} \Pi(\ell_b, \ell_a, L). \tag{2.24}$$

The parity factor $\Pi(\ell_b, \ell_a, L)$ in this equation is defined by

$$\Pi(\ell_1, \ell_2, \ell_3) = \begin{cases} 1 & \text{if } \ell_1 + \ell_2 + \ell_3 \text{ is even} \\ 0 & \text{otherwise} \end{cases}. \tag{2.25}$$

Combining the expressions, (2.19) and (2.20), we obtain

$$V_{\mathrm{HF}} R_a(r) = \sum_{n_b \kappa_b} [j_b] \left\{ v_0(b, b, r) R_a(r) - \sum_L \Lambda_{\kappa_b L \kappa_a} v_L(b, a, r) R_b(r) \right\}, \tag{2.26}$$

where the sum extends over the quantum numbers n_b, κ_b of the closed subshells of the atom. The radial Dirac equations (2.13) with

$$V R_a(r) = [V_{\mathrm{nuc}}(r) + V_{\mathrm{HF}}] R_a(r), \tag{2.27}$$

are solved self-consistently to give the radial functions $R_a(r)$ for each occupied level a.

If we wish to consider an atom with one valence electron outside closed shells in the Hartree-Fock approximation, we first solve the HF equations for the core orbitals ignoring the valence electron and then solve Eq. (2.13) for the valence electron in the fixed HF potential of the core electrons. The resulting *frozen-core* HF approximation leads to simplifications in the formulas of MBPT.

As a specific example of the frozen-core HF approximation, let us consider singly ionized zinc. This is an ion of zinc (nuclear charge $Z = 30$) with 28 core electrons and one valence electron. The core electrons are arranged in closed subshells according to the scheme:

$$(1s_{1/2})^2 (2s_{1/2})^2 (2p_{1/2})^2 (2p_{3/2})^4 (3s_{1/2})^2 (3p_{1/2})^2 (3p_{3/2})^4 (3d_{3/2})^4 (3d_{5/2})^6.$$

Here $(nl_j)^{2j+1}$ represents a subshell in which single-particle states with quantum numbers n, ℓ, j are occupied for all $2j + 1$ magnetic substates. We choose $U(r)$ for this ion as the frozen-core Hartree-Fock potential defined in the previous paragraph. The energy eigenvalues of the $n =$

Table 2.1. Single-particle energies (a.u.) for low-lying states of singly ionized zinc, $Z = 30$. The $n = 1, 2$ and 3 core states are found numerically by solving the radial Dirac Hartree-Fock equations self-consistently. The $n = 4$ valence states are then determined by solving the Dirac Hartree-Fock equations in the field of the frozen core. Experimental comparison values, designated by E_{exp}, are from Moore[91]

n	$s_{1/2}$	$p_{1/2}$	$p_{3/2}$	$d_{3/2}$	$d_{5/2}$
1	-358.51081				
2	-46.09351	-40.49198	-39.60300		
3	-6.55068	-4.71722	-4.59692	-1.53089	-1.51496
4	-0.61572	-0.41246	-0.40910		
E_{exp}	-0.66017	-0.43928	-0.43530		
E_{corr}	-0.04445	-0.02682	-0.02620		

$1, 2$ and 3 states are determined by solving the core HF equations self-consistently. The resulting energy eigenvalues are listed in the first three rows of Table 2.1. Eigenvalues for the $4s$ and $4p$ valence states, obtained in frozen core HF approximation are listed in the following row and compared with experimental energies.[91] The differences between the experimental energies and the HF energies, which are referred to as *correlation corrections*, and designated by E_{corr} in Table 2.1, are seen to be about 6%–7% for the $n = 4$ states. One of the aims of MBPT is to predict such correlation corrections from first principles.

2.3 No-pair Hamiltonian

The nonrelativistic Hamiltonian for a system of electrons moving independently in an average field $U(r)$ is expressed in second-quantized form as

$$H_0 = \sum_k \varepsilon_k\, a_k^\dagger a_k, \tag{2.28}$$

where ε_k is an eigenvalue of the one-electron Dirac equation (2.1). The quantities $a_k^\dagger$ and a_k are creation and annihilation operators for one-electron states k. To ensure that the Pauli exclusion principle is satisfied for many-electron states, these operators are required to satisfy the anti-commutation relations

$$\{a_k, a_\ell\} = 0, \quad \{a_k^\dagger, a_\ell^\dagger\} = 0, \quad \{a_k^\dagger, a_\ell\} = \delta_{k\ell}, \tag{2.29}$$

where $\{x, y\} = xy + yx$. In nonrelativistic many-body theory, the independent-particle Hamiltonian H_0 is taken as the lowest approximation to the actual many-body Hamiltonian. This same Hamiltonian with ε_k replaced by Dirac eigenvalues is also the lowest approximation to the relativistic no-pair Hamiltonian, provided positron states are excluded from the sum over states. The sum in Eq. (2.28), therefore, extends over electron bound and scattering states only; states k associated with positrons (those for which $\varepsilon_k < -2mc^2$) are excluded.

We designate the vacuum state by $|0\rangle$. This state has the property

$$a_k|0\rangle = 0, \quad \text{for all } k. \tag{2.30}$$

It follows that the vacuum state is a zero energy eigenstate of H_0. The state vector for a one-electron atom with quantum numbers $v = (n_v, \kappa_v, m_v)$ is

$$|v\rangle = a_v^\dagger|0\rangle. \tag{2.31}$$

With the aid of the anticommutation relations (2.29), one finds

$$H_0|v\rangle = \varepsilon_v|v\rangle. \tag{2.32}$$

The state vector for an N-electron atom with electrons in single-particle orbitals $a, b, c, \cdots$ is

$$|a, b, c, \cdots\rangle = a_a^\dagger a_b^\dagger a_c^\dagger \cdots |0\rangle. \tag{2.33}$$

This vector, which describes the independent motion of N electrons in the potential $U(r)$, satisfies

$$H_0|a, b, c, \cdots\rangle = W_0|a, b, c, \cdots\rangle, \tag{2.34}$$

with

$$W_0 = \varepsilon_a + \varepsilon_b + \varepsilon_c + \cdots. \tag{2.35}$$

2.3.1   Coulomb interaction

Rules for describing physical operators in second quantization and for calculating matrix elements of such operators are given in books on many-body theory such as Ref. [92]. One very important operator is that describing the residual electron-electron Coulomb interaction; *i.e.*, the difference between the Coulomb interaction and the average interaction $U(r)$. In second quantization, the residual interaction is given by

$$V_{\mathrm{I}} = \frac{1}{2} \sum_{ijk\ell} g_{ijk\ell} a_i^\dagger a_j^\dagger a_\ell a_k - \sum_{ij} U_{ij} a_i^\dagger a_j, \tag{2.36}$$

where $g_{ijk\ell}$ is a two-particle matrix element of the Coulomb interaction[†]

$$g_{ijk\ell} = \int d^3r \int d^3r' \, \phi_i^\dagger(\mathbf{r})\phi_j^\dagger(\mathbf{r}') \frac{e^2}{|\mathbf{r} - \mathbf{r}'|} \phi_k(\mathbf{r})\phi_\ell(\mathbf{r}'), \qquad (2.37)$$

and U_{ij} is a one-particle matrix element of the background potential $U(r)$:

$$U_{ij} = \int d^3r \, \phi_i^\dagger(\mathbf{r})U(r)\phi_j(\mathbf{r}). \qquad (2.38)$$

The Coulomb part of the no-pair Hamiltonian is just the sum of the independent particle Hamiltonian H_0 and the residual interaction V_I,

$$H = H_0 + V_I. \qquad (2.39)$$

Let us designate the eigenstates of H_0 by Φ_n,

$$H_0\Phi_n = W_n\Phi_n, \qquad (2.40)$$

and assume that Φ_0 is the state vector for the nondegenerate ground-state of a closed-shell atom. The unperturbed energy W_0 is given by the sum over all occupied orbitals of the Hartree-Fock eigenvalues ε_a

$$W_0 = \sum_a \varepsilon_a \qquad (2.41)$$

First-order perturbation theory leads to a correction to the energy of this state given by

$$E^{(1)} = \langle \Phi_0 | V_I | \Phi_0 \rangle. \qquad (2.42)$$

The matrix element of V_I can be expressed in terms of matrix elements of the ground-state HF potential as

$$E^{(1)} = \sum_a \left\{ \frac{1}{2}(V_{\text{HF}})_{aa} - U_{aa} \right\}, \qquad (2.43)$$

where we have used the fact, evident from Eq. (2.15), that

$$(V_{\text{HF}})_{ij} = \sum_b (g_{ibjb} - g_{ibbj}). \qquad (2.44)$$

It should be noted that if $U = V_{HF}$, then

$$\begin{aligned} W_0 + E^{(1)} &= \sum_a [\varepsilon_a - \frac{1}{2}(V_{HF})_{aa}] \\ &= \sum_a (T)_{aa} + \frac{1}{2}\sum_{ab} (g_{abab} - g_{abba}), \qquad (2.45) \end{aligned}$$

where T is the sum of the electron's kinetic energy and V_{nuc}. The expression on the second line of Eq. (2.45) is just the expectation value

---

[†] Cf. with Eqs. 1.79 and 1.81 where the variable "g" was a general two-body operator.

of H evaluated using HF wave functions; *i.e.*, $W_0 + E^{(1)} = \langle \Phi_0 | H | \Phi_0 \rangle$. The radial HF equations discussed previously could have been derived by applying a variational principle to this expression. Stated differently, the HF energy is the energy of a closed-shell system obtained in a first-order perturbation theory calculation using HF orbitals.

As an elementary example of these remarks, let us consider helium in the $(1s_{\frac{1}{2}})^2$ ground state. Taking $U(r)$ to be the HF potential, one solves the radial Dirac equations to find $\varepsilon_{1s} = -0.918 \cdots$ a.u. for the $1s$ eigenvalue. From Eq. (2.35), the energy W_0 of the helium atom in the state $|1s_{m=\frac{1}{2}}, 1s_{m=-\frac{1}{2}}\rangle$ is $W_0 = 2\varepsilon_{1s} = -1.836 \cdots$ a.u. The residual interaction energy, obtained from Eq. (2.43), is $E^{(1)} = -1.026 \cdots$ a.u. Combining these two results, one obtains $W_0 + E^{(1)} = -2.862 \cdots$ a.u. as a first approximation to the energy of the ground state of helium. This approximate value differs from the highly-accurate value $E = -2.90386 \cdots$ a.u., obtained from an all-order relativistic many-body calculation [93] by only 1.4%.

2.3.2 Breit interaction

The Breit interaction is the interaction that arises from the exchange of transverse photons between electrons. The form of this interaction can be inferred from QED using second-order perturbation theory. If we consider two electrons in single-particle states a and b, with energies ε_a and ε_b, respectively, then the interaction energy is given by the difference between direct and exchange matrix elements of the potential, [94]

$$V_{\text{tr}} = -\frac{e^2}{2\pi^2} \int \frac{d^3 k}{k^2 - \Gamma^2} e^{i\mathbf{k}\cdot\mathbf{r}_{12}} \left[\delta_{ij} - \frac{k_i k_j}{k^2} \right] \alpha_{1i} \alpha_{2j}. \qquad (2.46)$$

In this equation, α_{1i} is the ith component of the Dirac $\boldsymbol{\alpha}$-matrix for the electron at $\mathbf{r}_1$. The quantity Γ in the denominator of Eq. (2.46) is the momentum of the exchanged photon,

$$\Gamma = \begin{cases} 0 & \text{in direct matrix elements} \\ |\varepsilon_a - \varepsilon_b|/\hbar c & \text{in exchange matrix elements} \end{cases} . \qquad (2.47)$$

The exchange matrix element of V_{tr} is complex. The real part gives the energy shift and the imaginary part gives the decay rate for $a \to b$ when $\varepsilon_a > \varepsilon_b$, or for $b \to a$ when $\varepsilon_b > \varepsilon_a$. The $\mathbf{k}$ integration in Eq. (2.46) can be evaluated explicitly, leading to the following expression for the real part of the interaction

$$V_{\text{tr}} = -e^2 \left\{ \frac{\cos \Gamma r_{12}}{r_{12}} \delta_{ij} - \frac{\partial^2}{\partial_{1i}\partial_{2j}} \frac{\cos \Gamma r_{12} - 1}{\Gamma^2 r_{12}} \right\} \alpha_{1i} \alpha_{2j}. \qquad (2.48)$$

In direct matrix elements, the limiting form of Eq. (2.48) with $\Gamma \to 0$ is to be used. In exchange matrix elements, $\Gamma r_{12} \approx \alpha Z$, so the limiting form of V_{tr} can also be used, to the neglect of terms of order $\alpha^4 Z^3$ a.u. The limiting form of the transverse interaction is the Breit interaction; it is given by

$$V_{Br} = -\frac{e^2}{r_{12}} \left[\boldsymbol{\alpha}_1 \cdot \boldsymbol{\alpha}_2 - \frac{1}{2} (\boldsymbol{\alpha}_1 \cdot \boldsymbol{\alpha}_2 - \boldsymbol{\alpha}_1 \cdot \hat{r}_{12} \, \boldsymbol{\alpha}_2 \cdot \hat{r}_{12}) \right], \quad (2.49)$$

$$= -\frac{e^2}{2r_{12}} [\boldsymbol{\alpha}_1 \cdot \boldsymbol{\alpha}_2 + \boldsymbol{\alpha}_1 \cdot \hat{r}_{12} \, \boldsymbol{\alpha}_2 \cdot \hat{r}_{12}]. \quad (2.50)$$

It is this interaction that we will use in the applications to follow. The nonrelativistic reduction of the Breit interaction, leading to spin-orbit, spin-other-orbit, and spin-spin corrections to the electron-electron Coulomb interaction is given on page 199 in Ref. [94]. The first term in Eq. (2.49) is referred to as the Gaunt interaction; it was introduced by Gaunt in his 1929 study of the helium fine structure. [95] The second term in Eq. (2.50), referred to as the retardation interaction, was introduced later by Breit and used in his study of the helium fine structure. [96] In calculations of inner-shell energies for closed-shell atoms, it has been found that the Gaunt interaction contributes about 90% of the Breit interaction energy shift and the retardation interaction contributes the remaining 10%. [97]

In second-quantization, the Breit interaction is given by the two-particle operator

$$B_I = \frac{1}{2} \sum_{ijk\ell} b_{ijk\ell} a_i^\dagger a_j^\dagger a_\ell a_k, \quad (2.51)$$

where $b_{ijk\ell}$ is an unsymmetrized two-electron matrix element of V_{Br}. The first-order Breit correction to the energy of a closed-shell atom is

$$\begin{aligned} B^{(1)} &= \langle \Phi_0 | B_I | \Phi_0 \rangle \\ &= \frac{1}{2} \sum_{ac} (b_{acac} - b_{acca}) = -\frac{1}{2} \sum_{ac} b_{acca}, \end{aligned} \quad (2.52)$$

where the sums on the second line are over occupied core orbitals a and c. The direct interaction term in the first sum on the second line vanishes for closed-shell systems by symmetry considerations.

In the applications considered later in this chapter, the the Breit corrections are small compared to the Coulomb correlation corrections; therefore, we evaluate the Breit correction in lowest order only. Nevertheless, it is perfectly correct (although cumbersome) to treat the Breit interaction on the same footing as the Coulomb interaction, and to include second- and higher-order Breit corrections along with the corresponding Coulomb corrections.

To summarize this section, the no-pair Hamiltonian is the sum of the independent-particle Hamiltonian, the Coulomb interaction, and the Breit interaction, $H = H_0 + V_I + B_I$, where the corresponding operators are constructed from electron creation and annihilation operators only. This approximate Hamiltonian, which has been derived heuristically here, can also be obtained from the field-theoretic Hamiltonian of QED by a contact transformation. [65, 83] It should be emphasized that the no-pair Hamiltonian provides only a starting point for relativistic calculations. The effects of virtual pairs, leading to self-energy and vacuum-polarization corrections, must be included in calculations of the energy to achieve agreement with precise spectroscopic measurements. A discussion of these radiative corrections is given in Ref. [85].

2.4   Perturbation theory

We consider the no-pair Hamiltonian with Coulomb interactions only, $H = H_0 + V_I$, and seek a solution to the Schrödinger equation

$$H\Psi = E\Psi \,, \tag{2.53}$$

for an N-electron atom that reduces to the independent-particle state vector Φ_0 (assumed to be nondegenerate) when V_I vanishes. We use nondegenerate perturbation theory to expand Ψ and E in powers of V_I:

$$\Psi = \Phi_0 + \Psi^{(1)} + \Psi^{(2)} + \cdots \,,$$
$$E = W_0 + E^{(1)} + E^{(2)} + \cdots \,. \tag{2.54}$$

If we adopt the intermediate normalization scheme, $\langle \Phi_0 | \Psi \rangle = 1$, then Ψ and E are given by the equations:

$$\Delta E = E - W_0 = \langle \Phi_0 | V_I | \Psi \rangle \,, \tag{2.55}$$

and

$$\Psi = \Phi_0 + \sum_{k=1}^{\infty} \left[\frac{\mathbf{Q}(\Delta E - V_I)}{H_0 - W_0} \right]^k \Phi_0 \,. \tag{2.56}$$

The operator $\mathbf{Q} = 1 - \mathbf{P}$ in Eq. (2.56) is a projection operator onto states orthogonal to Φ_0;

$$\mathbf{P} = |\Phi_0\rangle\langle\Phi_0| \,, \quad \mathbf{Q} = \sum_{n \neq 0} |\Phi_n\rangle\langle\Phi_n| \,. \tag{2.57}$$

The Rayleigh-Schrödinger perturbation series are obtained by expanding Eqs. (2.55) and (2.56) systematically in powers of V_I; the first few terms in the expansion are:

$$E^{(1)} = \langle \Phi_0 | V_I | \Phi_0 \rangle$$

$$\Psi^{(1)} = \sum_{n \neq 0} \frac{\langle \Phi_n | V_I | \Phi_0 \rangle}{W_0 - W_n} \Phi_n , \qquad (2.58)$$

and

$$E^{(2)} = \sum_{n \neq 0} \frac{\langle \Phi_n | V_I | \Phi_0 \rangle \langle \Phi_0 | V_I | \Phi_n \rangle}{W_0 - W_n} ,$$

$$\Psi^{(2)} = \sum_{n,m \neq 0} \frac{\langle \Phi_n | V_I | \Phi_m \rangle \langle \Phi_m | V_I | \Phi_0 \rangle}{(W_0 - W_n)(W_0 - W_m)} \Phi_n$$

$$-E^{(1)} \sum_{n \neq 0} \frac{\langle \Phi_n | V_I | \Phi_0 \rangle}{(W_0 - W_n)^2} \Phi_n . \qquad (2.59)$$

When written in terms of single-particle states, these formulas lead to precisely the same expressions as the linked-cluster expansion described in the first chapter.[‡]

Assuming that the background potential is chosen as the HF potential, $U(r) = V_{HF}$, we find from Eq. (2.58) for a closed-shell atom that

$$E^{(1)} = -\frac{1}{2} \sum_{ab} (g_{abab} - g_{abba}) = -\frac{1}{2} \sum_a (V_{HF})_{aa} , \qquad (2.60)$$

where the indices a, b range over orbitals occupied in the ground state. From Eq. (2.59) one obtains

$$E^{(2)} = \frac{1}{2} \sum_{abmn} \frac{g_{abnm}(g_{nmab} - g_{nmba})}{\varepsilon_a + \varepsilon_b - \varepsilon_m - \varepsilon_n} \qquad (2.61)$$

for the second-order energy of a closed-shell atom, where the indices a, b again range over orbitals occupied in the atomic ground state and the indices m, n over unoccupied orbitals. The Brueckner-Goldstone (BG) diagrams associated with the two terms in Eq. (2.61) are given in parts (a) and (b) of Fig. 2.1.

An atom with one valence electron outside a closed core is described in lowest-order by the state vector $\Phi_0 = a_v^\dagger |0_c\rangle$, where $|0_c\rangle$ is the lowest-order state vector of the closed core. The nth order energy for such an atom can be decomposed into a sum of the nth order core energy $E_c^{(n)}$ (which is independent of the quantum numbers of the valence electron) and an nth order valence energy $E_v^{(n)}$. The first- and second-order core energies are given by the formulas derived in the previous paragraph, provided the frozen core HF potential is used to define the one-electron orbitals. In this case, the first-order valence energy vanishes

$$E_v^{(1)} = 0 . \qquad (2.62)$$

[‡] cf. Eqs.(1.64,1.65, 1.70-1.72).

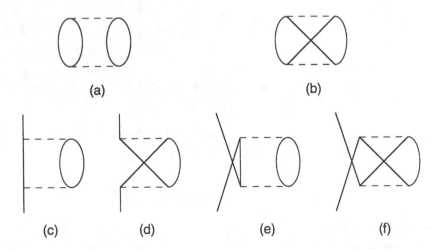

(a)   (b)

(c)   (d)   (e)   (f)

Fig. 2.1.   Parts (a) and (b) give the second-order energy for a closed-shell atom. Parts (c)–(f) give the second-order correction to the valence energy for ions with one valence electron.

This fact has the consequence that the negative of the frozen orbital Dirac-Fock eigenvalue for a valence orbital $-\varepsilon_v$ is an approximation to the ionization energy, accurate through first-order. The small residual differences between theory and experiment, such as those shown in Table 2.1, are dominated by the second-order correction to the valence energy

$$E_v^{(2)} = -\sum_{bmn} \frac{g_{mnvb}(g_{vbmn} - g_{vbnm})}{\varepsilon_m + \varepsilon_n - \varepsilon_v - \varepsilon_b} + \sum_{abn} \frac{g_{vnab}(g_{abvn} - g_{abnv})}{\varepsilon_v + \varepsilon_n - \varepsilon_a - \varepsilon_b} , \qquad (2.63)$$

which are shown in the BG diagrams in parts (c)–(f) of Fig. 2.1.

The methods described in Chapter 1 can be used to obtain expressions such as those derived above for the third- and higher-order corrections to the energy in terms of single-particle orbitals.

2.5   Angular reduction

To reduce the expressions for the higher-order corrections to the energy to forms suitable for numerical evaluation, one first carries out a summation over magnetic quantum numbers of the one-electron orbitals. To that end, we decompose the Coulomb integrals $g_{ijk\ell}$ defined in Eq. (2.37) into a sum of products of radial and angular factors. We then carry out the integrations over angles in the expression for $g_{ijk\ell}$ using the expansion of $1/|\mathbf{r} - \mathbf{r}'|$ in terms of spherical harmonics to obtain

$$g_{ijk\ell} = \sum_L J_L(ijk\ell) \, X_L(ijk\ell) , \qquad (2.64)$$

where

$$
J_L(ijk\ell) = \sum_M (-1)^{j_i+j_j+L-m_i-m_j-M} \begin{pmatrix} j_i & L & j_k \\ -m_i & M & m_k \end{pmatrix}
$$

$$
\times \begin{pmatrix} j_j & L & j_\ell \\ -m_j & -M & m_\ell \end{pmatrix} , \tag{2.65}
$$

and

$$
X_L(ijk\ell) = (-1)^L \langle i \| C_L \| k \rangle \langle j \| C_L \| \ell \rangle R_L(ijk\ell) . \tag{2.66}
$$

The Slater integral $R_L(ijk\ell)$ in Eq. (2.66) is defined by

$$
R_L(ijk\ell) = e^2 \int_0^\infty \int_0^\infty dr dr' \frac{r_<^L}{r_>^{L+1}} [P_i(r)P_k(r) + Q_i(r)Q_k(r)]
$$

$$
\times [P_j(r')P_\ell(r') + Q_j(r')Q_\ell(r')] . \tag{2.67}
$$

The dependence on magnetic quantum numbers in Eq. (2.64) is contained in the factor $J_L(ijk\ell)$. The factor $X_L(ijk\ell)$ depends only on the principal and angular quantum numbers n and κ of its arguments. Using the identity

$$
J_{L'}(ij\ell k) = -\sum_L [L] \begin{Bmatrix} j_i & j_\ell & L' \\ j_j & j_k & L \end{Bmatrix} J_L(ijk\ell) , \tag{2.68}
$$

where the expression in braces is a Wigner $6j$ symbol, we may express the antisymmetrized Coulomb integral $\tilde{g}_{ijk\ell} = g_{ijk\ell} - g_{ij\ell k}$ as

$$
\tilde{g}_{ijk\ell} = \sum_L J_L(ijk\ell)Z_L(ijk\ell) , \tag{2.69}
$$

where

$$
Z_L(ijk\ell) = X_L(ijk\ell) + \sum_{L'} [L] \begin{Bmatrix} j_i & j_\ell & L' \\ j_j & j_k & L \end{Bmatrix} X_{L'}(ij\ell k) . \tag{2.70}
$$

Taking advantage of the expansions Eq. (2.64) and Eq. (2.69), one can carry out the sums over magnetic quantum numbers in the expressions for $E^{(2)}$ given in the previous section. To accomplish this, we make use of the identity

$$
\sum_{m_a m_b m_m m_n} J_L(mnab)J_{L'}(abmn) = (-1)^{j_m+j_n-j_a-j_b} \frac{1}{[L]^2}\delta_{LL'} . \tag{2.71}
$$

With the aid of Eq. (2.71), we reduce the expression for the second-order correlation energy of a closed-shell atom to the following sum over radial integrals:

$$
E^{(2)} = -\frac{1}{2} \sum_{Labmn} (-1)^{j_m+j_n-j_a-j_b} \frac{1}{[L]^2} \frac{X_L(mnab)Z_L(abmn)}{\varepsilon_m + \varepsilon_n - \varepsilon_a - \varepsilon_b} . \tag{2.72}
$$

In a similar way, the second-order energy for an atom with one valence electron is found from Eq. (2.63) to be

$$
\begin{aligned}
E_v^{(2)} = &- \sum_{Lbmn} (-1)^{j_m+j_n-j_v-j_b} \frac{1}{[j_v][L]^2} \frac{X_L(mnvb)Z_L(vbmn)}{\varepsilon_m + \varepsilon_n - \varepsilon_v - \varepsilon_b} \\
&+ \sum_{Labn} (-1)^{j_v+j_n-j_a-j_b} \frac{1}{[j_v][L]^2} \frac{X_L(vnab)Z_L(abvn)}{\varepsilon_v + \varepsilon_n - \varepsilon_a - \varepsilon_b} .
\end{aligned} \tag{2.73}
$$

The expressions above for the second-order energies must be summed over the principal and angular quantum numbers n and κ of occupied orbitals $(a, b, \cdots)$ as well as the principal and angular quantum numbers of excited orbitals $(m, n, \cdots)$. These latter sums include integrations over the continuous spectrum of electron scattering states. Finite basis sets are introduced in the following section to aid in the evaluation of these sums.

2.6   Finite basis sets

The sums over excited states m, n in Eqs. (2.72) and (2.73) are difficult to evaluate accurately since there are an infinite number of possible angular momentum states κ_m for each orbital $\phi_m(\mathbf{r})$, and an infinite number of discrete states n_m as well as a positive-energy continuum for each κ_m.

A method that has been widely used to evaluate such sums in nonrelativistic calculations is a generalization of the Dalgarno-Lewis method, [99] in which a sum over m is converted into an integral over the solution to an inhomogeneous differential equation. Double sums over two excited states m and n, such as those occurring in the expressions for second-order MBPT, are converted into double integrals over the solutions of partial differential equations (the *pair-equations*) in this way. This method requires that the sum over excited states range over a complete set of single-particle orbitals $\phi_m(\mathbf{r})$. In the relativistic case, it is difficult to implement the pair-equation approach since the sums are restricted to electron states only and the completeness property is lost. The alternative used here is to replace the orbitals $\phi_m(\mathbf{r})$ by a finite basis set, and thereby reduce the infinite sums and integrals over the real spectrum to sums over a finite pseudospectrum.

Since the correlation corrections have finite range, we restrict our attention to the Dirac equation in a finite (but large) cavity of radius R. To study the ground-state or low-lying excited states of ions, the radius of this cavity is chosen to be $R \approx 40/Z_{\mathrm{ion}}$ a.u., where $Z_{\mathrm{ion}} = Z - N + 1$ is the ionic charge. Care is taken to insure that the results of calculations are independent of the cavity radius.

To avoid problems with the Dirac equation in a finite cavity associated with the Klein paradox (Ref. [100] p. 120), we impose MIT bag model [101] boundary conditions $P_m(r) = Q_m(r)$ at $r = R$. The spectrum in the cavity is discrete but infinite.

Next, we expand the solutions to the Dirac equation in a finite basis. Our basis is chosen to be a set of n B-splines of order k. Following deBoor, [102] we divide the interval from $r = 0$ to $r = R$ into segments. The endpoints of these segments are given by the knot sequence $\{t_i\}, i = 1, 2, \cdots, n + k$. The B-splines of order k, $B_{i,k}(r)$, on this knot sequence are defined recursively by the relations

$$B_{i,1}(r) = \begin{cases} 1, & t_i \leq r < t_{i+1} \\ 0, & \text{otherwise} \end{cases} \tag{2.74}$$

and

$$B_{i,k}(r) = \frac{r - t_i}{t_{i+k-1} - t_i} B_{i,k-1}(r) + \frac{t_{i+k} - r}{t_{i+k} - t_{i+1}} B_{i+1,k-1}(r) . \tag{2.75}$$

One sees that $B_{i,k}(r)$ is a piecewise polynomial of degree $k - 1$ inside the interval $t_i \leq r < t_{i+k}$ and $B_{i,k}(r)$ vanishes outside this interval. The knots defining our grid have k-fold multiplicity at the endpoints 0 and R; i.e. $t_1 = t_2 = \cdots = t_k = 0$ and $t_{n+1} = t_{n+2} = \cdots = t_{n+k} = R$. The knots $t_{k+1}, t_{k+2}, \cdots, t_n$ are distributed on an exponential scale between 0 and R.

The set of B-splines of order k on the knot sequence $\{t_i\}$ forms a complete basis for piecewise polynomials of degree $k - 1$ on the interval spanned by the knot sequence. We represent the solution to the Dirac equation as a linear combination of these B-splines and we work with the B-spline representation of the Dirac wave functions rather than the Dirac wave functions themselves. To obtain the desired representation, we start from the least action principle $\delta S = 0$ with

$$
\begin{aligned}
S = {} & \frac{1}{2} \int_0^R \Big\{ cP_\kappa(r)(d/dr - \kappa/r)Q_\kappa(r) - cQ_\kappa(r)(d/dr + \kappa/r)P_\kappa(r) \\
& + V(r)[Q_\kappa(r)^2 + P_\kappa(r)^2] - 2mc^2 Q_\kappa(r)^2 \Big\} dr \\
& - \frac{1}{2}\varepsilon \int_0^R [P_\kappa(r)^2 + Q_\kappa(r)^2] dr .
\end{aligned}
\tag{2.76}
$$

The parameter ε is a Lagrange multiplier introduced to insure that the normalization constraint

$$\int_0^R [P_\kappa(r)^2 + Q_\kappa(r)^2] dr = 1 \tag{2.77}$$

is satisfied. The variational principle $\delta S = 0$, together with the constraints $\delta P_\kappa(0) = \delta Q_\kappa(0) = 0$ and $\delta P_\kappa(R) = \delta Q_\kappa(R) = 0$, leads to the radial Dirac equations (2.13). The boundary conditions $P_\kappa(0) = 0$ and $P_\kappa(R) = Q_\kappa(R)$ can be imposed on the solution by adding a suitable boundary term to the action functional. [67]

We expand $P_\kappa(r)$ and $Q_\kappa(r)$ in terms of B-splines of order k as

$$P(r) = \sum_{i=1}^{n} p_i B_i(r) ,$$

$$Q(r) = \sum_{i=1}^{n} q_i B_i(r) . \tag{2.78}$$

The subscript κ has been omitted from $P_\kappa(r)$ and $Q_\kappa(r)$ and the subscript k has been omitted from $B_{i,k}(r)$ for notational simplicity. The action S becomes a quadratic function of the coefficients p_i and q_i when the expansions (2.78) are substituted into Eq. (2.76). The variational principle then leads to a system of linear equations for the expansion coefficients. These equations can be written in the form of a $2n \times 2n$ symmetric generalized eigenvalue equation

$$\underline{A}v = \varepsilon \underline{B}v , \tag{2.79}$$

where v is given by

$$v = (p_1, p_2, \cdots, p_n, q_1, q_2, \cdots, q_n) . \tag{2.80}$$

The matrices $\underline{A}$ and $\underline{B}$ are given explicitly in Ref. [67]. From Eq. (2.79) one obtains $2n$ real eigenvalues ε^λ and $2n$ eigenvectors v^λ. The eigenvectors satisfy the orthogonality relations

$$\sum_{i,j} v_i^\lambda B_{ij} v_j^\mu = \delta_{\lambda\mu} , \tag{2.81}$$

which leads to the orthogonality relations

$$\int [P^\lambda(r) P^\mu(r) + Q^\lambda(r) Q^\mu(r)] dr = \delta_{\lambda\mu} \tag{2.82}$$

for the corresponding radial Dirac functions.

The spectrum consists of two equal branches, each containing n eigenvalues. The lower branch contains terms with $\varepsilon^\lambda < -2mc^2$. These eigenvalues belong to the positron states and play no role in the present many-body calculations. The upper branch of n eigenvalues corresponds to the electron states; these terms are used to carry out the sums over excited states in the formulas from the previous section. The first few electron-state eigenvalues and eigenvectors in the cavity agree precisely with the first few bound-state eigenvalues and eigenvectors obtained by numerically integrating the radial Dirac equations, but as the principal quantum number increases, the cavity spectrum departs more and more from the real spectrum. The cavity spectrum is complete (in the space of piecewise polynomials of degree $k-1$) and, therefore, can be used instead of the real spectrum to evaluate correlation corrections to states confined to the cavity.

Table 2.2. The second-order energy for copper-like zinc, $Z = 30$, from Eq. (2.73) is compared with the experimental correlation energy

Term	$4s_{1/2}$	$4p_{1/2}$	$4p_{3/2}$
$E^{(2)}$	-0.04044	-0.02233	-0.02152
E_{corr}	-0.04445	-0.02682	-0.02620

The quality of the numerically generated B-spline spectrum can be tested by using it to evaluate various energy-weighted sum rules. A complete discussion of such tests is given in Ref. [67]. It suffices to say that calculations carried out using $n = 40$ splines with $k = 7$ satisfied the relativistic Thomas-Reiche-Kuhn sum rule to a few parts in 10^7, while calculations made with $n = 50$ splines of order $k = 9$ satisfied the same sum rule to a few parts in 10^9.

As an application, consider the second-order correction to the energy of the helium ground-state. The indices a and b in Eq. (2.72) are both restricted to the 1s ground state: $n_a = n_b = 1$, and $\kappa_a = \kappa_b = -1$. Angular momentum selection rules force the orbital angular momentum of the states n and m to equal L in the sum; moreover, there is an exact cancellation between the direct and exchange contributions to $Z_L(abmn)$ for $j_m \neq j_n$ so that only states with $\kappa_m = \kappa_n = L$ or $\kappa_m = \kappa_n = -L - 1$ contribute to the sum. One is finally left with a sum over $L = 0, 1, 2, \cdots$. For each L, there are double sums to be done over the basis sets for those excited states m, n permitted by the selection rules. The L sum converges as $1/L^4$, and can be extrapolated from $L = 0, 1, \cdots, 10$ to $L = \infty$ giving a value of $E^{(2)}$ accurate to a few parts in 10^6. The result of an explicit calculation for helium is $E^{(2)} = -.03737$ a.u. Adding this value to the first-order result discussed in Section 2.3, gives $W_0 + E^{(1)} + E^{(2)} = -2.89918$ a.u., which is within 0.1% of the all-order value from Ref. [93].

As a second example, we show in Table 2.2 the values of the second-order energy for the $4s$, $4p_{1/2}$ and $4p_{3/2}$ states of copper-like zinc, and compare the resulting values with the experimental correlation energies determined in Section 2.2. The first row of the table contains values determined from Eq. (2.73). These values are compared with the experimental values of the correlation energy given by the difference between the experimental energy and the Dirac-Hartree-Fock eigenvalues and listed in Table 2.1. We find that the second-order energy accounts for 80%–90% of the observed correlation energy for this ion. Higher-order contributions to the Coulomb energy, the Breit interaction, and QED corrections are expected to make up the remaining 10%–20% of the correlation energy.

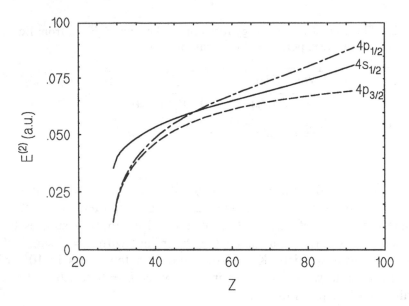

Fig. 2.2.   Second-order energies for $4s_{1/2}$, $4p_{1/2}$ and $4p_{3/2}$ states of copper-like ions given as functions of the nuclear charge Z.

In Fig. 2.2, we show the variation with Z of the second-order energy of the $4s$ and $4p$ states along the copper isoelectronic sequence. Nonrelativistically, one expects the second-order energy to become independent of Z, for large Z. The variation seen in Fig. 2.2 at large Z is a relativistic effect.

2.7   Higher-order corrections

Expressions for the third-order correction to the energy are given in Ref. [103]. These expressions may be reduced to sums over products of Slater integrals and angular factors using the methods outlined in Section 2.5, and evaluated numerically using the B-spline basis sets introduced in the previous section.

The successive corrections to the Coulomb energy are expected to decrease along an isoelectronic sequence approximately as Z^{2-n}, where n is the order of perturbation theory. This rule follows from the fact that the expressions in nth-order perturbation theory are sums of ratios which have products of n Slater integrals (which scale as Z) in their numerators and products of $n-1$ single-particle energies (which scale as Z^2) in the denominators. One therefore expects that predictions made using a fixed number of terms $E^{(n)}$ in the perturbation expansion will become more and more accurate as Z increases along a given isoelectronic sequence. In the present discussion of the copper sequence, we omit terms of fourth and

Table 2.3. Higher-order corrections to the energies of the $4s$ and $4p$ states of copper-like zinc, $Z = 30$

Term	$4s_{1/2}$	$4p_{1/2}$	$4p_{3/2}$
$E^{(2)}$	-0.04044	-0.02233	-0.02152
$E^{(3)}$	-0.00427	-0.00372	-0.00360
$B^{(1)}$	0.00038	0.00027	0.00020
$B^{(2)}$	-0.00019	-0.00009	-0.00009
$B^{(3)}$	0.00007	0.00006	0.00004
RM+MP	0.00001	0.00000	0.00000
Total	-0.04443	-0.02581	-0.02497
E_{corr}	-0.04445	-0.02682	-0.02620

higher order. These omitted terms are the dominant source of theoretical error near the neutral end of the isoelectronic sequence. For high Z, the theoretical error is due primarily to the omitted QED corrections.

For neutral helium, the third-order energy can be readily evaluated to give $E^{(3)} = -0.00368$ a.u., leading to a value $W_0 + E^{(1)} + E^{(2)} + E^{(3)} = -2.90286$ a.u., for the ground-state energy compared with the all-order value -2.90386 a.u., given in Ref. [93]. Contributions of the third-order energy to the $4s$ and $4p$ states in copper-like zinc are listed in the second row of Table 2.3. The calculations of these terms is similar to the corresponding second-order energies but it is far more time consuming since one must evaluate quadruple sums over excited states in third order compared to double sums in second order.

Further important contributions to the energy arise from the Breit interaction and from terms in perturbation theory corresponding to a single Breit interaction and one or more Coulomb interactions. The first-order Breit interaction $B_v^{(1)}$ grows as $\alpha^2 Z^3$ with increasing nuclear charge Z. This term is given by

$$B_v^{(1)} = -\sum_c b_{vccv} . \tag{2.83}$$

The corresponding contributions to the $4s$ and $4p$ energies in copperlike zinc are listed in the third row of Table 2.3.

The term in perturbation theory corresponding to one Breit interaction and one Coulomb interaction, $B_v^{(2)}$, grows as $\alpha^2 Z^2$. This term is given by

$$B_v^{(2)} = -\sum_{cmn} \frac{b_{mnvc}(g_{vcmn} - g_{vcnm})}{\varepsilon_m + \varepsilon_n - \varepsilon_v - \varepsilon_c} - \sum_{cmn} \frac{g_{mnvc}(b_{vcmn} - b_{vcnm})}{\varepsilon_m + \varepsilon_n - \varepsilon_v - \varepsilon_c}$$

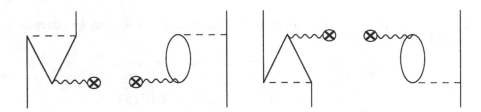

Fig. 2.3. Dominant second-order contributions to the Breit energy for ions with one valence electron.

$$+ \sum_{acn} \frac{b_{vnac}(g_{acvn} - g_{acnv})}{\varepsilon_v + \varepsilon_n - \varepsilon_a - \varepsilon_c} + \sum_{acn} \frac{g_{vnac}(b_{acvn} - b_{acnv})}{\varepsilon_v + \varepsilon_n - \varepsilon_a - \varepsilon_c}$$

$$+ \; B_{\mathrm{RPA}}^{(2)} \,, \tag{2.84}$$

where the RPA (random phase approximation) term on the third line is

$$B_{\mathrm{RPA}}^{(2)} = -\sum_{an} \frac{b_{an}(g_{nvav} - g_{nvva})}{\varepsilon_n - \varepsilon_a} + \text{c.c.}\,, \tag{2.85}$$

with

$$b_{an} = -\sum_{c} b_{accn}\,, \tag{2.86}$$

and where "c.c." denotes the complex conjugate of the preceding term. The dominant contribution to Eq. (2.84) is the RPA term given by Eq. (2.85) and is shown in the BG diagrams in Fig. 2.3. Because of the relatively large size of these contributions, they are iterated, as shown in Fig. 2.4 to include all third- and higher-order RPA corrections. The resulting second-order Breit corrections to the $4s$ and $4p$ states of copper-like zinc are listed in the fourth row of Table 2.3.

The third-order correction corresponding to one Breit interaction and two Coulomb interactions, $B_v^{(3)}$, grows as $\alpha^2 Z$. The two largest contributions to this term are the third-order RPA correction, which is already included with the second-order RPA corrections in $B_v^{(2)}$, and the Brueckner-orbital corrections $B_{\mathrm{BO}}^{(3)}$ shown in Fig. 2.5. These BO corrections are listed in the fifth row of Table 2.3.

In the sixth row of the Table 2.3 we give the reduced-mass and mass-polarization corrections to the energies of the $4s$ and $4p$ states. The reduced mass corrections are given by

$$\Delta E_{\mathrm{RM}} = -\frac{1}{M}(W_0 + E_v^{(2)} + E_v^{(3)} + \cdots + B_v^{(1)} + B_v^{(2)} + B_v^{(3)} + \cdots)\,, \tag{2.87}$$

where M is the nuclear mass expressed in atomic units. The mass polar-

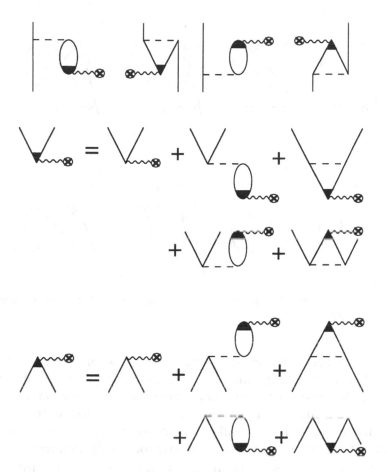

Fig. 2.4. Iterative solution to the RPA equations for the dominant contributions to the Breit energy for ions with one valence electron.

ization correction is determined by the matrix element of the operator

$$H_{\text{MP}} = \frac{1}{M} \sum_{i>j} \mathbf{p}_i \cdot \mathbf{p}_j . \tag{2.88}$$

In lowest order, the mass-polarization correction to the energy is given by

$$P_v^{(1)} = -\frac{1}{M} \sum_b |\langle v|\mathbf{p}|b\rangle|^2 . \tag{2.89}$$

Correlation corrections through third order to the mass-polarization energy can be worked out using the scheme described above for evaluating correlation corrections to the Breit interaction. The mass-polarization corrections listed in Table 2.3 include second- and third-order correlation corrections.

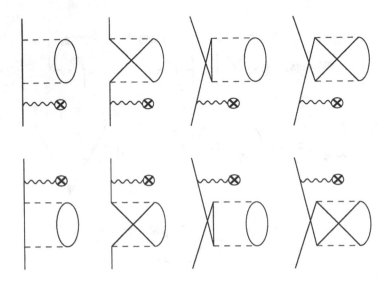

Fig. 2.5. Dominant third-order contributions to the Breit energy for ions with one valence electron.

The final theoretical energies listed in the row of Table 2.3 labeled "Total" differ from the experimental correlation energy because of omitted fourth- and higher-order correlation corrections and the omitted QED corrections. As mentioned previously, the correlation corrections are expected to dominate the the difference for nearly neutral ions, while the QED corrections are expected to dominate for highly charged ions.

In Table 2.4, we compare the $4p_{3/2} - 4s_{1/2}$ energy intervals for copper-like ions, calculated as described above, with measured intervals [104] throughout the copper isoelectronic sequence. For large Z, the difference between theory and experiment scales as Z^4. The difference is shown graphically in Fig. 2.6, where it is compared with phenomenological calculations of the Lamb shift. In the solid line, we show values of the hydrogenic $n = 2$ state self-energy and vacuum polarization corrections, scaled to $n = 4$, but with the nuclear charge reduced to $Z_{\text{eff}} = Z - 13.24$. This value of Z_{eff} was obtained by a one-parameter least-square fit to the energy differences in Table 2.4.

In the dashed curve, we show values of the $4p_{3/2} - 4s_{1/2}$ Lamb-shift obtained using the semi-empirical procedure contained in Grant's MCDF code. [88] Both estimates of the Lamb-shift contributions to the energy interval are in general agreement with the observed difference between theory and experiment.

The Lamb-shift energy difference for copper-like gold $Z = 79$ has been calculated from first principles using propagators constructed in a local

Table 2.4.   The $4p_{3/2} - 4s_{1/2}$ energy interval for copper-like ions (a.u.).   The theoretical energies are from a third-order relativistic MBPT calculation

Z	Theory	Exp.	Th.-Exp.	Z	Theory	Exp.	Th.-Exp.
29	0.1439(3)	0.1403	0.0036(3)	50	2.0882(3)	2.0807(1)	0.0075(3)
30	0.2261(2)	0.2249	0.0012(2)	53	2.4865(3)	2.4776(27)	0.0088(27)
31	0.3054(2)	0.3048	0.0006(2)	54	2.6035(3)	2.6195(1)	0.0109(3)
32	0.3836(2)	0.3832	0.0004(2)	56	2.9373(3)	2.9250(3)	0.0122(4)
33	0.4618(2)	0.4613	0.0005(2)	57	3.1008(3)	3.0877(3)	0.0131(4)
34	0.5403(2)	0.5398	0.0006(2)	60	3.6363(3)	3.6191(4)	0.0171(5)
35	0.6197(2)	0.6190	0.0007(2)	62	4.0347(3)	4.0143(5)	0.0204(6)
36	0.7003(2)	0.6993	0.0010(2)	63	4.2476(3)	4.2255(6)	0.0221(7)
37	0.7822(2)	0.7810	0.0012(2)	64	4.4704(3)	4.4470(7)	0.0234(7)
38	0.8659(2)	0.8644	0.0015(2)	66	4.9473(3)	4.9177(8)	0.0296(8)
39	0.9515(2)	0.9497	0.0018(2)	68	5.4695(3)	5.4363(10)	0.0332(10)
40	1.0393(2)	1.0371	0.0022(2)	70	6.0412(3)	6.0097(12)	0.0314(12)
41	1.1295(2)	1.1270	0.0025(2)	73	7.0029(3)	6.9690(16)	0.0339(16)
42	1.2224(2)	1.2194	0.0029(2)	74	7.3538(3)	7.3131(18)	0.0407(18)
44	1.4168(2)	1.4131	0.0038(2)	75	7.7212(3)	7.6803(19)	0.0409(20)
45	1.5190(2)	1.5144(1)	0.0046(3)	79	9.3704(3)	9.3123(29)	0.0580(29)
46	1.6246(2)	1.6198(1)	0.0049(3)	82	10.8199(3)	10.7550(38)	0.0650(38)
47	1.7341(3)	1.7287(1)	0.0055(3)	83	11.3487(3)	11.2797(42)	0.0689(42)
48	1.8477(3)	1.8416(1)	0.0061(3)	90	15.8041(3)	15.7169(81)	0.0871(81)
49	1.9657(3)	1.9586(1)	0.0071(3)	92	17.3557(3)	17.2589(98)	0.0969(98)

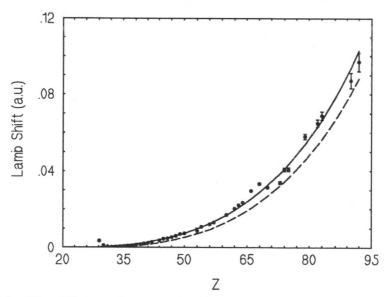

Fig. 2.6.   The difference between experimental and theoretical values of the $4p_{3/2} - 4s_{1/2}$ energy interval is compared with estimates of the $4p_{3/2} - 4s_{1/2}$ Lamb shift.   The solid curve gives scaled Coulomb values of the Lamb shift with $Z_{\text{eff}} = Z - 13.24$.   The dashed curve gives values of the Lamb shift calculated using the algorithms in Grant's MCDF code[88].

potential that closely approximates the HF potential. [105] One obtains the value 0.072(1) a.u. for the sum of the vacuum-polarization and self-energy corrections for the $4s$ state and 0.013(1) a.u. for the $4p_{3/2}$ state. The difference is 0.059(2) a.u., which can be compared with the value 0.058(3) a.u. inferred from Table 2.4.

As mentioned in the introduction, similar calculations have been carried out for the lithium and sodium isoelectronic sequences. The close agreement between such calculations (once low-order radiative corrections have been taken into account) and measurements is an illustration of the predictive power of atomic structure calculations based on the no-pair Hamiltonian.

3

Parity nonconservation in atoms

S. A. Blundell, W. R. Johnson, and J. Sapirstein

3.1 Introduction

The role of atomic physics in precision tests of QED is well known. However, atomic physics has not until recently played much of a role in testing the weak interactions owing to the extremely small ratio $\sim 10^{-11}$ of the energy scale of atoms to the mass of the $W^{\pm}$ and Z^0 bosons. Indeed, the first discussion of parity violating effects in atoms arising from weak-neutral-current interactions between the nucleus and electrons by Zel'dovich [106] concluded that these effects were probably too small to observe. However, after the existence of neutral-current weak interactions was established experimentally, the Bouchiats [107] showed in 1974 that parity nonconserving (PNC) transitions in heavy atoms with atomic number Z were enhanced by a factor of Z^3. While still very small, this effect has been observed in a variety of heavy atoms, specifically cesium ($Z = 55$), [108, 109] thallium ($Z = 81$), [110, 111] lead ($Z = 82$), [112] and bismuth ($Z = 83$). [113] These experiments deal with neutral atoms, and for this reason the electronic structure of the atoms must be understood with some precision before the experiments can be interpreted in terms of particle-physics concepts. This problem is not present for hydrogen, but PNC experiments in that atom have not proved successful.

Unfortunately, practical calculations that are accurate to the 1% level, the level of interest for particle physics, are quite difficult to carry out in heavy atoms. It is the purpose of this chapter to review the methods used in the recent highly accurate calculations of atomic PNC. The subject forms an interesting application for relativistic many-body techniques in their own right, but also has important implications for high-energy physics if the calculations can be performed convincingly to high precision. Since the most complete calculations to date are for Cs, we shall use this atom to illustrate the calculational techniques. The $6S$-$7S$ transition in

Cs presently offers the most accurate test of the standard model by an atomic measurement.

In the next section (Sec. 3.2) we review the particle-physics origin of atomic PNC. We then give a brief review of relativistic many-body perturbation theory (MBPT) in Sec. 3.3, to show how well the techniques of earlier chapters fare in heavy neutral atoms for standard properties such as ionization energies, dipole amplitudes, and hyperfine constants. The extension of these techniques to calculate PNC amplitudes is discussed in Sec. 3.4. We conclude with a comparison with the standard model and a discussion of the prospects for greater accuracy in Sec. 3.5.

3.2   Overview of PNC effects in atoms

The modern experiments to detect PNC in atoms were first proposed in 1974 [107] with a view to detecting PNC in electron-hadron interactions at a level consistent with the standard model. Nowadays the phenomenology of weak interactions is established, and the role of atomic measurements, along with accelerator-based experiments, is to make a *precision* test of the standard model. A new phase of precise tests of electroweak phenomena has been initiated by the recent accurate measurement of the mass of the Z^0 boson, [114] which can replace the Weinberg angle as an input parameter in electroweak calculations. As a result, the standard-model prediction for electroweak phenomena is now generally well-defined, with only small remaining uncertainties due mainly to the unknown masses of the Higg's particle and the top quark. Precise measurements of electroweak phenomena can be used to test the standard model prediction, and to constrain new physics beyond the standard model. It is hoped that atomic PNC can participate as one of such a set of measurements.

In a stable atom, the leading effects are neutral-current interactions, since these are diagonal in particle type. The dominant effect is the exchange of a virtual Z^0 boson between a quark in the nucleus and an atomic electron. This is a contact interaction, nonzero only where the electron wavefunction overlaps the nucleus. PNC arises in two ways: when the Z^0 current-current interaction is vector on the quark and axial on the electron ($V_q A_e$):

$$H_V = C_{1q} \frac{G_F}{\sqrt{2}} \int d\mathbf{x} \left[\bar{e}(\mathbf{x}) \gamma_\mu \gamma_5 e(\mathbf{x}) \right] \left[\bar{q}(\mathbf{x}) \gamma^\mu q(\mathbf{x}) \right], \qquad (3.1)$$

or vice versa ($A_q V_e$):

$$H_A = C_{2q} \frac{G_F}{\sqrt{2}} \int d\mathbf{x} \left[\bar{e}(\mathbf{x}) \gamma_\mu e(\mathbf{x}) \right] \left[\bar{q}(\mathbf{x}) \gamma^\mu \gamma_5 q(\mathbf{x}) \right]. \qquad (3.2)$$

Here G_F is the Fermi constant, and C_{1q} and C_{2q} are electron-quark coupling constants.* The dominant PNC contribution in an atom comes from H_V because all the quarks in the nucleus contribute coherently, while only the quarks from unpaired valence nucleons contribute to H_A. In addition, $C_{1q} \gg C_{2q}$ in the standard model. The quark vector current for H_V has a related conserved charge, the *weak nuclear charge* Q_W, on account of the "conserved vector current" hypothesis, where

$$Q_W = 2Z(2C_{1u} + C_{1d}) + 2N(C_{1u} + 2C_{1d}). \tag{3.3}$$

Here Z is the number of protons, N the number of neutrons, and $C_{1u,d}$ the vector part of the Z^0-quark vertex for the up and down quarks. The coupling constant Q_W forms the bridge between the atomic measurements and the particle theory: the aim is to extract Q_W as accurately as possible from the atomic experiment to compare with the standard-model prediction. Inserting the lowest-order values of C_{1u} and C_{1d} for the standard model gives:

$$Q_W = Z(1 - 4\sin^2\theta_W) - N . \tag{3.4}$$

Since $\sin^2\theta_W$ is close to $1/4$, the PNC effect is dominated by the neutrons, and $Q_W \approx -N$. Radiative corrections to this value are on the order of several percent. For Cs ($N = 78$), an analysis of radiative corrections that uses the very precise measurement of the Z^0 mass gives a standard-model prediction, [115]

$$Q_W = -73.2(0.2) , \tag{3.5}$$

that is remarkably independent of the unknown top-quark mass, as will be discussed in more detail in Sec. 3.5. Since the nucleus is quite nonrelativistic, the time-like contribution of the $(V_q A_e)$ current-current interaction dominates the spatial part. The time-like part can be described by the effective atomic Hamiltonian

$$h_W = \frac{G_F}{2\sqrt{2}} Q_W \rho_{\text{nuc}}(\mathbf{r})\gamma_5, \tag{3.6}$$

where $\rho_{\text{nuc}}(\mathbf{r})$ is a weighted average of the neutron and proton distributions in the nucleus, which leads to nuclear structure uncertainties that will be discussed in Sec. 3.4.5. Only the spherically averaged part of the density $\rho_{\text{nuc}}(\mathbf{r})$ makes a significant contribution.

The smaller $(A_q V_e)$ interaction H_A leads to a nuclear-spin-dependent effect that will be discussed in Sec. 3.4.6. Also discussed there is an

---

* In Eqs. (3.1) and (3.2) the functions $e(\mathbf{x})$ and $q(\mathbf{x})$ represent the electron and quark spinor functions, $\mathbf{x}$ represents the space-time four-vector, and the γ matrices are related to the α and β matrices of Eq. (2.3). See the Appendix.

interesting nuclear physics source of PNC known as the *anapole* effect, [116] which arises from photon exchange between the atomic electrons and the nucleus with weak radiative corrections on the nuclear vertex and nuclear wavefunctions. These nuclear weak-interaction effects can be parametrized in terms of an odd-parity vector moment of the nucleus known as the anapole moment. This gives an effective PNC contact interaction with the same tensor structure as the $(A_q V_e)$ term, but generally dominating that term. The anapole term enters at the several percent level, but in a way that can be subtracted out with experimental information, as described in Sec. 3.4.6.

A final PNC interaction in an atom is Z^0 exchange between electrons. This turns out to be very small, as is discussed in Sec. 3.4.7, and is at present undetectable in atomic experiments.

The weak interactions in an atom are dominated by the electromagnetic interaction, the scale of weak interactions being set by the Fermi constant, which in atomic units has the value $G_F = 2.22 \times 10^{-14}$. To distiguish the tiny weak-neutral-current interaction from electromagnetic effects, one makes use of its parity nonconserving signature. An immediate consequence of a small parity nonconserving term in the Hamiltonian is that the atomic states are no longer eigenstates of parity but contain small opposite-parity admixtures (typically with admixture coefficients on the order of 10^{-11}). A magnetic-dipole (M1) transition therefore acquires a small PNC electric-dipole (E1) component—this is the basis for practical experiments. By looking for a manifestly PNC signal, one tries to detect the interference term between the PNC-E1 amplitude and a parity-conserving electromagnetic amplitude, such as the M1 amplitude or a Stark-induced E1 amplitude. Examples of PNC signals include small asymmetries in fluorescence radiation relative to externally applied fields, and the rotation of the plane of polarization of light passing through a vapor of an element. From such signals one can extract the ratio of the PNC-E1 amplitude to the electromagnetic amplitude with which it interferes. The task for atomic theory is to relate this ratio of amplitudes to the weak nuclear charge.

For an M1 transition $I \rightarrow F$, the admixed PNC-E1 component is given by:

$$E_{PNC} = \sum_X \frac{\langle F|H_W|X\rangle\langle X|\mathbf{D}|I\rangle}{E_F - E_X} + \sum_X \frac{\langle F|\mathbf{D}|X\rangle\langle X|H_W|I\rangle}{E_I - E_X}, \qquad (3.7)$$

where X spans all N-body states with opposite parity to I and F, and where $\mathbf{D}$ is the dipole operator. While some experimental information is available for energies and dipole matrix elements, this expression also involves an unusual operator, H_W, on which we have no direct experimental handle.

The PNC electric-dipole amplitude $6S_{1/2} \rightarrow 7S_{1/2}$ has been observed in atomic cesium by groups in Paris [108] and Boulder. [109] Cesium has a nucleus consisting of 78 neutrons and 55 protons with nuclear spin $I = 7/2$. (The experiment is insensitive to other isotopes.) The total angular momentum of atomic S-states is therefore $F = 3$ or $F = 4$. Both of the transitions $6S(F = 4) \rightarrow 7S(F = 3)$ and $6S(F = 3) \rightarrow 7S(F = 4)$, have been measured, allowing the isolation of the nuclear-spin-dependent PNC effects (Sec. 3.4.6). The Cs experiments detect the interference between the PNC-E1 amplitude and a Stark-induced E1 amplitude, and yield the ratio of the PNC-E1 amplitude to the vector transition polarizability β_{pol} for the 6S-7S transition. The calculation of β_{pol} is discussed in Sec. 3.4.8. The structure of the Cs atom is the simplest of those in which PNC has been measured, since it can be described as a single valence electron outside a closed xenonlike core which is relatively unpolarizable.

The calculation that will be described in detail here leads from the Hamiltonian (3.6) to the prediction for the nuclear-spin-independent part of the PNC-E1 amplitude

$$E_{PNC} = -0.905(9) \times 10^{-11} i|e|a_0(-Q_W/N) . \qquad (3.8)$$

Here the unknown Q_W has been factored out, divided its approximate value $-N$. When this is compared with the experimental measurement

$$E_{PNC}^{exp} = -0.8252(184)[61] \times 10^{-11} i|e|a_0 , \qquad (3.9)$$

there results a determination of Q_W as

$$Q_W = -71.04(1.58)[0.88] \qquad (3.10)$$

where the first error is experimental and the second theoretical.

3.3   Relativistic MBPT in heavy neutral atoms

The PNC interactions discussed above are highly relativistic: they occur in the nucleus where the potential is many times the rest mass energy of an electron, and vanish in the nonrelativistic limit. The dominant PNC term, for instance, is proportional to γ_5, which couples large and small components of the Dirac wavefunction. The modern calculations described here use the "no-virtual-pair" approximation, with a relativistic finite basis set to restrict summations to positive-energy solutions to the Dirac equation, as described in earlier chapters.[†] The small Breit effects have been estimated by including the Breit interaction in the self-consistent Dirac-Fock equations (see Sec. 3.4.4).

[†] Cf. Eq. (1.19) and Sec. 2.3.

Table 3.1. Ionization energies (a.u.) for valence states of cesium calculated in second-order perturbation theory. The quantity ε_v^{HF} is the HF energy and $\varepsilon_v^{(2)}$ is the correction from second-order perturbation theory. See Ref. [117]

Orbital	ε_v^{HF}	$\varepsilon_v^{(2)}$	Theory	Expt.
$6s_{1/2}$	-0.12737	-0.01775	-0.14512	-0.14310
$6p_{1/2}$	-0.08562	-0.00691	-0.09253	-0.09217
$6p_{3/2}$	-0.08378	-0.00618	-0.08997	-0.08964

Some of the main issues of MBPT can be illustrated by considering the second-order many-body self-energy $\Sigma^{(2)}(\varepsilon)$, which in the Hartree-Fock (HF) potential is defined by its matrix elements

$$\langle i|\Sigma^{(2)}(\varepsilon)|j\rangle = -\sum_{abm}\frac{g_{abjm}\tilde{g}_{imab}}{\varepsilon_a+\varepsilon_b-\varepsilon_m-\varepsilon} + \sum_{amn}\frac{g_{aimn}\tilde{g}_{mnaj}}{\varepsilon_a+\varepsilon-\varepsilon_m-\varepsilon_n}. \qquad (3.11)$$

Here and subsequently, a, b, c, etc. denote core states, m, n, r, etc. denote excited states (which include the valence states), and i and j denote general states; $\tilde{g}_{abcd} = g_{abcd} - g_{bacd}$ denotes an antisymmetrized Coulomb matrix element, $g_{abcd} = \langle ab|1/r_{12}|cd\rangle$. ‡ The second-order ionization energy for a one-valence-electron system is given by $E^{(2)}(v) = \langle v|\Sigma^{(2)}(\varepsilon_v)|v\rangle$, where v is the valence electron state. Table 3.1 shows the effect of applying MBPT through second order for Cs ionization energies.

The lowest order results disagree with experiment at the 10% level, and the first order corrections vanish for the HF potential. The agreement is seen to improve substantially, however, with the inclusion of $E^{(2)}$, to the 1% level. Second-order energies such as these can be evaluated in a few minutes on modern workstations.

While calculations of energies are a useful monitor of the behavior of MBPT, we wish to accurately predict a parity violating transition amplitude. For this reason it is important to calculate standard parity-conserving amplitudes and compare them to experiment. We use hyperfine splittings and oscillator strengths for this purpose, since the former test the quality of the wavefunction at the origin, where the PNC interaction takes place, and the latter test the wavefunction at large distances and also enter directly in the expression for the PNC dipole matrix element. Here we illustrate the calculations with dipole transition amplitudes. They are determined by evaluating matrix elements of the dipole operator,

$$eZ_{op} = \sum_{ij}\langle i|ez|j\rangle a_i^\dagger a_j \qquad (3.12)$$

---

‡ Cf. with Eq. (2.37).

in perturbation theory. Working in the framework of the V^{N-1} Dirac-Fock potential, the first-order correction for a transition $v \to w$ is $d_{wv}^{(1)} = \langle w|ez|v\rangle$. The second-order correction is:

$$d_{wv}^{(2)} = \sum_{an} \frac{d_{an}^{(1)}\tilde{g}_{wnva}}{\varepsilon_v + \varepsilon_a - \varepsilon_n - \varepsilon_w} + \sum_{an} \frac{d_{na}^{(1)}\tilde{g}_{wavn}}{\varepsilon_w + \varepsilon_a - \varepsilon_n - \varepsilon_v}. \qquad (3.13)$$

While there are a large number of corrections from third order, two particularly important ones are given by

$$d_{wv}^{(3)}(\text{RPA}) = \sum_{abmn} \left[\frac{\tilde{g}_{wnva}d_{bm}^{(1)}\tilde{g}_{amnb}}{(\varepsilon_{mw} - \varepsilon_{bv})(\varepsilon_{nw} - \varepsilon_{av})} + \text{c.c.} \right] +$$
$$\sum_{abmn} \left[\frac{\tilde{g}_{mnab}d_{bm}^{(1)}\tilde{g}_{awvn}}{(\varepsilon_{nv} - \varepsilon_{aw})(\varepsilon_{mw} - \varepsilon_{bv})} + \text{c.c.} \right], \qquad (3.14)$$

and

$$d_{wv}^{(3)}(\text{BO}) = \sum_{abmi} \left[\frac{g_{abmv}d_{wi}^{(1)}\tilde{g}_{miba}}{(\varepsilon_i - \varepsilon_v)(\varepsilon_{mv} - \varepsilon_{ab})} + \text{c.c.} \right] +$$
$$\sum_{amni} \left[\frac{g_{aimn}d_{wi}^{(1)}\tilde{g}_{mnav}}{(\varepsilon_i - \varepsilon_v)(\varepsilon_{mn} - \varepsilon_{av})} + \text{c.c.} \right]. \qquad (3.15)$$

Here $\varepsilon_{ab} = \varepsilon_a + \varepsilon_b$, and "c.c." denotes the complex conjugate of the preceding term with v and w interchanged. The terms $d_{wv}^{(2)}$ and $d_{wv}^{(3)}(\text{RPA})$ are parts of the random-phase approximation (RPA), which describes the shielding effect of the core on an externally applied field. The RPA is our first example of an all-order method. The RPA amplitudes $t_{ij}(\omega)$ are obtained from the self-consistent equations:

$$t_{ij}(\omega) = d_{ij}^{(1)} + \sum_{bm} \frac{t_{mb}(\omega)\tilde{g}_{ibjm}}{\varepsilon_b - \varepsilon_m + \omega} + \sum_{bm} \frac{t_{bm}(\omega)\tilde{g}_{imjb}}{\varepsilon_b - \varepsilon_m - \omega}. \qquad (3.16)$$

These equations are first solved iteratively for the core-excited amplitudes $t_{an}(\omega)$ and $t_{na}(\omega)$; these amplitudes are then used to compute the RPA matrix element for the transition, $t_{wv}(\omega)$. The frequency ω takes its Dirac-Fock value in the simplest treatment, $\omega = \varepsilon_w - \varepsilon_v$. The contributions from $d_{wv}^{(1)}$, $d_{wv}^{(2)}$, and $d_{wv}^{(3)}(\text{RPA})$ are thereby picked up automatically together with the higher-order RPA terms. Equation (3.16) can be solved by converting it to a first-order differential equation that is solved by finite-difference techniques. Alternatively, it may be solved directly by use of a finite basis set.

The second contribution $d_{wv}^{(3)}(\text{BO})$ is relatively large because the energy denominator $\varepsilon_i - \varepsilon_v$ can be small when i is another valence state of similar energy. It can be rewritten as

$$d_{wv}^{(3)}(\text{BO}) = \langle \phi_w|ez|\delta\phi_v\rangle + \langle \delta\phi_w|ez|\phi_v\rangle, \qquad (3.17)$$

where $\delta\phi_{w,v}$ are lowest-order Brueckner orbital (BO) corrections to the HF valence states, defined by

$$\delta\phi_v = \sum_{i \neq v} \phi_i \frac{\langle i|\Sigma^{(2)}(\varepsilon_v)|v\rangle}{\varepsilon_v - \varepsilon_i}. \tag{3.18}$$

In this form we see clearly the role of the many-body self-energy operator as the formal representation of the polarization potential. Below we will see how the BO effect can be included to higher order in a simple way.

Also entering in the third-order matrix element are "structural radiation" terms, in which the dipole operator appears inside the self-energy unit (rather than to one side of a self-energy unit, as with the BO corrections). These terms are quite small for alkali systems, because they involve core excitations. To complete the third-order matrix element one must add small terms that correct the normalization of the many-body wavefunction as correlation effects are added. For a detailed discussion of these smaller third-order terms see Refs. [118, 119, 120].

At this point we will examine the behavior of MBPT for matrix elements in Cs when $d_{wv}^{(3)}$(BO) and all orders of RPA are included. In Table 3.2, we present these terms for hyperfine splittings and oscillator strengths. In this case, the lowest order results differ from experiment by up to 40%. However, both the RPA and BO corrections are substantial, and bring theory and experiment into agreement at the few-percent level. We can take as an informal error estimate the largest remaining discrepancy, the 7s hyperfine constant, which disagrees with experiment by 5%. As we shall see in Sec. 3.4.2, it is possible to carry out an analogous calculation for the PNC transition. When this is done, [121] one finds

$$E_{\text{PNC}} = -0.95(5) \times 10^{-11} i|e|a_0(-Q_{\text{W}}/N). \tag{3.19}$$

This result will change by 4% when higher-order terms are included, so in this case the crude error estimate is seen to be reliable.

The results of Table 3.2 indicate that while the frozen-core Hartree Fock potential gives relatively inaccurate results, MBPT through third order improves agreement with experiment to the few-percent level. Unfortunately, to be useful in precision tests of the standard model, this accuracy must be improved to the order of 1%. One possible approach is to include the next order of MBPT. However, when this is done for energies, it is found that the predictions actually worsen. For example, the 6s removal energy starts off 11% too low, becomes 1.4% too high after second order is included, but then becomes 2.5% too low when the third order is calculated. The reason for this is the very large second-order correction: this correction can be considered to be the first term of an infinite set, the next element of which enters in fourth order. In fact, when that fourth order term is added, agreement with experiment improves to

Table 3.2. Convergence of MBPT for ordinary matrix elements of cesium. A designates the hyperfine splitting in MHz, and D represents the $6p_{3/2} - 6s_{1/2}$ reduced dipole matrix element in units $|e|a_0$. See Ref. [117]

Order	$A\left(6s_{\frac{1}{2}}\right)$	$A\left(7s_{\frac{1}{2}}\right)$	D
(1)	5740	1576	-7.426
(2) RPA	1156	320	0.413
(3) BO	2568	392	0.842
Total	9464	2288	-6.171
Expt	9192	2184	-6.32(8)

the few tenths of a percent level. [122] It is, however, a somewhat dangerous procedure to include selected diagrams from higher order, since it is always possible that neglected diagrams may be larger than expected. It is clearly desirable to find an efficient way to include as many higher order diagrams as possible. Rather than going to the next higher order of MBPT, an alternative approach is to use all-orders methods, similar to the RPA but including a much wider class of higher-order MBPT diagrams. If chosen properly, these approaches can automatically pick up the first few orders of perturbation theory completely, at the same time accounting for infinite orders of diagrams built up as iterations of the dominant low-order diagrams.

Let us first consider the BO corrections. The valence-removal energy of a one-valence-electron system is given exactly as the eigenvalue of the quasiparticle equation: [123]

$$(h + V_{\mathrm{HF}} + \Sigma(\varepsilon_v))\psi_v = \varepsilon_v \psi_v . \tag{3.20}$$

Unfortunately, the exact proper self-energy operator $\Sigma(\varepsilon)$ here is rather hard to construct, but it is quite feasible to solve (3.20) for the second-order self-energy, $\Sigma(\varepsilon) = \Sigma^{(2)}(\varepsilon)$. When this is done, there results an infinite sum of second-order, fourth-order, sixth-order, etc. contributions to the energy, of which the fourth-order component contributes at the several percent level for heavy atoms, as mentioned above.

The quasiparticle wavefunctions ψ_v are also useful, since they contain the all-order modification of the HF wavefunction by the self-energy operator. They can be used to modify the external legs of any matrix-element diagram. For example, the first-order matrix element becomes modified to $-\langle\psi_w|ez|\psi_v\rangle$. This expression includes the first-order matrix element, the term $d^{(3)}_{wv}(\mathrm{BO})$ of Table 3.2, and higher-order BO effects.

A more complete set of terms including both BO and RPA effects together can be evaluated by replacing w and v in the RPA matrix element $t_{wv}(\omega)$ by the quasiparticle wavefunctions ψ_w and ψ_v. The frequency ω should then properly be the difference of quasiparticle eigenvalues. Schematically, we write this approximation as:

$$d_{wv}(\text{BO-RPA}) = \langle \psi_w | t(\omega) | \psi_v \rangle . \qquad (3.21)$$

It contains all the effects in Table 3.2, plus higher-order BO terms, and cross terms between the BO and RPA effects. A study of this approximation in Cs for the case $\Sigma(\varepsilon) = \Sigma^{(2)}(\varepsilon)$ is given in Ref. [120]. The agreement shown in Table 3.2 is worsened somewhat: the $6s$ and $6p_{\frac{1}{2}}$ hyperfine constants are about 7% greater than experiment, while the $6p_{\frac{1}{2}}$-$6s$ dipole amplitude is about 4% less than experiment.

The reason for this substantial disagreement is the use of the second-order self-energy. We have already seen that the third-order self-energy is large in Cs from the estimate of the third-order ionization energy quoted earlier, in which $E^{(3)}$ was -4.0% of experiment, while $E^{(2)}$ was 12.5%. A simple phenomenological way of estimating the effect of the higher-order self-energy on matrix elements is to scale the second-order self-energy, $\Sigma(\varepsilon) \rightarrow \lambda\Sigma^{(2)}(\varepsilon)$, with λ chosen to reproduce the experimental ionization energies. The scaling retains the correct physical $1/r^4$ dependence of the self-energy in the important large-r region. The $6s$, $7s$, $6p_{\frac{1}{2}}$, and $7p_{\frac{1}{2}}$ ionization energies are fitted very well by choosing $\lambda = 0.80$ for s states, and $\lambda = 0.84$ for $p_{\frac{1}{2}}$ states, showing that the second-order self-energy overestimates the exact self-energy "by about 20%." The significant overestimate provided by the second-order self-energy is typical for heavy elements. In Ref. [120], hyperfine constants and dipole amplitudes for Cs are plotted as a function of a continuous scaling λ. All properties improve systematically, and agree with experiment at the 1%–2% level when λ has its energy-fitted values.

It is therefore desirable to find an *ab initio* method for calculating $\Sigma(\varepsilon)$ to higher order. The group from Novosibirsk have used a Green's function technique [124] to add to the self-energy all-order terms corresponding to the "screening of the Coulomb interaction". These are the same terms that are known to be important in the correlation energy of the infinite electron gas (for example, see Ref. [125]). Using the approach for the matrix element in Eq. (3.21), they found agreement with experiment at the 1% level for matrix elements in Cs. This approximation works well for Cs because of the dominance of two of the fourteen Feynman diagrams that describe the third-order self-energy. [122] Ref. [122] also shows that this dominance is not as marked for Tl, so that an approach that summed a larger class of diagrams would perhaps be desirable.

Table 3.3. Valence removal energies for cesium from an all-order calculation. Units: a.u. See Ref. [126]

State	ε_v	$\delta\varepsilon_v$	Sum	Experiment
$6s_{\frac{1}{2}}$	0.12737	0.01521	0.14257	0.14310
$6p_{\frac{1}{2}}$	0.08562	0.00636	0.09198	0.09217
$6p_{\frac{3}{2}}$	0.08379	0.00572	0.08951	0.08964
$7s_{\frac{1}{2}}$	0.05519	0.00326	0.05845	0.05865
$7p_{\frac{1}{2}}$	0.04202	0.00183	0.04385	0.04393
$7p_{\frac{3}{2}}$	0.04137	0.00166	0.04303	0.04310

Such an approach can be constructed by using the all-orders techniques discussed in previous chapters, which culminate in the coupled-cluster (CC) approach.[§] Workable relativistic procedures of this general type have now been implemented by several groups; [93, 126, 30, 81] we consider here in some detail that of Ref. [126]. There an all-orders equation up to double excitations was constructed through the approximation of dropping the nonlinear terms in the CC approach for singles and doubles (see Ref. [29] for an extended discussion of CC methods for atoms). These terms were dropped principally because of the expense of their evaluation, but also because simple estimates suggest that they enter at or below the 1% level aimed for at this stage. However, a special set of terms that are related to triple excitations in the standard CC approach had to be added to enable the method to pick up the third-order ionization energy completely. The final approach, then, was complete through third-order for ionization energies, and contained subsets of fourth- and higher-order terms which were iterations of the structures in second and third order. Estimates of omitted fourth-order triples, and fourth-order nonlinear CC terms, suggest that they enter at the 1% level or less. Indeed, an explicit comparison of energies with experiment in Table 3.3 shows the method to be accurate to about 0.5% or better. Similar results are found for other alkali-metal systems such as K and Na, while for Li energies are accurate to the 0.1% level (10 μhartree). [127]

Matrix elements were evaluated in these all-order calculations through a generalization of Eq. (3.21), in which the all-order information about the wavefunction was combined consistently with the RPA. Some results of this method for Cs are shown in Table 3.4. In general, hyperfine constants

---

[§] See Sec. (1.7).

Table 3.4. Hyperfine constants (MHz) for ^{133}Cs with $I = 7/2$ and $g_I = 0.7377208$ from an all-order calculation. See Ref. [126]

State	A^{HF}	δA	Sum	Experiment
$6s_{\frac{1}{2}}$	1426.81	864.19	2291.00	2298.16
$7s_{\frac{1}{2}}$	392.05	151.99	544.04	545.90(9)
$6p_{\frac{1}{2}}$	161.09	131.58	292.67	291.90(13)
$7p_{\frac{1}{2}}$	57.68	35.53	94.21	94.35(4)
$6p_{\frac{3}{2}}$	23.944	25.841	49.785	50.275(3)
$7p_{\frac{3}{2}}$	8.650	7.605	16.255	16.605(6)

for $j = \frac{1}{2}$ states in Cs were accurate to the 1% level, and dipole matrix elements were slightly more accurate than this. Comparable accuracies to these can also be obtained for Tl with the same methods. [128]

A word is in order about the computational requirements for these relativistic all-order calculations in heavy atoms. For s and p states, a basis set with 25 positive-energy states was included; for d and higher states, 20 basis functions were used. All states with an angular momentum up to $l = 7$ (*i.e.* k states) were included, and all core states were allowed to be excited. The latter consideration is important, since excitations of the $1s$ shell make contributions of order 1% to hyperfine constants and matrix elements of the PNC Hamiltonian (although these enter primarily through RPA-like excitations). This then involved 16000 double-excitation channels, where a channel is defined to be an excitation $ab \to \kappa_n \kappa_m(L)$, L being the multipolarity of the radial excitation amplitude, and κ_n and κ_m the Dirac angular quantum numbers of the excited states n and m. Each channel is described by a 25×25 matrix of excitation coefficients, or by a 20×20 matrix for orbital angular momenta $\ell \geq 2$. The coefficients required about 80 megabytes of storage, and could be performed in core memory on a Cray-2 supercomputer. About eight hours of CPU time were required to converge the core-core excitation coefficients to the accuracy required, and about two hours were required for the core-valence channels per valence state. The method is quite close to a complete configuration-interaction (CI) calculation truncated at double excitations and solved iteratively, except that spurious unlinked terms have been carefully removed by the construction of the all-orders equations. [29]

The point is that in fact these computational requirements are not excessive. The large number of channels was used principally with a view to converging the doubles calculation to below the discrepancy with experiment, so that the residual discrepancy could be traced to perturbation terms omitted in the formalism and not to numerical effects. There is therefore hope that an extension of these methods to include the omitted fourth-order terms is feasible, perhaps involving a truncated basis set and a reduced set of allowed channels for the new terms. If this is indeed feasible, one could hope that *ab initio* calculations could approach the 0.1% level of precision (at which stage one might need to include QED effects).

3.4 PNC calculations

We now turn our attention to the calculation of the PNC-E1 amplitude. This is much harder than the simple matrix elements discussed above because it involves a crossed second-order term between two perturbations, H_W and the dipole operator **D**, and a sum over virtual many-body states X, as in Eq. (3.7). One can attempt to calculate E_{PNC} by saturating this sum, but this is difficult in general because in a complex open-shell atom many states X are involved. The special simplicity of Cs makes this an attractive approach, however, and we discuss it in Sec. 3.4.3. An alternative approach is to use mixed-parity MBPT. This method has been applied in some form to all transitions of experimental interest. We discuss it first, using the very complete calculations for the Cs transition as illustration. Our discussion follows mainly Refs. [129, 130, 120]. For more references to earlier calculations, and more details, the reader is also referred to the review article. [131]

3.4.1 Mixed-parity HF

In mixed-parity MBPT, one modifies the single-particle states by adding the weak-interaction Hamiltonian h_W to the HF potential. This approach leads to a generalization of the single-particle states in which each state acquires an opposite-parity admixture,

$$\phi_k \rightarrow \phi_k + \tilde{\phi}_k . \tag{3.22}$$

Thus, for example, each $s_{\frac{1}{2}}$ orbital will pick up a small $p_{\frac{1}{2}}$ state admixture, and vice-versa. Since h_W is a pseudoscalar, the states remain eigenstates of angular momentum, however. Treating the weak interaction to lowest order (which is certainly justified), the induced correction $\tilde{\phi}_v$ to the Dirac-Fock state ϕ_v satisfies the equation

$$(h + V_{HF} - \varepsilon_v)\tilde{\phi}_v = -h_W\phi_v , \tag{3.23}$$

in a first approximation. This equation may be solved for the $6s$ and $7s$ perturbed orbitals by either finite-difference techniques or basis-set techniques. In the latter case one evaluates:

$$\tilde{\phi}_v = \sum_{i \neq v} \phi_i \frac{\langle i | h_W | v \rangle}{\varepsilon_v - \varepsilon_i} . \tag{3.24}$$

The sum over i here is an example of a sum that can be properly extended to include negative-energy states, and indeed should include negative-energy states if the velocity form of the dipole operator is to be used. This is because the negative-energy states (virtual e^+–e^- pairs) make an enhanced contribution of the order of several percent when used with the velocity form, although their contribution is negligible if the length form ez of the dipole operator is used. [132] For the expression (3.24) to be accurate, the basis-set should represent small distances well; in the B-spline method discussed in Sec. 2.6, for example, one should place many knots inside the nucleus. One obtains in this first approximation:

$$\begin{aligned} E_{\mathrm{PNC}} &= \langle \phi_{7s} | ez | \tilde{\phi}_{6s} \rangle + \langle \tilde{\phi}_{7s} | ez | \phi_{6s} \rangle \\ &= -0.740 \times 10^{-11} i |e| a_0 (-Q_W/N) . \end{aligned} \tag{3.25}$$

The approximation made in Eq. (3.23) ignores the dependence of V_{HF} on the core orbitals, which themselves acquire small opposite-parity admixtures. If we take this dependence into account, then the HF potential is also modified, $V_{\mathrm{HF}} \rightarrow V_{\mathrm{HF}} + \tilde{V}_{\mathrm{HF}}$, where

$$\langle i | \tilde{V}_{\mathrm{HF}} | j \rangle = \sum_b (g_{i\tilde{b}jb} - g_{\tilde{b}ijb}) + (g_{ib j\tilde{b}} - g_{bij\tilde{b}}) . \tag{3.26}$$

Equation (3.23) is then replaced by the mixed-parity Hartree-Fock (PNC-HF) equations:

$$(h + V_{\mathrm{HF}} - \varepsilon_a) \tilde{\phi}_a = -(h_W + \tilde{V}_{\mathrm{HF}}) \phi_a , \tag{3.27}$$

$$(h + V_{\mathrm{HF}} - \varepsilon_v) \tilde{\phi}_v = -(h_W + \tilde{V}_{\mathrm{HF}}) \phi_v . \tag{3.28}$$

One solves Eqs. (3.27) self-consistently to obtain the perturbation to each core orbital $\tilde{\phi}_a$ and the perturbed HF potential $\tilde{V}_{\mathrm{HF}}$. Equation (3.28) is then solved to obtain the perturbed valence orbitals $\tilde{\phi}_v$. The corrections to the results of (3.25) obtained in this way are referred to as weak RPA corrections, since they can be shown to be equivalent to replacing the bare h_W with the RPA form given by Eq. (3.16) with $\omega = 0$. One obtains in this approximation: [133, 134, 129, 130, 135]

$$E_{\mathrm{PNC}}^{\mathrm{HF}} = -0.927 \times 10^{-11} i |e| a_0 (-Q_W/N) . \tag{3.29}$$

This value differs by only 2%–3% from the final value in Eq. (3.8), showing that the dominant many-body effects are the weak RPA terms already built in to the PNC-HF equations.

In practice one solves radial forms of Eqs. (3.27) and (3.28), which are closely related to the radial HF equation. A mixed-parity single-particle state in a spherically symmetric potential has the form:

$$\phi_{n\kappa m}(\mathbf{r}) + \tilde{\phi}_{n\kappa m}(\mathbf{r}) = \frac{1}{r}\begin{pmatrix} iP_{n\kappa}(r)\Omega_{\kappa m}(\hat{\mathbf{r}}) \\ Q_{n\kappa}(r)\Omega_{-\kappa m}(\hat{\mathbf{r}}) \end{pmatrix} + i\frac{1}{r}\begin{pmatrix} i\tilde{P}_{n\kappa}(r)\Omega_{-\kappa m}(\hat{\mathbf{r}}) \\ \tilde{Q}_{n\kappa}(r)\Omega_{\kappa m}(\hat{\mathbf{r}}) \end{pmatrix}, \quad (3.30)$$

where $\Omega_{\kappa m}(\hat{\mathbf{r}})$ is the spherical spinor discussed in Chapter 2.[¶] The radial functions are real, the phase factor i for the weakly perturbed part having been factored explicitly. The solution is an eigenstate of total angular momentum $\mathbf{j}$, but the weakly perturbed part has opposite parity because of the change in sign of κ. Introducing the two-component radial functions,

$$R_a(r) = \begin{pmatrix} P_{n_a\kappa_a}(r) \\ Q_{n_a\kappa_a}(r) \end{pmatrix}, \quad \tilde{R}_a(r) = \begin{pmatrix} \tilde{P}_{n_a\kappa_a}(r) \\ \tilde{Q}_{n_a\kappa_a}(r) \end{pmatrix}, \quad (3.31)$$

and following the same radial-reduction procedure as for the relativistic HF equations, one finds that (3.27) becomes:

$$(h_{-\kappa_a} + (V_{\mathrm{HF}})_{-\kappa_a} - \varepsilon_a)\tilde{R}_a(r) = -(h_{\mathrm{W}} + (\tilde{V}_{\mathrm{HF}})_{\kappa_a})R_a(r). \quad (3.32)$$

The radial Dirac Hamiltonian for a nuclear potential $V_{\mathrm{nuc}}(r)$ satisfies:

$$h_\kappa R(r) = \begin{pmatrix} V_n(r) & \hbar c\left(\dfrac{d}{dr} - \dfrac{\kappa}{r}\right) \\ -\hbar c\left(\dfrac{d}{dr} + \dfrac{\kappa}{r}\right) & V_{\mathrm{nuc}}(r) - 2mc^2 \end{pmatrix} R(r), \quad (3.33)$$

while the radial weak Hamiltonian is:

$$h_{\mathrm{W}} R(r) = \frac{G_{\mathrm{F}}}{2\sqrt{2}} Q_{\mathrm{W}}\rho_{\mathrm{nuc}}(r)\begin{pmatrix} 0 & -1 \\ 1 & 0 \end{pmatrix} R(r). \quad (3.34)$$

The radial relativistic HF potential $(V_{\mathrm{HF}})_\kappa$ has the form given in Eq. (2.26). Since the weakly perturbed orbitals are eigenstates of $\mathbf{j}$, $(\tilde{V}_{\mathrm{HF}})_\kappa$ has exactly the same j-dependent angular factors as $(V_{\mathrm{HF}})_\mathbf{j}$. However, the κ quantum numbers referring to the weakly perturbed core orbitals $\tilde{b}$ in Eq. (3.26) will have reversed sign, $-\kappa_b$. This has the immediate consequence that the direct parts of $(\tilde{V}_{\mathrm{HF}})_\kappa$ vanish, since they are proportional to $\langle \kappa_b \| C_0 \| - \kappa_b \rangle = 0$. One obtains finally:

$$(\tilde{V}_{\mathrm{HF}})_{\kappa_a} R_a(r) = -\frac{1}{[j_a]}\sum_{n_b\kappa_b,L}\Big[-\langle -\kappa_b \| C_L \| \kappa_a \rangle^2 v_L(\tilde{b},a;r)R_b(r)$$
$$+ \langle \kappa_b \| C_L \| \kappa_a \rangle^2 v_L(b,a;r)\tilde{R}_b(r)\Big], \quad (3.35)$$

where $[j_a] \equiv 2j_a + 1$, and the function v_L is defined in Eq. (2.18). Note the minus sign in front of the first term in Eq. (3.35), which originates from

---

[¶] Cf. Eq. (2.7).

the phase factor i for the weakly perturbed orbitals coupled with the fact that $\tilde{b}$ here appears in the bra position. The self-consistent solution of these radial equations by finite-difference techniques is now routine, and the value (3.29) is known to high numerical accuracy. In the next section, we will see how one can improve on the PNC-HF value systematically by generalizing the MBPT approach for a matrix element discussed in the previous section.

3.4.2 Mixed-parity MBPT

A convenient and systematic way to go beyond the PNC-HF result is to follow the MBPT procedure described in Sec. 3.3, but using mixed-parity single-particle states in place of the usual pure-parity states. The mixed-parity HF potential $V_{\mathrm{HF}} + \tilde{V}_{\mathrm{HF}}$ defines a complete set of mixed-parity single-particle states. Each state $|i\rangle + |\tilde{i}\rangle$ in this set has the usual HF energy ε_i because an odd-parity operator does not change energies in first order. For calculational purposes, it is convenient to construct a mixed-parity *finite* basis set. To do this, one first constructs a HF basis set $|i\rangle$ in a standard way, for example as described in Sec. 2.6, and then calculates the opposite-parity admixture $|\tilde{i}\rangle$ for each state by summation over the formal solution to Eq. (3.28):

$$|\tilde{i}\rangle = \sum_j \frac{|j\rangle\langle j|h_{\mathrm{W}} + \tilde{V}_{\mathrm{HF}}|i\rangle}{\varepsilon_i - \varepsilon_j} . \qquad (3.36)$$

It is assumed here that the potential $\tilde{V}_{\mathrm{HF}}$ has been constructed beforehand by a separate self-consistent solution of (3.27) for the weakly perturbed core states.

To illustrate the use of the mixed-parity finite basis set, let us consider the second-order matrix-element expression given in Eq. (3.13). Each state in this expression is now regarded as being of mixed parity, and the expression is linearized in the weak interaction. In practice, this linearization means replacing each of the states w, v, a, and n one at a time by its opposite-parity admixture, taking the appearances of a and n in the bra and ket positions separately. One thereby generates six terms in total.

As discussed in Sec. 3.3, one can generalize the second-order matrix element by using the all-order RPA amplitude $t_{wv}(\omega)$. In that case the RPA equations (3.16) must be evaluated with parity-mixed orbitals and linearized in the weak interaction, giving a rather complicated set of self-consistent equations. The RPA amplitude becomes modified, $t_{na}(\omega) \to t_{na}(\omega) + \tilde{t}_{na}(\omega)$, and the PNC-E1 matrix element becomes in the RPA approximation,

$$E_{PNC}(RPA) = \langle w|t(\omega)|\tilde{v}\rangle + \langle \tilde{w}|t(\omega)|v\rangle + \langle w|\tilde{t}(\omega)|v\rangle . \qquad (3.37)$$

Evaluating the first two terms of (3.37), and subtracting the implicitly contained PNC-HF result (3.29), one obtains the correction [129, 135, 120]

$$\delta E_{PNC}(RPA\text{-ext}) = 0.035 \times 10^{-11} i|e|a_0(-Q_W/N) . \qquad (3.38)$$

The third term in (3.37) evaluates to

$$\delta E_{PNC}(RPA\text{-int}) = 0.002 \times 10^{-11} i|e|a_0(-Q_W/N) . \qquad (3.39)$$

The first two terms in (3.37) are called *external* terms, since the PNC substitution is made on the external valence legs of the many-body diagram. The third term involves an *internal* substitution, corresponding to substitutions inside the bubbles of the RPA diagrams.

We here have our first encounter with an important simplifying result, that internal substitutions are small in alkali-like systems. This result has its origins in the tightly bound core of the alkali-metals: the internal substitutions involve core excitations by h_W, while the external substitutions involve valence excitations with a rather smaller associated energy denominator. The suppression of internal substitutions can be less pronounced in other systems. The Tl atom, for example, can be regarded as a one-valence-electron system with a valence $6p_{\frac{1}{2}}$ electron outside a closed $6s^2$ core. However, the internal terms are not then strongly suppressed because of the low excitation energy of the $6s^2$ shell and its strong interaction with the valence states.

Before solving the RPA equations, one must make a radial reduction. As we saw with the PNC-HF equations, the radial form for an MBPT term involving an opposite-parity substitution can be written down by inspection from the radial expression for the parity conserving case. Since the weakly perturbed states are eigenstates of **j**, the j-dependent angular factors remain unchanged, but the sign of the κ quantum number referring to a PNC substitution must reversed. Whenever a PNC substitution is made in a bra position, one must also remember to insert a minus sign to account for the phase factor i in (3.30). In this way one can write down the radial forms for the PNC-RPA equations directly from the radial RPA equations; they are written out explicitly in, for example, Ref. [134]. A useful overall check on numerics is provided by the equivalence of the length ($\mathbf{d}_L = e\mathbf{r}$) and velocity ($\mathbf{d}_V = -ic e\alpha/\omega$) forms of the dipole operator in the RPA approximation. The results (3.38) and (3.39) are known with high numerical accuracy.

The dominant third-order correction to the PNC amplitude arises from the BO terms. If we linearize Eq. (3.17) in the weak interaction, we obtain

$$
\begin{aligned}
\delta E_{\mathrm{PNC}}(\mathrm{BO}) &= \langle \delta\phi_{7s}|ez|\tilde{\phi}_{6s}\rangle + \langle \delta\tilde{\phi}_{7s}|ez|\phi_{6s}\rangle \\
&\quad + \langle \tilde{\phi}_{7s}|ez|\delta\phi_{6s}\rangle + \langle \phi_{7s}|ez|\delta\tilde{\phi}_{6s}\rangle \,,
\end{aligned} \tag{3.40}
$$

where $\delta\tilde{\phi}_v$ is the weak perturbation to the BO correction for the orbital ϕ_v. Each $\delta\tilde{\phi}_v$ consists of two parts, an external part arising from weak corrections to the valence orbital v and the state i in Eq. (3.18), and an internal part arising from weak corrections to $\Sigma^{(2)}(\varepsilon_v)$ itself,

$$
\delta\tilde{\phi}_v = \delta\tilde{\phi}_v^{\mathrm{ext}} + \delta\tilde{\phi}_v^{\mathrm{int}}\,. \tag{3.41}
$$

The external part leads to the dominant correction [121, 129, 135]

$$
\delta E_{\mathrm{PNC}}^{\mathrm{BO-ext}} = -0.058 \times 10^{-11} i|e|a_0(-Q_{\mathrm{W}}/N)\,, \tag{3.42}
$$

while the internal part gives the much smaller contribution

$$
\delta E_{\mathrm{PNC}}^{\mathrm{BO-int}} = -0.003 \times 10^{-11} i|e|a_0(-Q_{\mathrm{W}}/N)\,. \tag{3.43}
$$

To complete the third-order calculation, one must evaluate the small remaining contributions from "structural-radiation" and normalization. The former has the value, [130]

$$
\delta E_{\mathrm{PNC}}(\mathrm{StRad}) = -0.004 \times 10^{-11} i|e|a_0(-Q_{\mathrm{W}}/N)\,. \tag{3.44}
$$

This contribution is suppressed for alkali-metal atoms by energy-denominator considerations, much the same as the suppression of internal substitutions. Only the external valence PNC subsitutions have been made in evaluating (3.44). Normalization effects also enter in third order and give a correction [130]

$$
\delta E_{\mathrm{PNC}}(\mathrm{Norm}) = 0.008 \times 10^{-11} i|e|a_0(-Q_{\mathrm{W}}/N)\,. \tag{3.45}
$$

The results in Eqs. (3.44) and (3.45) both agree well with those of Ref. [129].

Now we turn to the evaluation of fourth- and higher-order corrections. A large correction arises when the approximate Brueckner orbitals obtained by solving Eq. (3.18) for $\delta\phi_v$ are replaced by the *chained* Brueckner orbitals determined by solving the second-order quasiparticle equation

$$
(h + V_{\mathrm{HF}} + \Sigma^{(2)}(\varepsilon_v))\psi_v = \varepsilon_v\psi_v \tag{3.46}
$$

exactly. Eq. (3.18) corresponds to a perturbative treatment of the above equation. Solving (3.46) to all orders in $\Sigma^{(2)}$ reduces the external BO corrections in Eq. (3.42) by a factor close to 2, leading to the value [130]

$$
\delta E_{\mathrm{PNC}}^{\mathrm{ChBO-ext}} = -0.029(9) \times 10^{-11} i|e|a_0(-Q_{\mathrm{W}}/N)\,. \tag{3.47}
$$

The error in this term was estimated by various methods. [120] In one, the self-energy was scaled, $\Sigma^{(2)}(\varepsilon_v) \to \lambda\Sigma^{(2)}(\varepsilon_v)$, with λ chosen to reproduce experimental removal energies, as described in Sec. 3.3. Although the values of $\lambda \sim 0.8$ thus obtained are quite different from unity, E_{PNC} was found to vary by only a few tenths of a percent. This behavior, which is in marked contrast to the behavior for hyperfine constants (Sec. 3.3), could be traced to accidental cancellations between numerator and denominator contributions in the implicit sum over states (see next section) in the present approach. The cancellations are fortunate, because they imply that corrections to the many-body self-energy beyond second order play a reduced role in the PNC effect.

We next replace the valence HF orbitals by chained Brueckner orbitals in an RPA calculation of the transition amplitude, accounting for a set of fourth-order corrections that correspond to core shielding of the Brueckner orbital corrections. At this stage we are using the mixed-parity equivalent of Eq. (3.21). One obtains a further modification of the amplitude [130]

$$\delta E_{PNC}^{RPA \times BO} = 0.014 \times 10^{-11} i|e|a_0(-Q_W/N) . \tag{3.48}$$

Putting together these different effects then gives a final prediction for the parity-mixed calculation [120]

$$\delta E_{PNC} = -0.904(9) \times 10^{-11} i|e|a_0(-Q_W/N) . \tag{3.49}$$

The uncalculated fourth-order terms are higher-order generalizations of the small structural radiation (3.44) and normalization (3.45) terms, or have already been included in error estimates such as the scaling of $\Sigma^{(2)}$. The result (3.49) is in good agreement with other calculations, as discussed in Ref. [120].

Dzuba *et al.* [129] have used a self-energy including higher-order terms corresponding to the shielding of the Coulomb interaction, giving results close to those in Eqs. (3.47)–(3.49). The final value is also close to that of Ref. [136], in which a semi-empirical polarization potential was used.

Because of the possible importance of atomic measurements of PNC, it is desirable to have as many different ways to calculate the PNC-E1 amplitude as possible. We now turn to a second method that can be used to calculate this amplitude, following Ref. [130].

3.4.3 *Sum-over-states approach for PNC amplitude*

An alternative and very direct way of calculating the PNC transition amplitude in Cs is to saturate the sum over exact many-body states X in

$$E_{PNC} = \sum_X \frac{\langle 7S|H_W|X\rangle\langle X|\mathbf{D}|6S\rangle}{E_{7S} - E_X} + \sum_X \frac{\langle 7S|\mathbf{D}|X\rangle\langle X|H_W|6S\rangle}{E_{6S} - E_X} . \tag{3.50}$$

One finds that about 98% of this sum is contributed by the states $X = |6P_{\frac{1}{2}}\rangle$ to $|9P_{\frac{1}{2}}\rangle$, and the remaining 2% by the states with $n \geq 10$, and by states in which X involves excitation of the core, which are autoionizing states. If the exact wavefunctions X are replaced by the corresponding lowest-order Slater determinants, this expression can be shown to reproduce the result (3.25). Here, however, we evaluate the contributions from the states with valence principle quantum numbers $n = 6 \cdots 9$ using our all-order wave functions and matrix element formalism; the remaining contributions are estimated using perturbation theory. We find: [120]

1. For $n = 6 - 9$

$$E_{\text{PNC}} = -0.893(7) \times 10^{-11} i |e| a_0 (-Q_{\text{W}}/N) \,,$$

2. For $n = 10 - \infty$

$$\delta E_{\text{PNC}} = -0.018(5) \times 10^{-11} i |e| a_0 (-Q_{\text{W}}/N) \,,$$

3. For autoionizing states

$$\delta E_{\text{PNC}} = 0.002(2) \times 10^{-11} i |e| a_0 (-Q_{\text{W}}/N) \,.$$

By using available experimental energies and dipole matrix elements, and by introducing scalings to account phenomenologically for terms omitted in the all-orders procedure, a set of values for the $n = 6$–9 contribution was determined whose scatter was used to define the error in that term. [120] The smaller errors in the other two terms were estimated by considering higher-order contributions. Adding these three contributions, we obtain

$$E_{\text{PNC}} = -0.909(9) \times 10^{-11} i |e| a_0 (-Q_{\text{W}}/N) \,. \qquad (3.51)$$

This result is consistent with the mixed-parity determination. The final result Eq. (3.8) is an average of the two methods taken together with the effect of the Breit interaction. We now turn to a discussion of that and other small PNC effects.

3.4.4   Breit interaction

The Breit interaction is taken into account by replacing the Coulomb interaction by the sum of the Coulomb and Breit interactions,[||]

$$g_{ijkl} \rightarrow g_{ijkl} + b_{ijkl} \,. \qquad (3.52)$$

With this replacement, the HF equations for the single-particle orbitals become

$$(h + V_{\text{HF}} + B_{\text{HF}}) \phi_k = \varepsilon_k \phi_k \,. \qquad (3.53)$$

[||] See Sec. 2.3.2.

Since the dominant contribution to the PNC amplitude is the PNC-HF contribution, it is sufficient to carry out a PNC-HF calculation including the Breit interaction in addition to the Coulomb interaction in order to evaluate the Breit correction. For this purpose, we solve the equations:

$$(h + V_{HF} + B_{HF} - \varepsilon_k)\tilde{\phi}_k = -(h_W + \tilde{V}_{HF})\phi_k , \qquad (3.54)$$

and use the resulting perturbed orbitals to evaluate the PNC amplitude as described in the section on mixed-parity calculations. This calculation leads to a 0.2% correction which has been included in Eq. (3.8).

3.4.5 Nuclear density

As mentioned in the introduction, the function $\rho_{nuo}(r)$ in the PNC Hamiltonian is a nuclear density function, close to the neutron density. Since there are no experimental values for the neutron density of ^{133}Cs, one can use instead an experimental proton density function. This proton density is taken to be a Fermi distribution

$$\rho_Z(r) = \frac{\rho_0}{1 + e^{-(r-c)/a}} , \qquad (3.55)$$

with parameters $a = 0.523$ fm and $c = 5.674(1)$ fm determined from muonic x-ray measurements. [90] The lowest-order PNC amplitude calculated using this distribution instead of the neutron density is

$$E_{PNC} = -0.7396 \times 10^{-11} i|e|a_0(-Q_W/N) . \qquad (3.56)$$

In the absence of an experimental neutron density, one can also use the theoretical neutron distribution function from a calculation that reproduces the experimental charge radius [137]

$$\rho_N(r) = \frac{\rho_0'}{(1 + e^{-(r-c')/a'})^{b'}} , \qquad (3.57)$$

with $a' = 0.6842$ fm, $b' = 1.589$, and $c' = 6.153$ fm. Calculating the lowest-order PNC amplitude with this distribution gives

$$E_{PNC} = -0.7390 \times 10^{-11} i|e|a_0(-Q_W/N) , \qquad (3.58)$$

a difference of only -0.08% from the value determined using the experimental proton distribution. At the 1% level of precision of interest at the present stage, we can obviously ignore the uncertainty in E_{PNC} caused by the lack of a precise understanding of the nuclear matter distribution. The uncertainty, however, does play a role when different isotopes are considered. The suggestion has been made to measure PNC in different isotopes, and to obtain information relatively free of electronic structure uncertainties by taking ratios. However, although the electronic structure is certainly almost unchanged, the neutron distribution in different nuclei

is more uncertain, and taking the ratio enhances the nuclear physics uncertainty. This issue has been addressed recently by Wilets *et al.*, [138] who
find significant effects when different nuclear models are used for the case
of lead $(Z = 82)$.

3.4.6 Nuclear spin-dependent effects

In addition to the dominant PNC interaction given in Eq. (3.6), there are
other smaller PNC interactions that must be considered. First, there is
the interaction between the nuclear axial-vector current and the electron
vector current from Z^0 exchange. In the limit of nonrelativistic nucleon
motion, this interaction is given approximately by the spin-dependent
Hamiltonian

$$h_W^{(2)} = -\frac{G}{\sqrt{2}} K_2 \frac{\kappa - 1/2}{I(I+1)} \boldsymbol{\alpha} \cdot \mathbf{I} \rho(r) . \tag{3.59}$$

Here, $\kappa = 4$, $I = 7/2$ and $K_2 \approx -0.05$ for the valence proton of
^{133}Cs. Additionally, parity nonconservation in the nucleus leads to to
a parity-nonconserving nuclear moment, the anapole moment mentioned
in Sec. 3.2, that couples electromagnetically to the atomic electrons. The
anapole-electron interaction is described by a Hamiltonian similar to (3.59)

$$h_W^a = \frac{G}{\sqrt{2}} K_a \frac{\kappa}{I(I+1)} \boldsymbol{\alpha} \cdot \mathbf{I} \rho(r) . \tag{3.60}$$

$K_a \approx 0.24-0.33$ is determined from nuclear model calculations. [139] These
two interactions can be treated together using (3.60) with $K_a \to K = K_a -
K_2(\kappa - 1/2)/\kappa$, if one makes the approximation that the nuclear densities
for each effect are the same. The resulting spin-dependent correction
was evaluated in the Dirac-Fock approximation including weak core-
polarization corrections. Combining that calculation with the previous
spin-independent result, we obtain [130]

$$E_{\text{PNC}} = -0.905(9) \times 10^{-11} i|e|a_0 \left[(-Q_W/N) + A(F', F)K \right] , \tag{3.61}$$

where the matrix $A(F', F)$ is found to be

$$\begin{pmatrix} A(3,3) & A(3,4) \\ A(4,3) & A(4,4) \end{pmatrix} = \begin{pmatrix} 0.029 & 0.048 \\ -0.041 & -0.022 \end{pmatrix} . \tag{3.62}$$

These values of $A(F', F)$ agree to within 10% with results of semi-
empirical [140] and MBPT [141] calculations. Linear combinations of the
amplitudes in (3.61) can be used to isolate either the spin-dependent or
spin-independent parts of the interaction as will be discussed below.

The interference between the hyperfine interaction and the spin-
independent PNC interaction leads to a tiny spin-dependent inter-

action [142, 143] that can also be included in the above analysis by adjusting the value of K in (3.61) slightly.

3.4.7 Electron-electron weak interaction

The effect of Z^0 exchange between electrons can be taken into account by adding a weak correction g_{ijkl}^W to the electron-electron Coulomb interaction. This correction takes the form of a contact interaction between a pair of electrons

$$g_{ijkl}^W = \sqrt{2}G \int \bar{\phi}_i(\gamma_\mu C_{1e} + \gamma_\mu \gamma_5 C_{2e})\phi_k \bar{\phi}_j(\gamma^\mu C_{1e} + \gamma^\mu \gamma_5 C_{2e})\phi_l \, d^3\mathbf{x}, \quad (3.63)$$

with $C_{1e} = -\frac{1}{2}(1 - 4\sin^2\theta_W)$, and $C_{2e} = \frac{1}{2}$ in lowest order in the standard model. Only the cross term proportional to $C_{1e}C_{2e}$ contributes to PNC. Treating this interaction in lowest-order perturbation theory leads to the following correction to the PNC amplitude:

$$E_{PNC}^{e-e} = \sum_{ai} \frac{d_{wi}^{(1)}\tilde{g}_{iava}^W}{\varepsilon_v - \varepsilon_a} + \sum_{ai} \frac{d_{iv}^{(1)}\tilde{g}_{wiai}^W}{\varepsilon_w - \varepsilon_a} + \sum_{am} \frac{d_{am}^{(1)}\tilde{g}_{wmva}^W}{\varepsilon_{av} - \varepsilon_{mW}} + \sum_{am} \frac{d_{am}^{(1)}\tilde{g}_{wavm}^W}{\varepsilon_{aW} - \varepsilon_{mv}}. \quad (3.64)$$

The last two terms in this expression are a generalization of the second-order matrix element (3.13); they turn out to be negligible. The most important effect comes from the first two terms, which correspond to a generalization of the HF potential to include the PNC interaction (3.63). Their evaluation is greatly simplified by a type of Fierz identity, $\tilde{g}_{ijkl}^W = 2g_{ijkl}^W$, which can be used to reduce the calculation to direct terms only. The radial reduction for the direct terms is then quite straightforward, and one obtains

$$\begin{aligned} E_{PNC}^{e-e} &= -0.0172\, C_{1e}C_{2e} \times 10^{-11}i|e|a_0(-Q_W/N) \\ &= -0.0003 \times 10^{-11}i|e|a_0(-Q_W/N). \end{aligned} \quad (3.65)$$

The bulk of this effect is contributed by the $1s$ state in the sum over a in (3.64). This small nuclear-spin-independent contribution is masked by the much larger uncertainty in the dominant term (3.6).

3.4.8 Vector transition polarizability

The interpretation of atomic PNC experiments in Cs also requires knowledge of the $6S$-$7S$ vector transition polarizability,

$$\beta_{pol} = \frac{1}{6}\sum_n \left[\langle 7S\|\mathbf{D}\|nP_{1/2}\rangle\langle nP_{1/2}\|\mathbf{D}\|6S\rangle \right.$$

$$\times \left(\frac{1}{E_{7S} - E_{nP_{1/2}}} - \frac{1}{E_{6S} - E_{nP_{1/2}}} \right)$$

$$+ \frac{1}{2} \langle 7S \| \mathbf{D} \| nP_{3/2} \rangle \langle nP_{3/2} \| \mathbf{D} \| 6S \rangle$$

$$\times \left(\frac{1}{E_{7S} - E_{nP_{3/2}}} - \frac{1}{E_{6S} - E_{nP_{3/2}}} \right) \Bigg]. \qquad (3.66)$$

This quantity has a double-perturbation structure similar to the PNC-E1 amplitude, but here experimentally derived values for the matrix elements are available and semi-empirical estimates already reach an accuracy of 1.5%. The oscillator strengths $f(6s \rightarrow 6p)$ and $f(6s \rightarrow 7p)$ can be taken from a direct measurement, while those for $7s \rightarrow 6p$ and $7s \rightarrow 7p$ can be inferred from the $7s$ lifetime and the $7s$ polarizability, respectively; the rapidly convergent "tail" for $n \geq 8$ can be estimated theoretically. Since the calculation of β_{pol} from (3.66) involves some delicate cancellations (β_{pol} vanishes nonrelativistically for a transition between two S states), it is preferable to calculate a related quantity α, the *scalar* transition polarizability, and then to estimate β_{pol} from measurements of $\alpha/\beta_{\mathrm{pol}}$. This procedure gives estimates $\beta_{\mathrm{pol}} = 27.2(4) \, a_0^3$ [144] and $\beta_{\mathrm{pol}} = 27.3(4) \, a_0^3$. [145] Alternatively, Bouchiat and Guena [146] have proposed inferring β_{pol} from a measurement of $M1^{\mathrm{hfs}}/\beta_{\mathrm{pol}}$, where $M1^{\mathrm{hfs}}$ is the component of the $6s \rightarrow 7s$ M1 amplitude that is induced via the hyperfine interaction with the nucleus. Using a theoretical estimate of $M1^{\mathrm{hfs}}$ they obtain $\beta_{\mathrm{pol}} = 27.17(35) \, a_0^3$.

Ab initio estimates of β_{pol} are more complicated than for the PNC-E1 amplitude: the additional perturbation $\mathbf{d} = e\mathbf{r}$ here is of tensor rank one so that the perturbed orbitals have a mixture of different angular momenta. Also, β_{pol} turns out to be rather sensitive to higher-order corrections to the self-energy because the cancellations found for E_{PNC} do not apply. However, the direct sum-over-states approach of Sec. 3.4.3 with intermediate states from an all-order approach can be used without difficulty and gives $\beta_{\mathrm{pol}} = 27.00(20) \, a_0^3$. [120] Future progress could result from a combination of improved measurements and more refined calculations.

3.5   Conclusions

We now make use of the above analysis to extract the value of the weak charge Q_{W} from the Cs experiment. The PNC amplitudes measured by Noecker, Masterson, and Wieman [109] in 1988 are

$$\Im(E_{\mathrm{PNC}})/\beta_{\mathrm{pol}} = \begin{cases} -1.639(47)(08) & 4 \rightarrow 3 \\ -1.513(49)(08) & 3 \rightarrow 4 \end{cases} \qquad (3.67)$$

in units of mV/cm. Eliminating the spin-dependence from (3.67) with the aid of Eq. (3.61) and using the theoretical value for $\beta_{\mathrm{pol}} = 27.00(20) \, a_0^3$

from Sec. 3.4.8, one finds

$$\Im(E_{\mathrm{PNC}}^{\mathrm{exp}}) = -0.8252(184)[61]10^{-11}|e|a_0 \,, \tag{3.68}$$

where the first error is from experiment and the second from theory. Combining this result with our calculation of the spin-independent amplitude given in Eq. (3.8), we obtain

$$Q_{\mathrm{W}} = -71.04(1.58)[0.88] \,. \tag{3.69}$$

Alternatively, if we use (3.61) to eliminate the spin-independent terms in (3.67), we obtain the value

$$K = 0.83 \pm 0.46 \tag{3.70}$$

for the constant governing the spin-dependent interaction.

Radiative corrections to the weak charge Q_{W} incorporating a parameterization of new physics beyond the standard model have been worked out by Marciano and Rosner, [115] who find

$$Q_{\mathrm{W}}(^{133}\mathrm{Cs}_{55}) = -73.20 - 0.8S - 0.005T \pm 0.13, \tag{3.71}$$

assuming the values $m_t = 140$ GeV for the top–quark mass and $m_H = 100$ GeV for the Higgs particle mass. The parameters S and T in Eq. (3.71) are associated partly with deviations of the top-quark and Higgs mass from their assumed values, and partly with new physics beyond the standard model. S represents physics that conserves weak isospin, while T represents physics that violates weak isospin conservation; deviations of the top-quark mass from $m_t = 140$ GeV enter through T. The small factor multiplying T makes this prediction very insensitive to the top-quark mass in the absence of new physics. Unfortunately, both the experimental and theoretical errors are presently too large to make atomic PNC in cesium a precision test of the standard model. However, there are two features of cesium PNC that even at the present accuracy lead to particle physics implications. The first is the possibility of a large positive values of S in technicolor theories. [147] A value of $S = 2$ moves the theoretical prediction for Q_{W} more than 2 experimental standard deviations away from the experimental value. The second is the effect of extra Z^0 bosons, which is not accounted for in Eq. (3.71). Exchange of new Z^0's can be shown to be strongly constrained by atomic PNC. [148] Of course there is also the possibility of having both a nonvanishing S parameter together with extra Z^0's, and in addition perhaps entirely new physics that has not been thought of. Since new physics affects different weak interaction tests differently, it is important to have as many such tests as possible. The value of atomic PNC tests will increase when the next stage of accuracy is reached.

Several possibilities exist for future improvements in the accuracy of both the experimental and the theoretical sides of atomic PNC. Work is in progress to reduce the experimental error in atomic Cs to several tenths of 1%, [149] at which point the anapole effect should become more evident. The incorporation of triple excitations into the all-order procedure could produce similar accuracies in the atomic theory. The all-order techniques could also be extended to systems with more than one valence electron, hopefully giving theoretical amplitudes accurate accurate to the 1% level or better for other transitions of experimental interest. A measurement on Bi using optical rotation is accurate to 1%, [113] and a similar accuracy has recently been reported for Pb. [150] It would be useful to combine measurements from many different transitions into a single model-independent analysis of atomic data. Finally, as mentioned in Sec. 3.4.5, a promising possibility is to measure atomic PNC in a pair of isotopes, the ratio of such measurements being essentially free of atomic structure uncertainty.

At present, in the most accurate case of atomic Cs, atomic PNC still plays only a qualitative role in testing electroweak radiative corrections, although it is on the threshold of making a more quantitative test. When the next stage of accuracy in the experimental and theoretical determination is reached, atomic PNC is likely to play a significant role as one of a set of precision tests of the standard model.

Part 2

PHOTOIONIZATION OF ATOMS

4

Many-body effects in single photoionization processes

J. J. Boyle and M. D. Kutzner

4.1 Introduction

The process of single photoionization occurs when an atom or molecule absorbs a photon and ejects a single electron. Photoionization studies of multi-electronic systems can provide excellent portraits of the many-body effects that lie within both the initial target state and the final state consisting of the ion plus the photoelectron. Important examples of many-body effects include autoionizing resonances, giant shape resonances, relaxation, and polarization. A common element of all of these effects is their prominence near ionization thresholds. In this chapter, we will examine the many-body effects that are present in atomic single photoionization problems within the framework of many-body perturbation theory (MBPT).

We will refer to the corrections to a one-electron approximation (such as a Hartree-Fock approximation) as correlation effects. Although a one-electron approximation is capable of describing many of the gross properties of photoionization in atoms, a scheme for including correlation effects will be necessary in order to describe many of the processes that are mentioned above.

There has been considerable development in experimental techniques to study photoionization over the past few decades. Synchrotron radiation has been used to measure total photoabsorption cross sections over a wide range of energies for many atomic species. Additionally, photoelectron spectroscopy has been used to partition total cross sections into channel cross sections. Methods for the measurement of the angular distribution asymmetry parameter, $\beta(\omega)$, and spin-polarization parameters have also been developed. (See chapter seven by S. Manson for a description of the β asymmetry parameter.) A detailed discussion of experimental advances has been given by Samson. [151] A recent review of the ex-

93

treme ultraviolet spectra of free metal atoms has been given by Sonntag and Zimmermann, [152] and a comprehensive review of the photoionization of rare gas atoms using synchrotron radiation has been given by Schmidt. [153]

Several theoretical methods have been developed in addition to MBPT in order to study correlation effects in photoionization. The random-phase approximation with exchange (RPAE) was introduced into atomic physics from nuclear theory by Altick and Glassgold. [154] It has been applied to the noble gases and to other closed-shell systems with success by Amusia, Cherepkov, and co-workers. [155] Extensions of the RPAE to open-shell systems have also been presented. [156] Johnson and Lin have developed a fully relativistic treatment, the relativistic random-phase approximation (RRPA), which also gives an excellent description of the process of photoionization in closed-shell atoms. [157, 158] Combining the RRPA with the multi-channel quantum defect theory has successfully accounted for autoionizing resonances in some closed-shell systems. [159] The *R*-matrix method introduced into atomic physics by Burke and Taylor has been applied to a wide variety of problems including photoionization and electron scattering in atoms, ions, and molecules. [160] The hyperspherical coordinates approach has been applied to the photoionization of helium [161] and to the valence shells of the alkaline earth metal atoms. [162] Configuration interaction and the multiconfiguration Hartree-Fock methods have also been utilized in photoionization calculations. [163] A detailed review of some of the theoretical approaches that are mentioned above, as well as other techniques has been given by Starace. [164]

In this section, we will review a theoretical framework for MBPT applied to photoionization problems. In Sec. 4.2, we will discuss methods and concerns of evaluating first-order MBPT diagrams. We will also demonstrate, in a low order calculation of platinum, how the choice of a potential can improve the convergence of a calculation. In Sec. 4.3, we will discuss how to evaluate an infinite order class of diagrams for special cases, and we will present the results of calculations performed on barium, xenon, and tungsten. In Sec. 4.4 we will present a summary.

4.1.1   *The photoionization cross section*

It is assumed that an *N*-electron atom with a nuclear charge *Z* may be described by the Hamiltonian

$$H = \sum_{i=1}^{N} \left(-\frac{\nabla_i^2}{2} - \frac{Z}{r_i} \right) + \sum_{i > j = 1}^{N} \frac{1}{r_{ij}} \qquad (4.1)$$

where the indices i and j represent atomic electrons. Atomic units ($\hbar = e = m = 1$) will be used throughout this chapter unless indicated otherwise. The Coulomb repulsion between the electrons is accounted for by the term $1/r_{ij}$.

The application of a harmonically varying external electric field $\mathbf{E} = \mathrm{Re}\left[F_0 e^{-i\omega t}\hat{\mathbf{n}}\right]$ introduces an additional term to Eq. (4.1),

$$H_{\mathrm{int}} = \mathbf{E} \cdot \mathbf{r} = \left[\frac{F_0}{2}e^{-i\omega t} + \frac{F_0^*}{2}e^{i\omega t}\right]\hat{\mathbf{n}} \cdot \mathbf{r} . \tag{4.2}$$

We will use the variable F_0 to denote the magnitude of the electric field. A variable that characterizes the process of photoionization is called the cross section $\sigma(\omega)$ and is defined by the relation

$$\sigma(\omega) = \frac{P(\omega)}{F} . \tag{4.3}$$

The variable $P(\omega)$ is the ionization probability per atom per unit time as a function of the photon frequency ω, and F is the flux of photons that are incident on the atom in units of photons per unit area per unit time. The flux F is the time averaged real part of the complex Poynting vector [165] divided by the photon energy ω

$$F = \frac{|F_0|^2 c}{8\pi\omega} . \tag{4.4}$$

The variable c is the speed of light. The application of Fermi's second Golden Rule [166, 167] for transition probabilities from the state $|0\rangle$ to the state $|f\rangle$ per unit time gives

$$P_{0 \to f}(\omega, k_f) = 2\pi \left|\left\langle f \left| \frac{F_0}{2}\hat{\mathbf{n}} \cdot \mathbf{r} \right| 0 \right\rangle\right|^2 \rho_f(E_f) , \tag{4.5}$$

where $|0\rangle$ and $|f\rangle$ are the exact solutions to Eq. (4.1) for the initial and final states of the atom respectively and are normalized to unity.* For a problem involving single photoionization, the momentum of the electron that makes the transition into the continuum and is detected will be denoted as k_f. The variable $\rho_f(E_f)$ is the density of states function for transitions into the continuum and is equivalent to a Dirac delta function:

$$\rho_f(E_f) = \delta(\omega + E_0 - E_f) = \delta\left(\omega - I_f - \frac{k_f^2}{2}\right) . \tag{4.6}$$

The variables E_0 and E_f are the exact eigenvalues of the states $|0\rangle$ and $|f\rangle$ respectively, with respect to the Hamiltonian H of Eq. (4.1). The

---

* Notice that the condition that the states $|0\rangle$ and $|f\rangle$ are normalized to unity is not the same as *intermediate normalization* used in perturbation theory, and shown in Eq. (1.31). This will be discussed later. Cf. Eqs. (4.33) and (4.34).

variable I_f is the exact single photoionization threshold for the final channel indicated by the subscript "f" (we will take $I_f > 0$). Formally, Eq. (4.5) yields a transition rate into the state $|f\rangle$ that is infinitely sharp at one point and is zero elsewhere. Since the energy eigenvalues of free electrons lie in the continuum, we are not interested in the transition rate into a particular energy state, which is infinitesimal, but the transition rate into small energy region centered around $\omega - I_f$. Following Dirac, [166] we define a total transition rate by summing over a small region where it is infinitely sharp:

$$P_{0\to f}(\omega) = \sum_{k_f} P_{0\to f}(\omega, k_f) \,. \tag{4.7}$$

This is the transition rate that will be used in Eq. (4.3).

We use the following normalization convention for the radial parts of our continuum electron orbitals: [94]

$$N_k \int_0^\infty r^2 dr R(kl)(r) \int_{k-\Delta k}^{k+\Delta k} R_{k'\ell}(r)dk' = 1 \,, \tag{4.8}$$

where N_k is a normalization constant to be determined and k is the electron momentum. The variable Δk is assumed small. We will take our continuum orbitals to vary asymptotically as Coulomb waves in the k-scale (or momentum scale) [94]

$$\lim_{r\to\infty} R_{k\ell}(r) \sim \frac{1}{r} \cos\left(kr + \frac{q}{k}\ln(2kr) - \frac{(\ell+1)\pi}{2} + \delta_\ell^C + \delta_\ell \right) \,, \tag{4.9}$$

where δ_ℓ^C is the coulomb phase shift, δ_ℓ is the non-coulomb phase shift, and q is the screened charge with which the continuum electron interacts with asymptotically.[†] (The variable q will be positive if, in the initial state, the atom is neutral in charge or a positive ion). From Eqs. (4.8) and (4.9) we have $N_k = 2/\pi$. Therefore, in order to extend the sum in Eq. (4.7) over the excited electron orbitals into a sum over the bound excited orbitals and an integration over the free (or continuum) orbitals, we use the following prescription:

$$\sum_{k_f} \to \frac{2}{\pi} \int dk_f \,. \tag{4.10}$$

The transformation indicated by Eq. (4.10) will be followed throughout this chapter. Equation (4.7) can be rewritten as

$$P_{0\to f}(\omega) = \frac{2}{\pi} \int P_{0\to f}(\omega, k_f)dk_f \,. \tag{4.11}$$

[†] Normalizing the continuum orbitals with respect to the *momentum* scale is only one convention that is used. Another convention is *energy scale* normalization which is used in chapter 5. Cf. Eq. (4.9) above with Eq. (5.43).

We will consider incident radiation polarized in the $\hat{z}$ direction and we will use the many-particle dipole operator $Z_{op} = \hat{z} \cdot \mathbf{r}$. Using the analytic properties of the Dirac delta function, we obtain

$$\sigma_{0 \to f}(\omega) = \int dk_f \frac{8\pi\omega}{k_f c} \delta \left(k_f - (2\omega - 2I_f)^{\frac{1}{2}} \right) |\langle f | Z_{op} | 0 \rangle|^2 . \qquad (4.12)$$

In this chapter, we will utilize two forms of the dipole operator Z_{op}. In order to express the photoionization cross section in terms of the length gauge, we take $Z_{op} = \sum_{i=1}^{N} z_i$, and obtain

$$\sigma_{0 \to f}^{(L)}(\omega) = \int dk_f \frac{8\pi\omega}{k_f c} \delta \left(k_f - (2\omega - 2I_f)^{\frac{1}{2}} \right) \left| \langle f | \sum_{i=1}^{N} z_i | 0 \rangle \right|^2 . \qquad (4.13)$$

If we use the following commutation relation, which is valid for exact eigenstates of the Hamiltonian H,

$$[\mathbf{r}, H] = i\mathbf{p} , \qquad (4.14)$$

then we obtain the velocity form of the cross section:

$$\sigma_{0 \to f}^{(V)}(\omega) = \int dk_f \frac{8\pi}{k_f c \omega} \delta \left(k_f - (2\omega - 2I_f)^{\frac{1}{2}} \right) \left| \langle f | \sum_{i=1}^{N} \frac{\partial}{\partial z_i} | 0 \rangle \right|^2 . \qquad (4.15)$$

Equations (4.13) and (4.15) will be the two forms of the photoionization cross section that we will utilize in this chapter. Other forms of the cross section can be defined by taking successively higher orders of the commutation relation shown in Eq. (4.14). [164]

4.1.2 The linked-cluster expansion

In order to proceed we need a suitable approximation for the states $|0\rangle$ and $|f\rangle$. To begin with, we choose a single-particle potential V that will approximate the Coulomb repulsion terms between the electrons in the system from Eq. (4.1): $\sum_{i>j=1}^{N} (r_{ij})^{-1}$. The explicit form of the potential V that will be used here will be discussed in more detail in Sec. 4.2.2. In the field of this potential we calculate a complete set of states $|\Phi_n\rangle$ satisfying

$$H_0 |\Phi_n\rangle = E_n^{(0)} |\Phi_n\rangle . \qquad (4.16)$$

Here H_0 is the single-particle approximate Hamiltonian

$$H_0 = \sum_{i=1}^{N} \left(-\frac{\nabla_i^2}{2} - \frac{Z}{r_i} + V_i \right) , \qquad (4.17)$$

and the state $|\Phi_n\rangle$ represents a linear combination of Slater determinants constructed from products of single-particle orbitals $|\phi_i\rangle$ that satisfy

$$\left(-\frac{\nabla_i^2}{2} - \frac{Z}{r_i} + V_i\right)|\phi_i\rangle = \varepsilon_i|\phi_i\rangle . \tag{4.18}$$

The perturbation interaction H_c is defined as the difference between the approximate single-particle potential from Eq. (4.17) and the electron-electron interaction from Eq. (4.1):

$$H_c = H - H_0 = \sum_{i>j=1}^{N} \frac{1}{r_{ij}} - \sum_{i=1}^{N} V_i . \tag{4.19}$$

Within the MBPT, we will denote the set of exact many-body solutions to the Hamiltonian H of Eq. (4.1) as $|\Psi_n\rangle$

$$H|\Psi_n\rangle = E_n|\Psi_n\rangle . \tag{4.20}$$

The states $|\Psi_0\rangle$ and $|\Psi_f\rangle$ will be identical to the states $|0\rangle$ and $|f\rangle$ introduced in Sec. 4.1.4.1.1 to within a normalization factor.

The perturbation expansion that gives the states $|\Psi_n\rangle$ of Eq. (4.20) in terms of the states $|\Phi_n\rangle$ of Eq. (4.16) may be obtained through the use of the time-evolution operator $U(t, t_0)$ in the interaction picture. [167] The operator $U(t, t_0)$ will be defined as the operator which transforms the state $|\Psi_n(t_0)\rangle$ at time t_0 to the state $|\Psi_n(t)\rangle$ at time t according to

$$|\Psi_n(t)\rangle = U(t, t_0)|\Psi_n(t_0)\rangle . \tag{4.21}$$

Although the problem of including the interaction represented by H_c is a time-independent one, we formally make it a time-dependent problem by incorporating an adiabatic damping factor $e^{\eta t}$, and transforming H_c to a time-dependent operator $H_c(t)$ in the interaction representation:

$$H_c(t) = e^{iH_0t}H_c e^{\eta t}e^{-iH_0t} . \tag{4.22}$$

The perturbation $H_c(t)$ is turned on adiabatically starting at $t_0 = -\infty$. According to Eqs. (4.16) and (4.17), the solutions at this time are given by $|\Psi_n(t_0 = -\infty)\rangle = |\Phi_n\rangle$. Once we have transformed to the time $t = 0$ and decided upon a normalization convention, we will take the limit as $\eta \to 0^+$.

The fundamental differential equation characterizing the time evolution of the system in the interaction picture is

$$i\frac{\partial}{\partial t}\left[e^{iH_0t}|\Psi_n(t)\rangle\right] = H_c(t)e^{iH_0t}|\Psi_n(t)\rangle . \tag{4.23}$$

From Eq. (4.21), we observe that Eq. (4.23) is equivalent to the operator differential equation

$$i\frac{d}{dt}\left[e^{iH_0t}U_\eta(t,t_0)e^{-iH_0t_0}\right] = H_c(t)e^{iH_0t}U_\eta(t,t_0)e^{-iH_0t_0}, \qquad (4.24)$$

subject to the condition that

$$U_\eta(t_0,t_0) = 1 \qquad (4.25)$$

is the unit operator. We mention that $U_\eta(t,t_0)$ is defined for the variable t over the interval $0 \geq t \geq t_0 = -\infty$. The operator differential equation (Eq. (4.24)) may now be converted to the following expression by integrating on both sides:

$$e^{iH_0t}U_\eta(t,t_0)e^{iH_0t_0} = 1 - i\int_{t_0}^{t} dt_1 H_c(t_1)e^{iH_0t_1}U_\eta(t_1,t_0)e^{-iH_0t_0}. \qquad (4.26)$$

The solution to Eq. (4.26) is obtained by iteration, which yields

$$
\begin{aligned}
e^{iH_0t}U_\eta(t,t_0)e^{-iH_0t_0} = &\ 1 - i\int_{t_0}^{t} dt_1 H_c(t_1) \\
&\times \left[1 - i\int_{t_0}^{t_1} dt_2 H_c(t_2)e^{iH_0t_2}U_\eta(t_2,t_0)e^{-iH_0t_0}\right],
\end{aligned}
$$

$$(4.27)$$

or

$$
e^{iH_0t}U_\eta(t,t_0)e^{-iH_0t_0} = 1 + \sum_{n=1}^{\infty}(-i)^n \int_{t_0}^{t} dt_1 \cdots \int_{t_0}^{t_{n-1}} dt_n H_c(t_1)\cdots H_c(t_n),
$$

$$(t_0 < t_n < t_{n-1} < \cdots < t_2 < t_1 < 0) \qquad (4.28)$$

which is known as the Dyson series. [3]

The time-evolution operator as expressed in Eq. (4.28) may now be applied to the unperturbed states in the interaction representation $e^{iH_0t_0}|\Psi_n(t_0 = -\infty)\rangle = e^{iH_0t_0}|\Phi_n\rangle$ and the integrations carried out from $t_0 = -\infty$ to $t = 0$ to yield

$$
\begin{aligned}
U_\eta(0,-\infty)|\Phi_n\rangle = &\ e^{i\theta_n}\left[1 + \frac{1}{E_n^{(0)} - H_0 + i\eta}H_c \right. \\
&\left. + \frac{1}{E_n^{(0)} - H_0 + 2i\eta}H_c\frac{1}{E_n^{(0)} - H_0 + i\eta}H_c + \cdots\right]|\Phi_n\rangle,
\end{aligned}
$$

$$(4.29)$$

which is a representation of our state function $|\Psi_n(t = 0)\rangle$, and where θ_n is an arbitrary phase factor. The phase factor appears, in part, due to the fact that we have not yet imposed any normalization constraints on $|\Psi_n\rangle$.

We adopt the intermediate normalization scheme (cf. Eq. (1.31)):

$$\langle \Phi_n | \Psi_n(t = 0) \rangle = 1 \, . \tag{4.30}$$

This leads to a normalization denominator for $U_\eta(0, -\infty)|\Phi_n\rangle$. We now have

$$|\Psi_n\rangle = |\Psi_n(t = 0)\rangle = \lim_{\eta \to 0^+} \frac{U_\eta(0, -\infty)|\Phi_n\rangle}{\langle \Phi_n | U_\eta(0, -\infty)|\Phi_n\rangle} \, . \tag{4.31}$$

Finally, in Eq. (4.31), we have taken the limit as $\eta \to 0^+$.

By the use of Wick's theorem, individual terms in the series of Eq. (4.31) for $U_\eta(0, -\infty)|\Phi_n\rangle$ may be represented by Goldstone diagrams or graphs. [17] The convention that will be used in this chapter is as follows: an occupied excited orbital ("particle" line) is represented by a line with an arrow directed upwards and an unoccupied unexcited orbital ("hole" line) is represented by a line with an arrow directed downwards; the direction of increasing time is upwards. Another convention which is frequently followed is to take the direction of increasing time to proceed from the left to the right, with the appropriate orientation of the "particle" (occupied excited orbital) and "hole" (unoccupied unexcited orbital) lines. [155, 168] The specific rules for constructing a mathematical expression from a Goldstone diagram will be presented in Sec. 4.2, where we consider the evaluation of low-order diagrams.

A distinction is made between the so-called "linked" and "unlinked" diagrams.‡ An *unlinked diagram* is defined as a diagram any part of which is completely disconnected from the rest and has no external lines attached. Two examples of unlinked diagrams are shown in Fig. 4.1. Both Fig. 4.1(a) and Fig. 4.1(c) are considered "unlinked." External lines are straight lines with arrows that continue off to infinity, such as the lines labelled by the indices *a, r, b,* and *s* in Figs. 4.1(b-d). *Linked diagrams* are defined as diagrams containing no unlinked parts. According to this definition, both Figs. 4.1(b) and 4.1(d) are considered "linked," although Fig. 4.1(d) would, in addition, be considered "disconnected." We mention that the horizontal dashed lines in Fig. 4.1 indicate matrix elements, and will be discussed in more detail in Sec. 4.2.

According to the linked-cluster theorem, [16] all of the unlinked terms in the numerator of Eq. (4.31) are exactly factored out by identical unlinked terms in the denominator, leaving a perturbation expansion for $|\Psi_n\rangle$ given

---

‡ For further discussion, see Sec. 1.5.1.

by (cf. Eq. (1.9))

$$|\Psi_n\rangle = \sum_{\substack{\text{linked} \\ \text{diagrams}}} \sum_{m=0}^{\infty} \left(\frac{1}{E_n^{(0)} - H_0} H_c\right)^m |\Phi_n\rangle . \tag{4.32}$$

Notice that the initial- and final-state wavefunctions obtained from the linked expansion are not normalized to unity and that matrix elements involving the wave functions of Eqs. (4.30) and (4.31) need to be modified in order to take the normalization correction into account. Therefore, the states $|0\rangle$ and $|f\rangle$ that were introduced in Eq. (4.5) may finally be represented as

$$|0\rangle = \frac{|\Psi_0\rangle}{\sqrt{\langle\Psi_0|\Psi_0\rangle}} , \tag{4.33}$$

and

$$|f\rangle = \frac{|\Psi_f\rangle}{\sqrt{\langle\Psi_f|\Psi_f\rangle}} . \tag{4.34}$$

In this way the MBPT builds correlation into wavefunctions. A photoionization cross section may be obtained by evaluating the dipole matrix elements of Eqs. (4.13) and (4.15) using the normalized wavefunctions shown in Eqs. (4.33) and (4.34), and obtained from the linked expansion of Eq. (4.32):

$$\sigma_{0\to f}^{(L)}(\omega) = N_{0\to f} \int dk_f \frac{8\pi\omega}{k_f c} \delta\left(k_f - (2\omega - 2I_f)^{\frac{1}{2}}\right) \left|\langle\Psi_f|\sum_{i=1}^{N} z_i|\Psi_0\rangle\right|^2 , \tag{4.35}$$

and

$$\sigma_{0\to f}^{(V)}(\omega) = N_{0\to f} \int dk_f \frac{8\pi}{k_f c\omega} \delta\left(k_f - (2\omega - 2I_f)^{\frac{1}{2}}\right) \left|\langle\Psi_f|\sum_{i=1}^{N} \frac{\partial}{\partial z_i}|\Psi_0\rangle\right|^2 , \tag{4.36}$$

where

$$N_{0\to f} = \left(\langle\Psi_0|\Psi_0\rangle\langle\Psi_f|\Psi_f\rangle\right)^{-1} , \tag{4.37}$$

is the normalization correction and is usually close to unity.

The amount of correlation included at a given order of perturbation theory may be inferred to some degree by comparing the length $\sigma^{(L)}$ and velocity $\sigma^{(V)}$ results. Fully-correlated wavefunctions are exact eigenstates of H and therefore satisfy the commutation relation (Eq. (4.14)) precisely. It is possible, however, for wavefunctions which are not well correlated to give good length and velocity agreement so that this criterion should be considered as necessary but not sufficient. In fact, the length and velocity

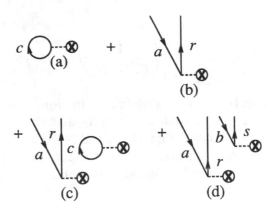

Fig. 4.1. A selection of Brueckner-Goldstone diagrams representative of the function $U_\eta(0, -\infty)|\Phi_n\rangle$ defined in Eq. (4.29). The function in Eq. (4.29) will include both linked and unlinked terms in its expansion. Figures 4.1(a) and 4.1(c) are considered unlinked and Figs. 4.1(b) and 4.1(d) are considered linked. Only the diagrams from Figs. 4.1(b) and 4.1(d) will contribute to the wavefunction $|\Psi_n\rangle$ defined in Eqs. (4.31) and (4.32). The horizontal dashed lines indicate matrix elements and will be discussed in more detail in Sec. 4.2.

cross sections will be equal even in the single-particle approximation if a local potential is chosen for the unperturbed states $|\Phi_n\rangle$. [164]

4.1.3   *An alternative expansion*

In the previous subsection, we defined a linked expansion for the wave-functions according to Eq. (4.32). An alternative diagrammatic expansion may be devoloped directly for the total photoionization cross section through an analysis of the frequency dependent polarizability $\alpha_{pol}(\omega)$. The relationship between the total photoionization cross section and the frequency-dependent polarizability is [169]

$$\sigma(\omega) = \frac{4\pi\omega}{c} \text{Im} \left[\alpha_{pol}(\omega) \right] . \tag{4.38}$$

The polarizability is a measure of the linear response of the atom to an external field. For example, if an atom is in the external electric field $\mathbf{E} = \text{Re}[F_0 e^{-i\omega t}\hat{z}]$, and if we use the variable $\mathbf{P}$ to denote the electric dipole moment of an atom, then the polarizability $\alpha_{pol}(\omega)$ is defined through the relation:

$$\mathbf{P} = \left(\alpha_{pol}(\omega)\frac{F_0}{2}e^{-i\omega t} + \alpha_{pol}^{*}(\omega)\frac{F_0^{*}}{2}e^{i\omega t} \right) \hat{z} . \tag{4.39}$$

Through an analysis of the dipole moment of the atom,

$$\mathbf{P} \cdot \hat{\mathbf{z}} = \frac{\langle \Psi'_N | - Z_{op} | \Psi'_N \rangle}{\langle \Psi'_N | \Psi'_N \rangle} , \tag{4.40}$$

where $|\Psi'_N\rangle$ is analogous to $|\Psi_N\rangle$ of Eq. (4.32), and includes the corrections due to the external field interaction ($H_c \rightarrow H_c + H_{int} \Rightarrow |\Psi_N\rangle \rightarrow |\Psi'_N\rangle$, here H_{int} is from Eq. (4.2)). One obtains for the polarizability $\alpha_{pol}(\omega)$ the series of all *closed* Goldstone diagrams that involve two interactions with the dipole perturbation Z_{op} and any number of interactions with H_c. Also, the energy denominators between the two dipole interactions have the additional terms $\pm\omega$. A linked expansion for the wavefunction $|\Psi'_N\rangle$ is obtained in the same manner that Eq. (4.32) was derived, but with a harmonic time-dependent perturbation $H_{int} = \mathbf{E} \cdot \mathbf{r}$ added to the correlation perturbation H_c of Eqs. (4.23) and (4.24).

A derivation of the polarizability series $\alpha_{pol}(\omega)$ has been given by Kelly. [170] The equality between the cross sections obtained from the polarizability series of Kelly [170] and the cross sections obtained in Eqs. (4.35) and (4.36), particularly with respect to the normalization factors in each series, has been demonstrated to first order by Carter. [171]

4.2 The evaluation of first-order diagrams

We will focus our attention on the diagrams which result from the dipole transition matrix element $\langle \Psi_f | Z_{op} | \Psi_0 \rangle$, where the two state functions $|\Psi_f\rangle$ and $|\Psi_0\rangle$ are defined by Eq. (4.32). Because the dipole operator Z_{op} is a tensor of rank one, it will always carry us out of the initial state of the atom, and the matrix elements $\langle \Psi_f | Z_{op} | \Psi_0 \rangle$ will involve a series of *open* diagrams which contain *one* interaction with the dipole operator, and any number of interactions with the correlation perturbation H_c in both the initial and final states.

The notation that we will use here is as follows. The N electron that are occupied in the initial state of the atom are referred to as *core* orbitals and are indicated by the use of the subindices a, b, and c The remaining solutions to Eq. (4.18) that are not occupied in the initial state of the atom are referred to as *excited* orbitals and are indicated by the use of the subindices r, s, t and u. Finally, the use of the subindices i and j will indicate a summation over both the *core* and *excited* orbitals. The use of these subindices follows that of Lindgren and Morrison. [29]

One of the important features of the MBPT is that corrections to the unperturbed one-electron matrix element can be evaluated term by term. Provided that the unperturbed states are a fairly close approximation to the exact wavefunctions, classes of diagrams in the perturbation expansion can be associated with certain classes of physical processes. In this section we present diagrams of first order in the correlation interaction H_c and discuss their evaluation.

4.2.1 Ground- and final-state correlation diagrams

The MBPT diagram representing the lowest order dipole matrix element $\langle \phi_r | Z_{op} | \phi_a \rangle$ contributing to single photoionization is shown in Fig. 4.2(a). Since time is assumed to be increasing upwards in these diagrams, the unperturbed ground state is represented by the empty region at the base of the figure. The dipole matrix element is represented by the dashed line terminating in a small open circle. Following the dipole matrix element in a time ordered sense, there occurs a "hole" orbital $|\phi_a\rangle$ and the "particle" orbital $|\phi_r\rangle$. This diagram represents the lowest order term that describes the absorption of a photon and the promotion of an electron from the orbital $|\phi_a\rangle$ to the orbital $|\phi_r\rangle$.

Diagrams that are first order in the perturbation interaction (Eq.(4.19)) are shown in Figs. 4.2(b-e) where the matrix element of the two-particle operator $1/r_{ij}$ is represented by a horizontal dashed line connecting two sets of particle/hole lines. The interaction with the potential $-V_i$ is indicated by a horizontal dashed line that ends in a circle enclosing a bold cross. There are energy denominators that are associated with the intermediate virtual excitations, according to the prescription of Eq. (4.32). We note that no energy denominator is associated with the final external lines since the amplitude associated with the dipole matrix element (and hence the diagrams) must have the dimensions of length. If the virtual excitation occurs "before" the dipole interaction, as in Fig. 4.2(b), then the general form of the energy denominator D is

$$D = \sum_{i=1}^{N_{ph}} \left(\varepsilon_{h_i} - \varepsilon_{p_i} \right) , \qquad (4.41)$$

where the variable N_{ph} represents the number of particle-hole pairs that occur in the intermediate state, ε_{h_i} is the single-particle energy of the ith hole orbital, and ε_{p_i} is the single-particle energy of the ith particle orbital. If the virtual excitation occurs "after" the dipole interaction, as shown in Fig. 4.2(c), then the general form of the energy denominator D is

$$D = \sum_{i=1}^{N_{ph}} \left(\varepsilon_{h_i} - \varepsilon_{p_i} \right) + \omega , \qquad (4.42)$$

where ω is the energy of the incident photon. An overall sign is attributed to each diagram and is given by $(-1)^{h+l}$ where h is the number of internal hole lines and l is the number of closed loops. Exchange diagrams although

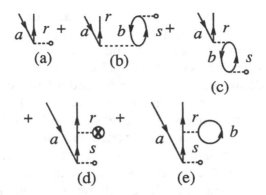

(a) (b)

(c)

(d) (e)

Fig. 4.2. A selection of low order Brueckner-Goldstone diagrams that contribute to the dipole matrix element for photoionization. The time ordering of these diagrams proceeds from the bottom to the top. The solid lines with arrows refer to the orbitals in the basis set, and the direction of the arrow signifies the occupation status of the repective orbital. In the initial state of the system, all of the core orbitals are occupied and none of the excited orbitals is. An arrow pointing down indicates an unoccupied core orbital and an arrow pointing up indicates an occupied excited orbital. A dashed line connected to a small circle indicates a dipole interaction. A dashed line that connects two sets of arrows indicates a Coulomb interaction. The dashed line that is connected to a circle enclosing a bold cross indicates an interaction with the potential $-V$. Diagrams where a Coulomb interaction occurs after the dipole interaction in a time ordered sense, as in Figs. 4.2(c-e), are referred to as "final state correlation" diagrams. Diagrams where a Coulomb interaction occurs before the dipole interaction in a time ordered sense, as in Fig. 4.2(b), are referred to as "ground state correlation" diagrams.

not explicitly shown, should also be included. All of the diagrams in Fig. 4.2 satisfy the condition $(-1)^{h+l} = 1$. The exchange versions of the diagrams shown in Figs. 4.2(b,c) and 4.2(e) will satisfy $(-1)^{h+l} = -1$. The are no "exchange" diagrams, as such, for the diagrams shown in Figs. 4.2(a) and 4.2(d).

The diagram represented in Fig. 4.2(b) is a ground state correlation diagram since the interaction with the perturbation $1/r_{ij}$ occurs earlier in time than the interaction with the dipole operator. The analytic form of the matrix element represented by the ground state correlation diagram in Fig. 4.2(b) is

$$= \sum_{b,s} \frac{d_{bs}^{(1)} g_{rsab}}{(\varepsilon_a + \varepsilon_b - \varepsilon_r - \varepsilon_s)}. \qquad (4.43)$$

Here $d_{bs}^{(1)} = \langle \phi_b | Z_{op} | \phi_s \rangle$ and $g_{rsab} = \langle \phi_r(1)\phi_s(2) | r_{12}^{-1} | \phi_a(1)\phi_b(2) \rangle$. The summation over the index "s" in Eq. (4.43) involves a discrete sum over bound excited states $\sum_s$(bound), and an integral over the continuum excited states $2/\pi \int dk_s$ according to Eq. (4.10). In practice, the summation over the index "b" in Eq. (4.43) can be truncated in order to explore the effects of coupling between individual channels. The energy denominator in Eq. (4.43) does not vanish and standard numerical integration techniques can be used in order to evaluate ground state correlation diagrams for single photoionization problems with minimal error. We also mention that, in practice, the excited orbitals $|\phi_s\rangle$ that lie in the continuum will be calculated for a discrete set of electron momentum $k_s^{(i)}$ values, where the index "i" corresponds to the index of this k-mesh. This implies that, in practice, there will be a maximum cutoff point $k_s^{(\max)}$ to the integration: the integration $2/\pi \int dk_s$ will not extend to infinity.

The diagram represented in Fig. 4.2(c) is a final state correlation diagram since the interaction with the perturbation occurs after the dipole interaction. In terms of matrix elements we write

$$\text{[diagram]} = \sum_{b,s} \frac{g_{rbas} d_{sb}^{(1)}}{(\varepsilon_b - \varepsilon_s + \omega)} . \qquad (4.44)$$

In the case of final state correlations, as is seen in Eqs. (4.42) and (4.44), the energy denominator may vanish whenever the photon energy ω equals the difference between the energy of the excited orbitals ε_{p_i} (ε_s in Eq. (4.44)) and the hole orbitals ε_{h_i} (ε_b in Eq. (4.44)). When the denominator vanishes in the region where $\varepsilon_s = k_s^2/2 > 0$, it is handled by adding to the denominator a small imaginary contribution $+i\eta$ consistent with the incoming wave boundary condition in the final state [164] (p.566 of Ref. [172]) and using the rule

$$\lim_{\eta \to 0^+} (D + i\eta)^{-1} = \mathsf{P}(D^{-1}) - i\pi\delta(D) , \qquad (4.45)$$

where P represents a principal value integration and $\delta(D)$ is a Dirac delta function. Due to the presence of the delta function $\delta(\varepsilon_b - \varepsilon_s + \omega)$ under the integral sign $\int dk_s$, the imaginary part of the diagram of Fig. 4.2(c) and Eq. (4.44) is readily obtained by interpolation.

The real principal value part of Eq. (4.44) can be evaluated by an analytic integration of interpolatory type. [173] Since the evaluation of the real principal value portion of Eq. (4.44) will be discussed later in this chapter, we will briefly outline a method of performing this integration. We assume that the numerator of Eq. (4.44) $g_{rbas} d_{sb}^{(1)}$ can be interpolated by a polynomial in k_s between sufficiently close k_s-mesh values:

$$g_{rbas}d_{sb}^{(1)} \approx \sum_{\kappa=0}^{3} \sum_{i=1}^{max} (k_s)^\kappa c^{(i)}(\kappa) g_{rbas(i)} d_{s(i)b}^{(1)} , \qquad (4.46)$$

where κ is assumed to take on integral values. The coefficients $c^{(i)}(\kappa)$ are interpolation coefficients and they will be functions of the integral power κ and of the index of the k_s-mesh i. The interpolation over the values of κ shown in Eq. (4.46) can be performed by using a four-point Lagrange interpolation. The principal value integrations may now be performed analytically between the successive values of the k_s-mesh:

$$\frac{2}{\pi} P \int dk_s \frac{g_{rbas}d_{sb}^{(1)}}{\left(\varepsilon_b - \dfrac{k_s^2}{2} + \omega \right)}$$

$$\approx \frac{2}{\pi} \sum_{\kappa=0}^{3} \sum_{i=1}^{max} P \int_{k=k_s^{(i-1)}}^{k=k_s^{(i)}} dk_s \frac{(k_s)^\kappa c^{(i)}(\kappa) g_{rbas(i)} d_{s(i)b}^{(1)}}{\left(\varepsilon_b - \dfrac{k_s^2}{2} + \omega \right)} . \qquad (4.47)$$

One analytically evaluates the terms

$$P \int_{k_s^{(i-1)}}^{k_s^{(i)}} dk_s \frac{(k_s)^\kappa}{\left(\varepsilon_b - \dfrac{k_s^2}{2} + \omega \right)} , \qquad (4.48)$$

and algebraically collects the coefficients of the product $g_{rbas(i)} d_{s(i)b}^{(1)}$ in Eq. (4.47). Equation (4.47), therefore, becomes

$$\frac{2}{\pi} P \int dk_s \frac{g_{rbas(i)} d_{s(i)b}^{(1)}}{\left(\varepsilon_b - \dfrac{k_s^2}{2} + \omega \right)} \approx \frac{2}{\pi} \sum_{i=1}^{max} a^{(i)}(b, \omega) g_{rbas(i)} d_{s(i)b}^{(1)} , \qquad (4.49)$$

defining coefficients $a^{(i)}(b, \omega)$.

In this way, we have discretized the integration over the continuum orbitals and we have analytically performed the principal value integrations.

Up to now, our discussion of the final state correlation diagrams has concerned only the case when $\varepsilon_s > 0$. The final state correlation diagram of Fig. 4.2(c) is also the lowest-order term accounting for autoionizing resonances when $|\phi_s\rangle$ occurs in a bound-excited state with an energy that is degenerate with a continuum state containing the orbital $|\phi_r\rangle$. However, higher-order diagrams must be included to give the autoionizing resonance the correct energy position and width. This will be discussed in Sec. 4.3.

Interactions with the potential are represented by the diagram of Fig. 4.2(d) where the circle enclosing the bold cross indicates the matrix element

$-\langle\phi_i|V|\phi_j\rangle$ relating to Eq. (4.19). Figure 4.2(e) and its exchange represent corrections to the final state by an interaction with an unexcited electronic orbital. Finally, we mention that all of the diagrams of Fig. 4.2 as well as higher order terms are automatically taken into account if one uses the RPAE technique. [168]

4.2.2   The choice of the potential V

In this section, we will discuss the implicit evaluation of low order diagrams through the choice of the potential V. In Eqs. (4.16) and (4.17), we introduced the single-particle approximate Hamiltonian H_0

$$H_0 = \sum_{i=1}^{N}\left(-\frac{\nabla_i^2}{2} - \frac{Z}{r_i} + V_i\right), \tag{4.50}$$

where,

$$\left(-\frac{\nabla_i^2}{2} - \frac{Z}{r_i} + V_i\right)|\phi_i\rangle = \varepsilon_i|\phi_i\rangle. \tag{4.51}$$

The operator V_i is a single-particle Hermitian potential, which should be chosen to approximate the Coulomb repulsion term $\sum_{i>j=1}^{N}(r_{ij})^{-1}$. as closely as possible in order to minimize the correlation correction of Eq. (4.19)

$$H_c = H - H_0 = \sum_{i>j=1}^{N}\frac{1}{r_{ij}} - \sum_{i=1}^{N}V_i. \tag{4.52}$$

We mention that Eq. (4.52) represents the entire perturbation that is under consideration. The presence of the potential V_i in Eq. (4.52) accentuates the fact that we are using approximate electron orbitals $|\phi_i\rangle$ calculated in the field of this potential. If we make a poor choice for V_i, then the magnitude of the correlation Hamiltonian H_c can become large and the convergence of the perturbation series will deteriorate.

   We will require our core orbitals to be defined in a potential such that all of the first order diagonal corrections vanish, or at least, very nearly vanish. (We mention that, for closed-shell atoms, the cancellation between the first order corrections and the potential can be effected as follows, but for open-shell atoms, this condition will be eased). After we have equated the energy denominators, this requirement for the electron orbitals can be expressed mathematically for a single-determinantal wavefunction as

$$\langle\phi_s|V|\phi_a\rangle = \sum_{b}(g_{sbab} - g_{bsab}). \tag{4.53}$$

Equation (4.53), combined with the condition that the spin-orbitals $|\phi_i\rangle$ are functions that are separable into radial, orbital angular momentum, and spin angular momentum functions of the form

$$\phi_i(r, \theta, \varphi, \chi) = R_i(r) Y_{\ell_i m \ell_i}(\theta, \varphi) \sigma(m s_i) \tag{4.54}$$

defines a Hartree-Fock potential for closed shell atoms. [174] (The variable σ on the right side of Eq. (4.54) represents the spin coordinates of the orbital ϕ_i.) For an open-shell atom, we will use what is referred to as a "restricted" Hartree-Fock potential and assume the functional form that is indicated in Eq. (4.54) for the basis orbitals. In this situation, the cancellation that is indicated in Eq. (4.53) will not be exact, although it may be satisfied to a good approximation. For the conditions associated with the definition of a restricted Hartree-Fock potential, see Lindgren and Morrison. [29] We will denote both the Hartree-Fock potential that is obtained from the algorithm of Eq. (4.53), and the restricted Hartree-Fock potential for open-shell atoms, by the operator V^{HF}.

We mention that in Eq. (4.53) we have carried out our analysis within the perturbation series for the wavefunctions. Specifically, we have defined a potential for the core orbitals of the atom. An important point to make is that there are an infinite number of potentials that can be defined over all of the orbitals (both the core and the excited orbitals) but that behave exactly as depicted in Eq. (4.53) for the core orbitals. We consider this type of Hermitian potential below: [175]-[178]

$$V = V^{HF} + \left[1 - \sum_a |\phi_a\rangle\langle\phi_a| \right] \left(V^{N-1} - V^{HF} \right) \left[1 - \sum_a |\phi_a\rangle\langle\phi_a| \right] , \tag{4.55}$$

This potential will satisfy the following identities:

$$\langle\phi_a|V|\phi_a\rangle = \langle\phi_a|V^{HF}|\phi_a\rangle , \tag{4.56}$$

$$\langle\phi_r|V|\phi_a\rangle = \langle\phi_a|V|\phi_r\rangle = \langle\phi_a|V^{HF}|\phi_r\rangle , \tag{4.57}$$

and

$$\langle\phi_r|V|\phi_r\rangle = \langle\phi_r|V^{N-1}|\phi_r\rangle . \tag{4.58}$$

In Eqs. (4.56-4.58) we have used $\langle\phi_a|\phi_r\rangle = 0$, which is an automatic result of the fact that V is a Hermitian potential, provided that $\varepsilon_a \neq \varepsilon_r$.

The definition of the potential shown in Eq. (4.55) allows us freedom in the choice of the V^{N-1} potential for the excited orbitals. This choice of the potential should be determined by the type of physical interaction that is under investigation. [179] In our case, it is photoionization, and so our analysis will focus on the perturbative expansion of the dipole matrix elements. A previous investigation into this problem was made by Qian, Carter, and Kelly. [180] The analysis that will be presented here is due to Boyle, and arrives at a slightly different potential and results from that of Qian *et al.* [181]

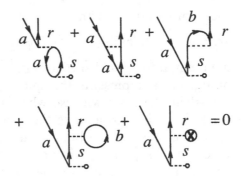

Fig. 4.3. The diagrams that contribute to the effective single-particle potential. The dashed line that is connected to a circle enclosing a bold cross indicates an interaction with this potential. We require that the first order final-state correlation terms that contribute to the potential cancel. The final averaging procedure is given in Eqs. (4.64) and (4.67).

We will consider a single-determinant initial state function for electron orbitals residing in subshell $n_a \ell_a$:

$$|(n_a \ell_a)^{q_a} : M_L\rangle = \frac{1}{\sqrt{q_a!}} \det \left[|\phi_{a:(1)}\rangle |\phi_{a:(2)}\rangle \cdots |\phi_{a:(q_a)}\rangle \right] . \tag{4.59}$$

where q_a is the occupation number of subshell $n_a \ell_a$. The additional indices "i" in the subscript "$a : (i)$" arate m_ℓ and m_s values that the electron orbitals can have within the subshell $n_a \ell_a$ and that are degenerate in energy. The sum of the m_ℓ values of the orbitals $|\phi_{a:(i)}\rangle$ should equal the value of M_L.

We define an averaged squared total dipole matrix element for transitions from the subshell $n_a \ell_a$ to the continuum orbital $k_r \ell_r$ $\left| \langle Z^{(a \to r)} \rangle \right|^2$ by

$$\left| \langle Z^{(a \to r)} \rangle \right|^2 = \sum_{\substack{i=i(M_L) \\ M_L}} \frac{|\langle \phi_{r:(i)} | Z_{\mathrm{op}} | \phi_{a:(i)} \rangle|^2}{(2L+1)} . \tag{4.60}$$

In Eq. (4.60) we have recognized that the m_ℓ values of the individual orbitals $\phi_{a:(i)}$ and $\phi_{r:(i)}$, indicated by the indices "(i)," will be dependent upon the total M_L value of the determinant. This is demonstrated by making the summation over the index "i" a function of M_L, and then averaging over the M_L values that are possible. Notice that we have ignored the single-particle m_s values and the total M_S values in the averaging procedure indicated in Eq. (4.60). This neglect is due to the fact that the dipole operator Z_{op} does not have any effect upon the spin.

In terms of MBPT diagrams, we consider, next, the first order diagram that represents the interaction with the potential $-V$ on the excited orbital:

$$a:(i) \quad \begin{matrix} r:(i) \\ \otimes \\ s:(i) \\ \circ \end{matrix} = \sum_s \frac{\langle \phi_{r:(i)}| - V_{(M_L)}^{N-1}|\phi_{s:(i)}\rangle \langle \phi_{s:(i)}|Z_{op}|\phi_{a:(i)}\rangle}{(\varepsilon_a - \varepsilon_s + \omega)}. \tag{4.61}$$

In Eq. (4.61), we have recognized that, in terms of the MBPT diagrams, the potential that we are considering will also be dependent upon the M_L value of the determinant of the initial state. This is indicated in Eq. (4.61) by the subscript M_L.

Analogous to Eq. (4.53), which was for the core orbitals and defined the Hartree-Fock potential V^{HF}, we will require our $V_{(M_L)}^{N-1}$ potential for the excited orbitals to satisfy the condition that the first order final state correlation corrections that fall into the class of potential corrections cancel with the potential. Diagrammatically, this is depicted in Fig. 4.3. Again, after equating the energy denominators, this will appear mathematically as

$$\langle \phi_{r:(i)}|V_{(M_L)}^{N-1}|\phi_{s:(i)}\rangle$$
$$= \sum_{b,j} \left(\langle \phi_{r:(i)}(1)\phi_{b:(j)}(2)|r_{12}^{-1}|\phi_{s:(i)}(1)\phi_{b:(j)}(2)\rangle \right.$$
$$\left. - \langle \phi_{b:(j)}(1)\phi_{r:(i)}(2)|r_{12}^{-1}|\phi_{s:(i)}(1)\phi_{b:(j)}(2)\rangle \right) \frac{\langle \phi_{s:(i)}|Z_{op}|\phi_{a:(i)}\rangle}{\langle \phi_{s:(i)}|Z_{op}|\phi_{a:(i)}\rangle}$$
$$+ \sum_j \left(\langle \phi_{r:(i)}(1)\phi_{a:(j)}(2)|r_{12}^{-1}|\phi_{a:(i)}(1)\phi_{s:(j)}(2)\rangle \right.$$
$$\left. - \langle \phi_{a:(j)}(1)\phi_{r:(i)}(2)|r_{12}^{-1}|\phi_{a:(i)}(1)\phi_{s:(j)}(2)\rangle \right) \frac{\langle \phi_{s:(j)}|Z_{op}|\phi_{a:(j)}\rangle}{\langle \phi_{s:(i)}|Z_{op}|\phi_{a:(i)}\rangle}. \tag{4.62}$$

To this point, we have followed the analysis given by Qian et al. [180] We will now deviate from that analysis. Utilizing Eq. (4.60), we define the following M_L independent V^{N-1} potential for excitations from the subshell $n_a\ell_a$, $\langle \phi_r|V^{N-1}|\phi_r\rangle$, as the first order correction to the M_L independent dipole matrix element $\langle Z^{(a\to r)}\rangle$,

$$\left| \langle Z^{(a\to r)}\rangle - \sum_s \frac{\langle \phi_r|V^{N-1}|\phi_s\rangle \langle Z^{(a\to s)}\rangle}{(\varepsilon_a - \varepsilon_s + \omega)} \right|^2$$
$$= \sum_{\substack{i=i(M_L) \\ M_L}} \frac{1}{(2L+1)} \left| \langle \phi_{r:(i)}|Z_{op}|\phi_{a:(i)}\rangle \right.$$
$$\left. - \sum_s \frac{\langle \phi_{r:(i)}|V_{(M_L)}^{N-1}|\phi_{s:(i)}\rangle \langle \phi_{s:(i)}|Z_{op}|\phi_{a:(i)}\rangle}{(\varepsilon_a - \varepsilon_s + \omega)} \right|^2. \tag{4.63}$$

Notice that we have introduced the M_L independent potential V^{N-1} to the left of the equals sign in Eq. (4.63). Notice, also, that it will be defined in terms of the M_L dependent potential included on the right side of the equals sign $V_{(M_L)}^{N-1}$ and defined in Eq. (4.62). Using Eq. (4.60), keeping terms to first order in the potential interaction, and setting $s = r$, we obtain from Eq. (4.63):

$$\langle \phi_r | V^{N-1} | \phi_r \rangle$$
$$= \frac{\displaystyle\sum_{\substack{i=i(M_L) \\ M_L}} |\langle \phi_{r:(i)} | Z_{op} | \phi_{a:(i)} \rangle|^2 \langle \phi_{r:(i)} | V_{(M_L)}^{N-1} | \phi_{r:(i)} \rangle}{\displaystyle\sum_{\substack{i'=i'(M_L') \\ M_L'}} |\langle \phi_{r:(i')} | Z_{op} | \phi_{a:(i')} \rangle|^2},$$

(4.64)

where $\langle \phi_{r:(i)} | V_{(M_L)}^{N-1} | \phi_{r:(i)} \rangle$ is given in Eq. (4.62). Equation (4.64) is given in terms of single determinants in the initial state of the atom, such as the determinant that is shown in Eq. (4.59). In terms of the multi-determinant LS-coupled initial state

$$|G(M_L)\rangle = |(n\ell)^q LS\gamma M_L M_S\rangle, \tag{4.65}$$

and an LS-coupled final channel

$$|F(I)\rangle = |((n\ell)^{q-1} L_I S_I \gamma_I ; \varepsilon\ell_F) L_F S_F \gamma_F M_L M_S\rangle, \tag{4.66}$$

Eq. (4.64) can be rewritten

$$\langle V^{N-1}(n\ell)^q : LS\gamma ; (n\ell, \varepsilon\ell_F)\rangle$$
$$= \frac{\displaystyle\sum_{F,I,J,M_L} \langle G(M_L)|C_{10}|F(I)\rangle \langle F(I)|r_{12}^{-1}|F(J)\rangle \langle F(J)|C_{10}|G(M_L)\rangle}{\displaystyle\sum_{F',I',M_L'} \langle G(M_L')|C_{10}|F'(I')\rangle \langle F'(I')|C_{10}|G(M_L')\rangle},$$

(4.67)

where C_{10} is the spherical tensor operator appropriate to the dipole interaction. The radial dependence and, therefore, the gauge dependence of the matrix element of the external field has been factored out. Notice that the potential depends only on the initial state coupling and on the interaction between $n\ell$ and $\varepsilon\ell_F$. There is no dependence on L_F, S_F, γ_F, or L_I, S_I, γ_I.

It has been demonstrated by Boyle, that the definition shown in Eq. (4.67) is also consistent with the requirement that first-order corrections in the perturbative series of the dipole polarizability $\alpha_{pol}(\omega)$ cancel with the potential for the excited orbitals. [182] The potential presented in

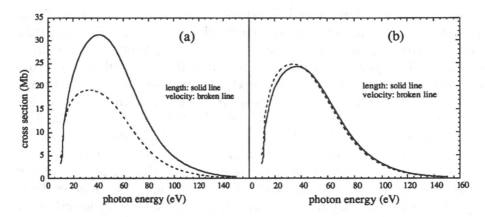

Fig. 4.4. (a) The lowest order $5d$ partial cross section of platinum $[\text{Xe}]4f^{14}5d^96s$ with the excited orbitals calculated with the potential defined in Eq. (4.67). (b) The $5d$ partial cross section of platinum that explicitly includes the amplitude associated with the first order ground state correlation diagram, and that corresponds to a $5d \rightarrow k\ell$ virtual excitation. Such a ground state correlation is in the same class of interactions defined in the potential of Eq. (4.67), but with a different time ordering.

Eq. (4.67) also possesses some very useful analytic properties. One property relates the V^{N-1} potential for transitions from an initial state that has a closed subshell to the corresponding V^{N-1} potential for transitions from an initial state that has one fewer electron than a closed subshell

$$\langle V^{N-1}(n\ell)^{4\ell+1} : L = \ell, S = 1/2; (n\ell, \varepsilon\ell_F)\rangle$$
$$= \frac{4\ell}{4\ell+1}\langle V^{N-1}(n\ell)^{4\ell+2} : L = 0, S = 0; (n\ell, \varepsilon\ell_F)\rangle. \quad (4.68)$$

One can also partition the angular coefficients associated with the potential $\langle V^{N-1}(n\ell)^q : LS\gamma; (n\ell, \varepsilon\ell_F)\rangle$ into an average term that is dependent only upon the angular momentum pair $(n\ell, \varepsilon\ell_F)$, and a correction to the average term that is dependent upon the angular momentum pair $(n\ell, \varepsilon\ell_F)$ and the total angular momentum quantum numbers of the initial state $LS\gamma$.[181] We mention that the Qian et al. potential does differ slightly from the potential shown in Eq. (4.67) for most open-shell atoms. A comparison with the Qian et al. potential and an itemization of the analytic properties of this potential, including those that are mentioned above, have been given by Boyle.[182]

In Fig. 4.4, we present an example of the use of the potential for the excited orbitals defined in Eq. (4.67), as well as the importance of ground-state correlation diagrams. In Fig. 4.4(a), we show the lowest order $5d$ partial cross section of platinum $[\text{Xe}]4f^{14}5d^96s(^3D_{J=3})$ $(Z = 78)$

using the potential for the excited orbitals that is defined in Eq. (4.67). In the lowest order cross section, we observe that there exists approximately 30% disagreement between the length and velocity versions of the cross section. It is important to keep in mind that the potential of Eq. (4.67) automatically includes final state correlation diagrams of the type shown in Fig. 4.2(c) with $(n_a\ell_a, k_r\ell_r) = (n_b\ell_b, k_s\ell_s) = (5d, k\ell)$. In the first order, the other time ordered version of Fig. 4.2(c) is Fig. 4.2(b). In Fig. 4.4(b), we have shown the $5d$ partial cross section of platinum explicitly including the correlation of Fig. 4.2(b) with $(n_a\ell_a, k_r\ell_r) = (n_b\ell_b, k_s\ell_s) = (5d, k\ell)$, and we observe that the length and velocity agreement of the cross sections has improved substantially. We mention that we have explicitly evaluated only those amplitudes associated with the lowest order dipole (Fig. 4.2(a)) and the first order ground state correlation diagram (Fig. 4.2(b) with $(n_a\ell_a, k_r\ell_r) = (n_b\ell_b, k_s\ell_s) = (5d, k\ell)$) in order to obtain the cross section that is shown in Fig. 4.4(b). It would have been necessary to evaluate more diagrams in order to obtain this result if we had used a different non-local potential from V defined in Eqs. (4.55) and (4.67).

4.3 The evaluation of an infinite-order series of diagrams

As mentioned in Sec. 4.2.1, the final state correlation diagram of Fig. 4.2(c) and its exchange have energy denominators which may vanish giving rise to a simple pole when the excited orbital $|\phi_s\rangle$ is a bound-excited state. This is the lowest-order term that accounts for the effect of autoionizing resonances. In order to improve the accuracy of a description of the resonances, including the resonance width and energy shift, one must include higher order iterations of the final state correlation diagrams. This may be done by summing a geometric series of diagram segments. In this section we will present two such summations. The first, the coupled-equations technique of Brown, Carter, and Kelly, [183] includes more correlations and utilizes the analytic principal value technique discussed in Sec. 4.2.1. The second technique, the generalized resonance technique of Garvin [184] is an approximation to the coupled-equations scheme, and is useful when there are many angular momentum channels that are available to an outgoing electron.

4.3.1   The coupled-equations technique

We define the correlated dipole matrix element $\bar{d}_{ar}$ as the sum of the uncorrelated dipole matrix element $d_{ar}^{(1)} = \langle \phi_r | Z_{\text{op}} | \phi_a \rangle$ plus all orders of

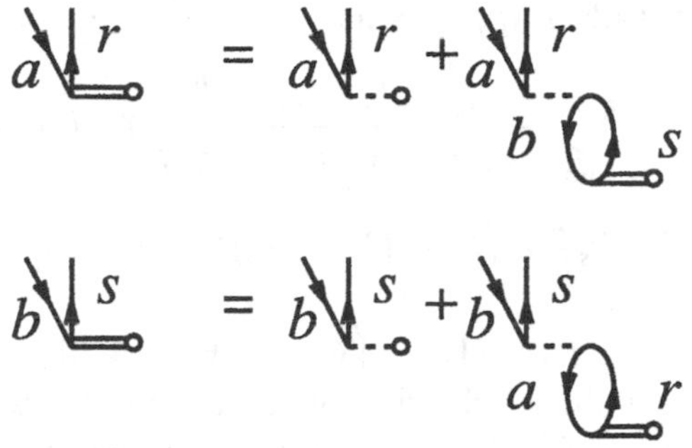

Fig. 4.5. An example of a two channeled set of coupled-equations. The solid double lines ending in an open-circle indicate an "all-orders" correlation. The technique as it has been presented here was introduced by Brown, Carter, and Kelly [183].

the time-forward final-state correlation diagrams coupling any number of channels with the total LS values that are the same in all cases. This definition is shown diagrammatically in Fig. 4.5 for the case of the two channels $|\phi_a\rangle \rightarrow |\phi_r\rangle$ and $|\phi_b\rangle \rightarrow |\phi_s\rangle$. Each solid double line in Fig. 4.5 that ends in an open circle represents an "all-orders" correlation matrix element $\bar{d}_{ij}$. We mention that in practice there will be other diagrams included in an actual calculation that couples the two channels $|\phi_a\rangle \rightarrow |\phi_r\rangle$ and $|\phi_b\rangle \rightarrow |\phi_s\rangle$ than those that are shown in Fig. 4.5. For example, potential correction terms and exchange terms would need to be considered as well. However, in order to simplify the presentation of the mathematical aspects, we will consider only the limited coupling that is shown in Fig. 4.5.

The first line of Fig. 4.5 may be written as an equation involving the correlated matrix element $\bar{d}_{ar}$ for channel $|\phi_a\rangle \rightarrow |\phi_r\rangle$, the uncorrelated matrix element $d_{ar}^{(1)}$ for the same channel, and the perturbation interaction matrix element $\langle H_c \rangle$ allowing for interactions originating from the channel $|\phi_b\rangle \rightarrow |\phi_s\rangle$ and decaying into the $|\phi_a\rangle \rightarrow |\phi_r\rangle$ channel:

$$\bar{d}_{ar} = d_{ar}^{(1)} + \sum_s \frac{g_{rbas}\bar{d}_{bs}}{(\varepsilon_b - \varepsilon_s + \omega)} . \tag{4.69}$$

As usual, the sum in Eq. (4.69) over the s index represents a summation over bound excited orbitals and an integration over continuum orbitals. The unknowns in this coupled set of integral equations are the correlated dipole matrix elements: $\bar{d}_{ar}$ and $\bar{d}_{bs}$.

We will now utilize the technique developed in Eqs. (4.45)–(4.49) as a practical means of evaluating the final state correlation diagrams shown in Fig. 4.5. Equation (4.69) appears as

$$\bar{d}_{ar} = d_{ar}^{(1)} + \sum_{\text{bound } s} \frac{g_{rbas}\bar{d}_{bs}}{(\varepsilon_b - \varepsilon_s + \omega)} + \frac{2}{\pi}\sum_{i=1}^{\text{max}} a^{(i)}(b,\omega)g_{rbak_s^{(i)}}\bar{d}_{bk_s^{(i)}}$$

$$- i\int_{k_s} dk_s \frac{2}{k_s}\delta\left(k_s - (2\omega + \varepsilon_b)^{\frac{1}{2}}\right)g_{rbak_s}\bar{d}_{bk_s}. \tag{4.70}$$

Notice that, in going from Eq. (4.69) to Eq. (4.70) we have assumed that the off-the-energy-shell behavior of the unknown function $\bar{d}_{jk_r}$ varies smoothly as a function of k_r for a given photon energy ω.

Let us suppose that we have N_B explicitly calculated bound orbitals and N_K explicitly calculated continuum orbitals. We can define a complex vector $V^{s(i)}(r,b,a,\omega)$ with a range $i = 1, (N_B + N_K)$ as follows

$$V^{s(i)}(r,b,a,\omega)\Big|_{i=1,N_B} = \frac{g_{rbas(i)}}{(\varepsilon_b - \varepsilon_{s(i)} + \omega)}\Bigg|_{\text{bound } s}, \tag{4.71}$$

$$V^{s(i)}(r,b,a,\omega)\Big|_{i=N_B+1,N_B+N_K} = \frac{2}{\pi}a^{(i-N_B)}(b,\omega)g_{rbak_s^{(i-N_B)}} - i\int_{k_s} dk_s \frac{2}{k_s}$$

$$\times \delta\left(k_s - (2\omega + \varepsilon_b)^{\frac{1}{2}}\right)\sum_{\kappa=0}^{3}\sum_{i=1}^{N_K}(k_s)^{\kappa}c^{(i)}(\kappa)g_{rbak_s^{(i)}}. \tag{4.72}$$

The variable $c^{(i)}(\kappa)$ used in Eq. (4.72) is the interpolation coefficient that was introduced in Eq. (4.46). Utilizing the coefficients defined above, Eq. (4.69) now becomes a linear system of $N_C \times (N_B + N_K)$ coupled algebraic equations, where N_C is the number of channels to be coupled. In the case described here $N_C = 2$. Figure 4.5 can be expressed mathematically as

$$\bar{d}_{ar(j)} = d_{ar(j)}^{(1)} + \sum_{i=1}^{N_B+N_K} V^{s(i)}(r(j),b,a,\omega)\bar{d}_{bs(i)} \tag{4.73}$$

$$\bar{d}_{bs(j)} = d_{bs(j)}^{(1)} + \sum_{i=1}^{N_B+N_K} V^{r(i)}(s(j),a,b,\omega)\bar{d}_{ar(i)}. \tag{4.74}$$

This can be rewritten in matrix form as

$$\begin{bmatrix} \delta_{il} & -V^{s(m)}(r(i),b,a,\omega) \\ -V^{r(l)}(s(j),a,b,\omega) & \delta_{jm} \end{bmatrix}\begin{bmatrix} \bar{d}_{ar(l)} \\ \bar{d}_{bs(m)} \end{bmatrix} = \begin{bmatrix} d_{ar(i)}^{(1)} \\ d_{bs(j)}^{(1)} \end{bmatrix}. \tag{4.75}$$

The correlated dipole matrix elements $\bar{d}_{ij}$ on the left side of Eq. (4.75) may then be solved by matrix inversion techniques for each photon energy value ω, and then interpolated to the on-the-energy-shell values. In practice a discrete set of continuum orbitals (usually 30–40) is used along with approximately 10 explicitly calculated bound excited orbitals.

Not only can the series of final state correlation diagrams be effectively summed in this way, but by inserting various combinations of diagrammatic virtual transitions in the diagonal entries to the matrix of the perturbation interaction, one can also include higher order effects. It is important to point out, however, that this technique is valid only if the condition expressed by Eq. (4.46) holds in the numerator of the Eq. (4.69). Stated another way, the technique as it is presented here will be able to describe only simple pole structures. In order to treat interactions that have higher pole structures, this technique would need further modification.

We will now discuss using the coupled equations technique (Eqs. (4.71)– (4.75)) in order to isolate particular many-body effects. Two effects which can be important in single-electron photoionization are relaxation and polarization. Relaxation is the rearrangement that takes place among the core electrons of the residual ion as a result of the removal of the photoelectron. Core polarization results from the perturbation of the ionic core orbitals by the outgoing photoelectron. The magnitudes of each of these effects should depend on the speed of the outgoing photoelectron since a slowly moving photoelectron spends more time in the vicinity of the ion.

The effects of relaxation have been carefully studied by Altun, Kutzner, and Kelly for the $4d$ photoemission of xenon [185] ($Z = 54$) and by Kutzner, Altun, and Kelly for the $4d$ photoemission of barium [186] ($Z = 56$). The inner-shell $4d$ photoionization spectra of these elements and the lanthanides are characterized by broad, delayed absorption peaks often called giant resonances. [187] In addition to the interesting shape of the total absorption curves, studies involving photoelectron spectroscopy have revealed significant photoionization-with-excitation and double photoionization cross sections in these regions.

One method of approximating the effects of relaxation is to calculate excited orbitals in the potential of the relaxed ion rather than in the usual frozen-core Hartree-Fock V^{N-1} approximation. This approach should be valid for low kinetic energies of the photoelectron where the relaxation process is completed before the photoelectron has left the potential of the ion. Since this method uses different sets of orbitals in the initial and final states, one must include overlap integrals multiplying the dipole matrix elements as outlined by Löwdin. [188] The effect of the overlap integrals is to reduce the single-electron cross section, with the oscillator strength going into "shake-up" and "shake-off" multiple-excitation channels. [164] Relaxation and polarization can also be included by the evaluation of appropriate many-body diagrams. The lowest-order contributions to re-

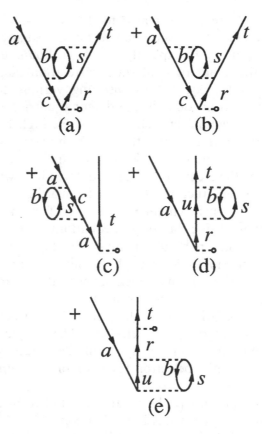

Fig. 4.6. A selection of diagrams that contribute to the dipole matrix element and that are representative of the effects of relaxation, ionization energy threshold shift, and polarization in lowest order. The diagrams in Figs. 4.6(a) and 4.6(b) represent the effects of relaxation when $a = c$. Figure 4.6(c) can be accounted for by using an experimentally determined threshold energy for the excitation $|\phi_a\rangle \rightarrow |\phi_t\rangle$. The diagrams in Figs. 4.6(d) and 4.6(e) represent the effects of polarization in the final and initial states respectively.

laxation are the second-order diagrams shown in Figs. 4.6(a) and 4.6(b) with $a = c$. The diagram of Fig. 4.6(c) can be effectively summed to all orders by using the experimental removal energy. [168] Diagrams which account for polarization in the final and initial states, respectively, are shown in Fig. 4.6(d) and 4.6(e). The diagrams can be evaluated order by order or the diagram segments above the first dipole interaction may be evaluated and included as diagonal entries to the coupled-equations matrix and effectively summed to all orders. The cross sections calculated from this second method are evaluated only for the on-the-energy-shell omega values appropriate to $4d \rightarrow kf$ transitions, *i.e.*, $\omega + \varepsilon_{4d} - \varepsilon_{kf} = 0$.

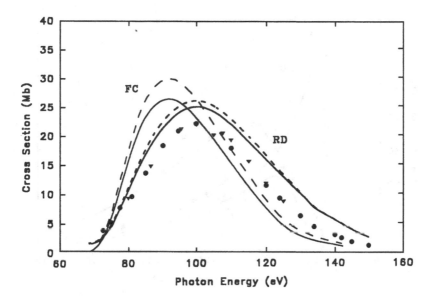

Fig. 4.7. The partial 4*d* photoionization cross sections of xenon as a function of photon energy. The label FC represents calculations without relaxation (frozen core) and the label RD represents calculations that include relaxation diagrams. Solid and dashed lines are the length and velocity versions of the cross sections, respectively. The solid dots are the measurements of Becker *et al.* [189] and the solid triangles are the measurements of Kämmerling *et al.* [190].

The importance of including these diagrams is illustrated in Fig. 4.7 for the 4*d* photoionization of xenon [185] and in Fig. 4.8 for the 4*d* photoionization of barium. [186] The length forms of the cross sections are shown as solid lines and the velocity forms of the cross section are shown as dashed lines. The curves labelled FC represent calculations performed without including relaxation (frozen core), the curves labelled RD represent the inclusion of relaxation diagrams using the coupled-equations technique, and the curves labelled RPD represent calculations that were performed including both the relaxation and polarization diagrams. The RPD curves are shown here only for barium in Fig. 4.8. The solid dots in Fig. 4.7 are the measurements of Becker *et al.* [189] and the solid triangles in Fig. 4.7 are the measurements of Kämmerling *et al.* [190] The solid dots in Fig. 4.8 are the measurements of Becker [191] and the solid triangles in Fig. 4.8 are the measurements of Bizau *et al.* [192] For both xenon and barium, the unrelaxed results (FC) peak too high and too early. The inclusion of relaxation diagrams (Figs. 4.6(a) and 4.6(b)) has the effect of lowering the cross section (RD) and moving the peak to higher photoelectron energies. In barium [186] when the polarization diagrams were also included (RPD), the agreement with experiment improved.

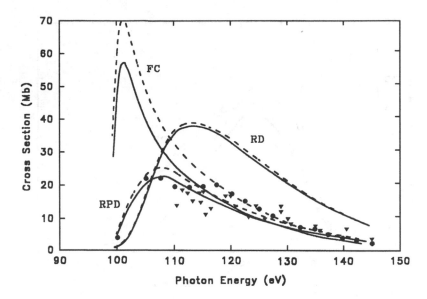

Fig. 4.8. The partial photoionization cross sections of barium as a function of photon energy. The label FC represents calculations performed without including relaxation (frozen core), the label RD represents calculations that include relaxation diagrams, and the label RPD represent calculations that were performed including both relaxation and polarization. The solid and dashed lines are the length and velocity versions of the cross sections respectively. The solid dots are the measurements of Becker *et al.* [191], and the solid triangles are the measurements of Bizau *et al.* [192].

Corrections to the potential may also be included as diagonal matrix elements in the coupled-equations matrix. In addition, the "uncorrelated" dipole of Eq. (4.69) $d_{ar}^{(1)}$ may have already been corrected for ground-state correlations. In principle any number of channels with the same L and S can be coupled. In practice, however, the demands on computer core memory sets the limits on the number of coupled channels that can be usefully considered.

4.3.2   *The generalized resonance technique*

Within the coupled equations technique as described above, the complete linear system of equations for including a class of final-state correlations to infinite order is $N_C \times (N_B + N_K)$, where N_C is the number of channels to be coupled, N_B is the number of bound-excited orbitals, and N_K is the number of continuum orbitals. In certain cases, there can be a large number of LS- or LSJ-coupled channels that interact strongly and need to be included in the set of equations indicated in Eq. (4.75). This will increase the value of N_C. In such cases, when the resonance-resonance and resonance-continuum channels interact strongly, an approximation to Eq. (4.75) may be appropriate. Notice that if we consider only the imaginary contributions that carry us from the channel $|\phi_b\rangle \to |\phi_s\rangle$ to the channel $|\phi_a\rangle \to |\phi_r\rangle$, and if we interpolate all of the matrix elements to their on-the-energy-shell values, then the effective size of the matrix to be inverted becomes N_C. This technique can become especially useful if an atom contains a subshell that is not completely filled in its initial state: $(n_a\ell_a)^{q_a}$ where $q_a < 4\ell_a + 2$. In such a situation, a resonant excitation from an inner subshell $(n_b\ell_b)^{4\ell_b+2}$ may occur into the open subshell, creating the resonant state $(n_b\ell_b)^{4\ell_b+1}(n_a\ell_a)^{q_a+1}$ and which can decay into a continuum channel $(n_b\ell_b)^{4\ell_b+2}(n_a\ell_a)^{q_a-1}k_r\ell_r$. An example of this process is depicted in Fig. 4.9. We have primed the excited a index in Fig. 4.9 in order to identify the orbital from which an electron is excited into the open $(n_a\ell_a)^{q_a}$ subshell. The term "(corrections)" in Fig. 4.9 refers to the infinite-order series of corrections that are depicted to the right of the equals sign.

If we identify the dipole matrix element correlated in this manner by $\bar{d}_{ar}$, then the mathematical counterpart to Fig. 4.9 is

$$\bar{d}_{ar} = d_{ra}^{(1)} + \sum_{b,a'} \frac{g_{rbaa'} d_{a'b}^{(1)}}{(\varepsilon_b - \varepsilon_{a'} + \omega)}$$

$$+ \sum_{s,b,a'} \left(\frac{-2i}{|k_s|}\right) \frac{g_{rbaa'} g_{a'abs}}{(\varepsilon_b - \varepsilon_{a'} + \omega)} \bar{d}_{as} \qquad (4.76)$$

Following Wendin, [193] we have denoted the contribution from the Dirac delta function $-i\pi\delta(D)$ in the diagrams in Fig. 4.9 by a solid horizontal line. The imaginary contributions arise from the boundary condition that is presented in Eq. (4.45). Equation (4.76) can be rewritten as

$$\sum_t \left(\delta_{rt} + \sum_{b,a'} \left(\frac{2i}{|k_t|}\right) \frac{g_{rbaa'} g_{a'abt}}{(\varepsilon_b - \varepsilon_{a'} + \omega)}\right) \bar{d}_{at}$$

$$= d_{ra}^{(1)} + \sum_{b,a'} \frac{g_{rbaa'} d_{a'b}^{(1)}}{(\varepsilon_b - \varepsilon_{a'} + \omega)} \qquad (4.77)$$

Again, in practice, there will be other terms in the perturbation expansion, such as exchange terms, that need to be included in the equation shown above and in Fig. 4.9. Equation (4.77) is known as the "generalized resonance" series, and was first discussed in the form presented here by Garvin. [184] Each term is known in Eq. (4.77) except for the correlated dipole $\bar{d}_{at}$. As with the coupled equations technique (Eq. (4.75)), Eq. (4.77) can be solved for the column vector $\bar{d}_{at}$ by matrix inversion techniques at each photon energy value ω. In practice, higher order corrections can

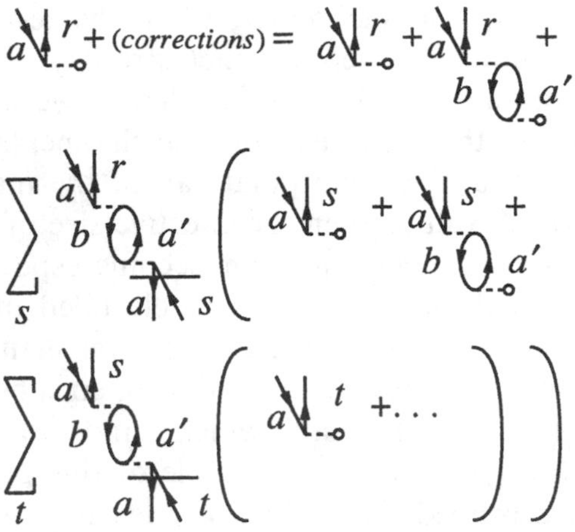

Fig. 4.9. The diagrams that contribute to the generalized resonance series and that are discussed in Sec. 4.3.2. The exchange versions of these diagrams, although they are not shown here, should be included. Following the example of Wendin [193], we use a solid horizontal line drawn through a segment of an MBPT diagram to indicate that we are including only the imaginary contribution $-i\pi\delta(D)$ from that segment of the diagram. The imaginary contribution $-i\pi\delta(D)$ arises from the boundary conditions that are presented in Eq. (4.45).

be included as simpler geometric sums within the basic equation that is given above.

The first application of this technique was by Boyle, Altun, and Kelly [194] in order to obtain a photoionization cross section of atomic tungsten [Xe]$4f^{14}5d^46s^2$ ($Z = 74$). Tungsten is complicated due to profusion of singly excited channels that are available to the outgoing electron. The $5d$ subshell in tungsten is an open subshell, and corresponds to the open $n_a\ell_a$ subshell depicted in Fig. 4.9. In terms of the variables N_C, N_B, and N_K discussed above and including the spin-orbit interaction: for the $5d \rightarrow (kl)$ channels, we take $N_C(kl) = 43$, $N_B(kl) = 0$, and a reasonable value for $N_K(kl)$ is ≈ 40; for the $5p \rightarrow 5d$ channels, we take $N_C(5d) = 32$, $N_B(5d) = 1$, and $N_K(5d) = 0$. In this situation, $N_C(kl) \times [N_B(kl) + N_K(kl)] + N_C(5d) \times [N_B(5d) + N_K(5d)] \approx 1752$. Therefore, the size of the coupled-equations matrix to be inverted would be 1752×1752 to include just these effects. Within the generalized resonance technique, however, the size of the matrix to be inverted will be $N_C(kl) = 43$, or 43×43.

The results of the calculation of Boyle, Altun, and Kelly, [194] using the generalized resonance technique are shown in Fig. 4.10. Figure 4.10 displays the total photoionization cross section of atomic tungsten over the region of the $5p^65d^4. \rightarrow 5p^55d^5$ resonant transitions compared with

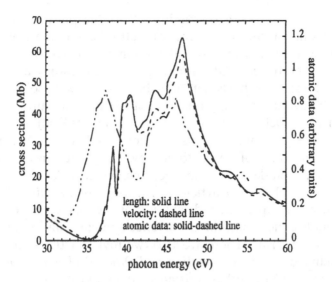

Fig. 4.10. The total resonant cross section in length form (solid line) and velocity form (dashed line) of tungsten $[Xe]4f^{14}5d^46s^2$ from 30 eV to 60 eV. The 5d partial cross section included in this figure contains the correlations that are associated with the generalized resonance technique. The 6s, 5p, 5s, and 4f partial cross sections are the lowest order non-resonant partial cross sections. The solid-dashed line represents the atomic data of Costello *et al.* [195].

a measurement of the photoabsorption of atomic tungsten performed recently by Costello *et al.* [195] The length result is given by the solid line, and the velocity result is given by the dashed line. The solid-dashed line represents the measurement of Costello *et al.* [195] The large peak in the experimental spectrum at 37 eV is not well reproduced by the calculation. However, there is general qualitative agreement between the atomic measurement and the generalized resonance calculation above 43 eV. We mention that some of the discrepancy might be due to the approximations made in obtaining Eq. (4.77).

4.4 Summary

In this chapter, we have reviewed some of the details behind MBPT calculations of single-photoionization cross sections, and have provided examples of explicit calculations performed for platinum, xenon, barium, and tungsten. We have not discussed such important topics as photoionization with excitation and double photoionization processes. Aspects of these processes will be considered in other chapters. (See chapter five by T.-N. Chang for a discussion of photoionization with excitation, and chapter six by Z. Liu for a discussion of double photoionization.) We mention that one of the important strengths behind the use of MBPT is the property that a complete calculation should converge, order by order, to the exact answer. Additionally, we have demonstrated that, with the proper choice of a basis set of orbitals, one can isolate classes of interactions by evaluating only a subset of MBPT diagrams.

The main force behind the development and application of the techniques that were presented in this chapter was Hugh Kelly. Useful review articles on some of these topics by Kelly that have not already been referenced include a 1968 article included in *Advances in Theoretical Physics* [196] and a 1969 article in *Advances in Chemical Physics*. [34] Both of these articles provide a discussion of the linked-cluster expansion and some details concerning the occurrence and cancellation of exclusion-principle violating terms. There are also two review articles, one written in 1976 and included in *Photoionization and Other Probes of Many-Electron Interactions* [197] and the other written in 1985 and included *Fundamental Processes in Atomic Collision Physics*, [198] that provide a discussion of Rayleigh Schrödinger perturbation theory as well as details of photoionization cross section calculations.

5

Photoionization dominated by double excitation in two-electron and divalent atoms

T.-N. Chang

5.1 Introduction

Photoionization of a two-electron or a divalent atom in the vicinity of a doubly excited resonance due to the strong interaction between two outer electrons represents perhaps one of the most direct and unambiguous atomic processes for a quantitative study of the many-electron effects in atomic transitions. Recent applications of high resolution intense lasers and elaborate experiments in shorter wavelength regions with more advanced applications of synchrotron radiation to highly correlated atomic systems have opened up more opportunities for such a study. For example, operating at high synchrotron radiation intensity, Domke, Remmers, and Kaindl [199] have recently observed the doubly-excited $2pnd$ ^1P series of He below the He$^+$ $N = 2$ threshold with an improved energy resolution of $\cong 4$ meV. Parallel to the fast growing experimental advances, significant progress have also been made in many of the existing theoretical approaches.

One of the best known doubly excited photoionization structures is perhaps the He $2s2p\,^1P$ resonance shown in Fig. 5.1. This resonance was first observed in detail by Madden and Codling [200] in He photoionization from its ground state and classified subsequently by Cooper, Fano, and Prats. [201] Theoretically, the photoionization structure of an isolated resonance located at resonant energy E_r is best described by the Fano formula [202] in terms of a set of resonant parameters, which includes the resonance width Γ, the asymmetry parameter q, and the nonresonant background cross section σ_b. The peak cross section $\sigma_{\max}$, located at $E_{\max} = E_r + \frac{1}{2}(\Gamma/q)$ according to the Fano formula, equals $\sigma_b(1 + q^2)$. Also, the cross section is expected to reach a zero at an energy $E_{\min} = E_r - \frac{1}{2}(\Gamma q)$.

Physically, the resonant profile, which is determined by the asymmetry

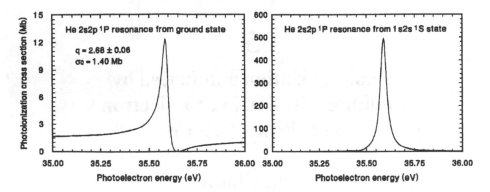

Fig. 5.1. Theoretical photoionization cross sections in the vicinity of $2s2p\,^1P$ resonance from both the ground state and the $1s2s\,^1S$ excited state of He. Data are taken from [207].

parameter q, measures qualitatively the interference between transitions from initial state to final bound and continuum components of the state wave function. If the contribution from the transition to the bound component is very small in comparison with the transition to the continuum background, q is very small and a zero cross section is expected either at or near E_r. For an intermediate q value, the resonant profile is generally asymmetric such as the one for the He $(1s^2\,^1S) \rightarrow$ He $(2s2p\,^1P)$ transition shown in Fig. 5.1. On the other hand, if the transition is strongly dominated by the contribution from the transition to the bound component of the state wave function, the q value is large, and a more symmetric photoionization structure, such as the one for the He $(1s2s\,^1S) \rightarrow$ He $(2s2p\,^1P)$ transition, also shown in Fig. 5.1, is expected. Unlike the q parameter, the resonant width Γ, which measures the interaction strength between the bound and continuum components of the state wave function of a doubly excited autoionization state, is independent of the transition process. The accuracy of an experimental determination of Γ is often limited by the difficulty in unraveling the asymmetric structure profile resulting from the simultaneous change in orbital of two electrons in a double excitation process. For a narrow resonance, the width measurement could also be hampered by the lack of adequate energy resolution. As a result, for a two-electron atom, the resonant widths are in general more readily available from the theoretical calculations than from the experimental measurements.

For a transition originated from a *bound* state confined in a finite volume to a continuum extending to infinity, a complete characterization of the asymptotically oscillating continuum wave functions is not necessarily required if the transition is dominated by the short-range interaction. In fact, the transition matrix can be evaluated accurately with an *effective* continuum wave function, which is a linear combination of an L^2 integrable basis confined in a finite volume and normalized with a proper boundary condition that is consistent with its asymptotic behavior at a large distance. Such an attempt was made earlier by Heller, Reinhardt, and co-workers, [203] which led to successful applications to electron-hydrogen scattering and photoejection of one and two electron from H⁻. Using an L^2 calculational procedure, Moccia and Spizzo [204] have also successfully represented the continuum spectrum by a set of elaborate one-electron orbitals including modified Slater-type orbitals (STOs) with an explicit $\cos(kr)$ dependence. In addition, Martin and co-workers [205] have employed the usual STO basis to determine the phase shifts of the continuum wave functions with an elaborate fitting procedure. The earlier multiconfiguration Hartree-Fock (MCHF) approach for the continuum has also been extended to study the resonance structure for the photoionization of He. [206] The purpose of this chapter is to review the basic theoretical features of an alternative L^2 method, *i.e.*, a B-spline-based configuration-interaction method for the *continuum spectrum* (CIC), [207] which has been applied recently with success to photoionization of a two-electron atom [208] and the single and multiphoton ionization of a divalent atom. [209]

In Section 5.2 we present the basic procedure of the B-spline-based configuration-interaction (CI) procedure, including the use of frozen-core Hartree-Fock (FCHF) orbitals for divalent atoms. In Section 5.3 we extend the basic procedure outlined in Section 5.2 to the *continuum spectrum*, including detailed discussion on calculational procedures for the scattering phase shifts, the energies and widths of the doubly excited resonant series, and the photoionization cross sections. Applications to two-electron atoms and divalent atoms are presented in Section 5.4.

5.2 B-spline-based configuration-interaction method

Within the Breit-Pauli approximation, [94] the total Hamiltonian consists of the N-electron nonrelativistic Hamiltonian H_{nr} and the sum of all relativistic contributions H_m. The N-electron nonrelativistic Hamiltonian is given in atomic units by

$$H_{nr}(\mathbf{r}_1, \mathbf{r}_2, \cdots) = \sum_{i=1}^{N} \left(-\frac{1}{2}\frac{d^2}{dr_i^2} - \frac{Z}{r_i} + \frac{1}{2}\frac{\mathbf{l}^2}{r_i^2} \right) + \sum_{i<j}^{N} \frac{1}{r_{ij}} \quad (5.1)$$

where Z is the nuclear charge and ℓ is the one-particle orbital angular momentum operator. In a straightforward configuration-interaction (CI) calculation, the energy eigenvalue and the state wave function corresponding to an energy eigenstate of a divalent or a two-electron atom can be obtained by diagonalizing a Hamiltonian matrix constructed from a basis set consisting of J-dependent *basis functions* $\Psi^{\Omega}_{n\ell n'\ell'}(\mathbf{r}_1, \mathbf{r}_2, \cdots)$, where $\Psi^{\Omega}_{n\ell n'\ell'}$ is characterized by a set of quantum numbers $\Omega \equiv (SLJM_J)$ and a two-electron configuration $(n\ell, n'\ell')$. S, L, J, and M_J are the total spin, the total orbital angular momentum, the total angular momentum and its corresponding magnetic quantum number, respectively. The basis functions $\Psi^{\Omega}_{n\ell n'\ell'}$ represents a sum of J-*independent* basis functions $\Psi^{\Lambda}_{n\ell n'\ell'}$ over all M_S and M, [210] *i.e.*,

$$\Psi^{\Omega}_{n\ell,n'\ell'} = \sum_{M_S,M} (-1)^{L-S} (2J+1)^{1/2} \begin{pmatrix} S & L & J \\ M_S & M & -M_J \end{pmatrix} \Psi^{\Lambda}_{n\ell,n'\ell'}, \quad (5.2)$$

where $\Lambda \equiv (SLM_S M)$ is a set of quantum numbers S, L, M_S, and M. M_S and M are the magnetic quantum numbers of S and L, respectively. Within the central field approximation, the basis function $\Psi^{\Lambda}_{n\ell,n'\ell'}$ is expressed as a sum of N-particle Slater determinant wave functions over all magnetic quantum numbers, *i.e.*, [211]

$$\Psi^{\Lambda}_{n\ell,n'\ell'}(\mathbf{r}_1, \mathbf{r}_2, \cdots) = \sum_{\text{all } m'_s} (-1)^{\ell'-\ell} [(2S+1)(2L+1)]^{1/2} \begin{pmatrix} \ell & \ell' & L \\ m & m' & -M \end{pmatrix}$$

$$\times \begin{pmatrix} \frac{1}{2} & \frac{1}{2} & S \\ m_s & m'_s & -M_S \end{pmatrix} \Phi^{mm_s m' m'_s}_{n\ell,n'\ell'}(\mathbf{r}_1, \mathbf{r}_2, \cdots), \quad (5.3)$$

where the Slater-determinant wave function is constructed from the one-electron orbitals u_β, *i.e.*,

$$\Phi^{mm_s m' m'_s}_{n\ell,n'\ell'}(\mathbf{r}_1, \mathbf{r}_2, \cdots) = (N!)^{-1/2} \det_N |u_\beta(\mathbf{r}_\mu)|, \quad (5.4)$$

and β represents a set of quantum numbers $n_\beta, \ell_\beta, m_\beta$, and m_{s_β}. The one-particle orbital function $u_{n\ell mm_s}$ is given by the product of its spatial and spin parts, *i.e.*,

$$u_{n\ell mm_s}(\mathbf{r}) = \frac{P_{n\ell}(r)}{r} Y_{\ell m}(\theta, \varphi) \sigma(m_s). \quad (5.5)$$

This CI procedure is essentially non-variational, and its success depends critically on the choice of $P_{n\ell}$ in the construction of the Hamiltonian matrix, since the only component in the basis functions that is not predetermined is the radial part $P_{n\ell}$ of the one-particle orbital function u.

For an N-electron divalent atom, the first $N - s2$ of the u_β in Eq. (5.4) represent the orbital functions for the N - 2 core electrons, and the remaining two orbital functions represent the orbitals of the two valence electrons denoted by a two-electron configuration $(n\ell, n'\ell')$. A factor of $2^{-1/2}$ is added to Eq. (5.3) when $n\ell = n'\ell'$ to ensure normalization. The nonrelativistic matrix element for H_{nr} between basis functions corresponding to a pair of two-electron configurations, $(n_\mu \ell_\mu, n_\nu \ell_\nu)$ and $(n_\delta \ell_\delta, n_\gamma \ell_\gamma)$, can be expressed as the sum of three energy terms, [211] *i.e.*,

$$\langle \Psi^\Lambda_{n_\mu \ell_\mu, n_\nu \ell_\nu} \mid H_{\mathrm{nr}} \mid \Psi^\Lambda_{n_\delta \ell_\delta, n_\gamma \ell_\gamma} \rangle = \sum_{i=0}^{2} E_i(n_\mu \ell_\mu \, n_\nu \ell_\nu, n_\delta \ell_\delta \, n_\gamma \ell_\gamma), \quad (5.6)$$

where

$$E_0(n_\mu \ell_\mu \, n_\nu \ell_\nu, n_\delta \ell_\delta \, n_\gamma \ell_\gamma)$$
$$= \left(\delta_{n_\mu n_\delta} \delta_{n_\nu n_\gamma} \delta_{\ell_\mu \ell_\delta} \delta_{\ell_\nu \ell_\gamma} + (-1)^{L+S+\ell_\delta+\ell_\gamma} \delta_{n_\mu n_\gamma} \delta_{n_\nu n_\delta} \delta_{\ell_\mu \ell_\gamma} \delta_{\ell_\nu \ell_\delta} \right) E_{\mathrm{core}}^{HF}, \quad (5.7)$$

$$E_1(n_\mu \ell_\mu \, n_\nu \ell_\nu, n_\delta \ell_\delta \, n_\gamma \ell_\gamma) = \delta_{\ell_\mu \ell_\delta} \delta_{\ell_\nu \ell_\gamma} (\delta_{n_\mu n_\delta} h_{\nu\gamma}^{\mathrm{HF}} + \delta_{n_\nu n_\gamma} h_{\mu\delta}^{\mathrm{HF}}) +$$
$$(-1)^{L+S+\ell_\delta+\ell_\gamma} \delta_{\ell_\mu \ell_\gamma} \delta_{\ell_\nu \ell_\delta} (\delta_{n_\mu n_\gamma} h_{\nu\delta}^{\mathrm{HF}} + \delta_{n_\nu n_\delta} h_{\mu\gamma}^{\mathrm{HF}}),$$
$$(5.8)$$

$$E_2(n_\mu \ell_\mu \, n_\nu \ell_\nu, n_\delta \ell_\delta \, n_\gamma \ell_\gamma)$$
$$= (-1)^{\ell_\nu - \ell_\delta} \left[\sum_k (-1)^L \left\{ \begin{matrix} \ell_\mu & \ell_\nu & L \\ \ell_\gamma & \ell_\delta & k \end{matrix} \right\} \langle n_\mu \ell_\mu, n_\nu \ell_\nu \parallel V^k \parallel n_\delta \ell_\delta, n_\gamma \ell_\gamma \rangle \right.$$
$$\left. + \sum_k (-1)^S \left\{ \begin{matrix} \ell_\mu & \ell_\nu & L \\ \ell_\delta & \ell_\gamma & k \end{matrix} \right\} \langle n_\mu \ell_\mu, n_\nu \ell_\nu \parallel V^k \parallel n_\gamma \ell_\gamma, n_\delta \ell_\delta \rangle \right], \quad (5.9)$$

and

$$h_{\nu\gamma}^{\mathrm{HF}} = \int dr \, P_{n_\nu \ell_\nu}(r) \, h_{\ell_\gamma}^{\mathrm{HF}}(r) \, P_{n_\gamma \ell_\gamma}(r). \quad (5.10)$$

The Hartree-Fock energy $E_{\mathrm{core}}^{\mathrm{HF}}$ for the $N-2$ core electrons is evaluated by using the radial functions $P_{n_0 \ell_0}$ of the occupied core orbitals, where $P_{n_0 \ell_0}$ satisfy the eigenequation

$$h_{\ell_0}^{\mathrm{HF}} P_{n_0 \ell_0} = \varepsilon_{n_0 \ell_0} P_{n_0 \ell_0}. \quad (5.11)$$

The one-particle Hartree-Fock Hamiltonian is given by

$$h_\ell^{\mathrm{HF}}(r) = \left(-\frac{1}{2} \frac{d^2}{dr^2} - \frac{Z}{r} + \frac{1}{2} \frac{\ell(\ell+1)}{r^2} \right) + V_\ell^{FCHF}(r), \quad (5.12)$$

and $V_\ell^{FCHF}(r)$ is the frozen-core Hartree-Fock (FCHF) potential defined explicitly by [211]

$$V_\ell^{FCHF}(r) f_\ell(r)$$
$$= \sum_{n_0 \ell_0}^{\mathrm{core}} 2 \left(\frac{2\ell_0 + 1}{2\ell + 1} \right)^{1/2} \langle \ell \parallel V^0(P_{n_0 \ell_0}, P_{n_0 \ell_0}; r) \parallel \ell \rangle f_\ell(r)$$
$$- \frac{1}{2\ell+1} \sum_{n_0 \ell_0}^{\mathrm{core}} \sum_\nu (-1)^\nu \langle \ell \parallel V^\nu(P_{n_0 \ell_0}, f_\ell; r) \parallel \ell_0 \rangle P_{n_0 \ell_0}(r), \quad (5.13)$$

where

$$\langle \ell \parallel V^{\nu}(a,b;r) \parallel \ell' \rangle = \langle \ell \parallel C^{[\nu]} \parallel \ell' \rangle \langle \ell_a \parallel C^{[\nu]} \parallel \ell_b \rangle$$
$$\times \int_0^{\infty} ds\, a(s)b(s) \frac{r_<^{\nu}}{r_>^{\nu+1}}. \qquad (5.14)$$

The Coulomb matrix in Eq. (5.9) is expressed in terms of the radial integral

$$\langle ab \parallel V^{\nu} \parallel cd \rangle = \int_0^{\infty} dr\, P_a(r) \langle \ell_a \parallel V^{\nu}(P_b, P_d; r) \parallel \ell_c \rangle P_c(r). \qquad (5.15)$$

and $\langle \ell \parallel C^{[\nu]} \parallel \ell' \rangle$ in Eq. (5.14) is the reduced matrix element of the tensor operator $C^{[\nu]}$ for spherical harmonics.

In a frozen-core Hartree-Fock (FCHF) approximation, all radial functions P included in the Hamiltonian matrix calculation are also generated from Eq. (5.11). The total Hartree-Fock core energy $E_{\text{core}}^{\text{HF}}$ is a constant, which may be set to zero for simplicity. As a result, the nonrelativistic Hamiltonian matrix element reduces to

$$\langle \Psi_{n_\mu \ell_\mu, n_\nu \ell_\nu}^{\Lambda} \mid H_{\text{nr}} \mid \Psi_{n_\delta \ell_\delta, n_\gamma \ell_\gamma}^{\Lambda} \rangle$$
$$= \left(\delta_{n_\mu n_\delta} \delta_{n_\nu n_\gamma} \delta_{\ell_\mu \ell_\delta} \delta_{\ell_\nu \ell_\gamma} + (-1)^{L+S+\ell_\delta+\ell_\gamma} \delta_{n_\mu n_\gamma} \delta_{n_\nu n_\delta} \delta_{\ell_\mu \ell_\gamma} \delta_{\ell_\nu \ell_\delta} \right) (\varepsilon_{n_\mu \ell_\mu} + \varepsilon_{n_\nu \ell_\nu})$$
$$+ E_2(n_\mu \ell_\mu \, n_\nu \ell_\nu, n_\delta \ell_\delta \, n_\gamma \ell_\gamma). \qquad (5.16)$$

Unlike other more elaborate CI methods, by using the predetermined one-particle orbital functions such as the FCHF radial functions generated by Eq. (5.11), the CI approach discussed above is carried out without the optimization procedure for each energy eigenstate. In spite of its simplicity, a direct application of this approach is often limited by its inability to include the positive-energy orbitals in the basis functions due to the numerical difficulty in the calculation of $\langle ab \parallel V^{\nu} \parallel cd \rangle$, resulting from the long-range behavior of the Coulomb interaction. In most of the earlier applications (*e.g.*, *truncated-diagonalization method or TDM approach*, [212, 213]) the basis functions are limited to the products of two negative-energy (bound) one-electron orbitals (*i.e.*, BB-type). The CI contribution from the positive-energy (continuum) orbitals, in terms of basis functions that include products of bound-continuum (BC-type) and continuum-continuum (CC-type) orbitals, is often excluded due to the numerical difficulties, even when considering highly correlated systems. This quantitative obstacle can be circumvented if one replaces the incomplete set of *bound-only* one-electron orbital functions by a nearly *complete* set of finite L^2 basis functions, which includes both bound and continuum one-electron orbital functions confined in a finite radius R. The radius R should be, in practice, larger than the estimated physical size of the energy eigenstates of interest.

The energy contribution from the intrashell core excitation and the intershell core-valence interaction for a divalent atom is represented by a parametrized long-range dipole core-polarization potential in the form of [214, 215, 216]

$$V_p = -\frac{\alpha_{pol}}{r^4} \left[1 - e^{-(r/r_0)^6} \right] \tag{5.17}$$

where α_{pol} is the static dipole polarizability and r_0 is a cut-off parameter for V_p as r approaches zero. For a more elaborate calculation, a short-range interaction [214, 215]

$$V_s = \sum_{\mu=0}^{2} a_\mu r^\mu e^{-\beta_1 r} + \sum_{\mu=0}^{2} b_\mu r^\mu e^{-\beta_2 r} \tag{5.18}$$

is also included to account for the additional interactions (*e.g.*, the relativistic effects) involving the core electrons. Recent detailed calculation by Chang and Chung [214] have concluded that the combined use of V_p and V_s to represent the core-related interactions is well supported except for those dynamical properties closely associated with the small-r behavior of the radial functions.

In a recent relativistic many-body perturbation calculation, Johnson *et al.* [67] have demonstrated *explicitly* the ability of the B-spline-based finite basis set to account for the many-body interactions in atomic process. Unlike the Slater-type orbitals, which favor the small-r region, the B-splines, with similar amplitude between $r = 0$ and $r = R$, are able to treat the entire physical region more uniformly. One of the key advantages in the application of B-splines is its independence of any *a priori* procedure in selecting the nonlinear parameters for the exponential functions. Detailed discussion of the basic properties of B-splines can be found elsewhere. [67, 102, 217] We will limit our discussion to those features related to the calculation of the one-particle radial functions P employed in the construction of the CI basis functions.

A set of B-splines of order K and total number n is often [67, 217] defined with an exponentially increasing knot sequence in a bound-state calculation. Such a knot sequence satisfies the need for, on the one hand, *densely* populated B-splines near the nucleus to accommodate the fast-rising inner s-orbitals at small r. On the other hand, more *evenly* populated B splines are required at larger r to represent the oscillating behavior of the positive energy orbitals at large distance. As a result, to take into account both the small- and large-r behavior of the orbital functions for transitions involving both a bound state and a continuum, we have chosen a sine-like knot sequence d_v defined by [207]

$$d_v = R \, \sin\left[\frac{\pi}{2} \left(\frac{(v-1)\Delta h}{R} \right)^y \right]; \quad v = 1, 2, \cdots, n-K+2, \tag{5.19}$$

where $\Delta h = R/(n - K + 1)$. The distribution of knot points can be adjusted by changing the position of the first non-zero knot d_2, which in turn determines the value of y according to Eq. (5.19). By employing such a knot sequence, we are able to limit the size of B-spline set to a modest n in our calculation.

The nonrelativistic one-particle radial functions P, subject to a non-local potential $V(r)$, satisfy an eigenequation, e.g., Eq. (5.11), in the form of

$$-\frac{1}{2}\frac{d^2 P}{dr^2} + V(r) P = \varepsilon P \tag{5.20}$$

The solution P can be expanded in terms of a set of B splines defined between $r = 0$ and $r = R$, i.e.,

$$P(r) = \sum_{i=1}^{n} c_i B_i(r). \tag{5.21}$$

The index K is dropped from the functions B_i for simplicity. All B splines equal zero except for B_1 and B_n at the endpoints $r = 0$ and $r = R$, i.e.,

$$B_1(r = 0) = 1 \quad \text{and} \quad B_n(r = R) = 1. \tag{5.22}$$

In a nonrelativistic calculation, the radial functions P are subject to the boundary conditions, $P(0) = P(R) = 0$, which can be satisfied if we set $c_1 = c_n = 0$. Substitution of Eq. (5.21) into Eq. (5.20) leads to a $(n - 2) \times (n - 2)$ generalized eigenvalue equation:

$$\underline{H}C = \varepsilon \underline{A}C, \tag{5.23}$$

where $\underline{H}$ and $\underline{A}$ are $(n - 2) \times (n - 2)$ matrices given by

$$H_{ij} = -\frac{1}{2}\langle B_i \mid \frac{d^2}{dr^2} \mid B_j \rangle + \langle B_i \mid V \mid B_j \rangle;$$
$$i \text{ and } j = 2, \ldots, (n - 1), \tag{5.24}$$

$$A_{ij} = \langle B_i \mid B_j \rangle; \qquad i \text{ and } j = 2, \ldots, (n - 1). \tag{5.25}$$

The radial eigenfunction P_v of Eq. (5.20) corresponding to an energy eigenvalue ε_v is given by

$$P_v(r) = \sum_{i=2}^{n-1} c_i B_i(r), \tag{5.26}$$

where the set of $n - 2$ coefficients c_i forms the eigenvector,

$$C_v = (c_2, c_3, \ldots, c_{n-1}), \tag{5.27}$$

of Eq. (5.23). The $n - 2$ radial eigenfunctions P_v of Eq. (5.20) form the one-particle finite basis set in the CI basis functions.

For each orbital angular momentum ℓ, the calculated $\varepsilon_{v\ell}$ of the first few lowest negative-energy eigenfunctions, which are completely confined in a radius R, should agree with the numerical results of Eq. (5.20) from direct integration. The positive-energy orbitals, with energy $\varepsilon_{v\ell}$ up to a few rydbergs, should exhibit an oscillatory behavior at large r. In practice, only those positive-energy orbitals with momentum $k_{v\ell} = (2\varepsilon_{v\ell})^{\frac{1}{2}}$, that satisfy the boundary conditions, $P_{v\ell}(r = 0) = P_{v\ell}(r = R) = 0$, or,

$$k_{v\ell}R + \frac{Z}{k_{v\ell}} \ln(2k_{v\ell}R) - \frac{\ell\pi}{2} + \delta_\ell^C + \delta_\ell = m\pi, \qquad (5.28)$$

appear in the nearly complete set of discretized radial functions $P_{v\ell}$. For a pure hydrogenic system, the short-range phase shift δ_ℓ vanishes and δ_ℓ^C equals the analytical Coulomb phase shift.

The CI basis set consists of a number of two-electron *configuration series* $n\ell\ell'$. Each $n\ell\ell'$ series includes a set of basis functions $\Psi_{n\ell,n'\ell'}^\Lambda$ corresponding to one of the valence electrons in a fixed orbital $n\ell$ and the other one with orbital angular momentum ℓ' but variable energy, both negative and positive, over an entire set of eigenfunctions P of Eq. (5.20). When the $n\ell$ orbital is bound, a configuration series is in theory equivalent to an open channel in close-coupling calculation. Such a series includes only the BB- and BC- types of configuration. The CC-type is included in the basis set when $n\ell$ represents a positive-energy orbital. Within the nonrelativistic FCHF approximation, the energy eigenvalue E_μ^Λ of a state $| \mu \rangle$ is calculated by diagonalizing the Hamiltonian matrix given by Eq. (5.16). The corresponding state function is given by

$$\Phi_\mu^\Lambda = \sum_{v\ell\ell'} \Xi_{\mu,v\ell\ell'}^\Lambda, \qquad (5.29)$$

where the *configuration series function*

$$\Xi_{\mu,v\ell\ell'}^\Lambda = \sum_{v'} C_\mu^\Lambda(v\ell, v'\ell') \, \Psi_{v\ell \, v'\ell'}^\Lambda \qquad (5.30)$$

represents the contribution to the state wave function from the $v\ell\ell'$ configuration series. A complete set of coefficients $C_\mu^\Lambda(v\ell, v'\ell')$ forms the eigenvector of the state $| \mu \rangle$.

For a divalent atom, the energy eigenvalue of an eigenstate dominated by a two-electron configuration $(n_i\ell_i n_o\ell_o)$ with the inner valence electron in an $n_i\ell_i$ orbital and the outer valence electron in an $n_o\ell_o$ orbital is denoted as $E_{n_i\ell_i n_o\ell_o}^\Lambda$. For simplicity, the energy corresponding to the ionization threshold with both valence electrons removed is set to zero. The calculated $E_{n_i\ell_i n_o\ell_o}^\Lambda$ corresponding to an energy eigenstate $n_i\ell_i n_o\ell_o \, (^{2S+1}L)$ can therefore be expressed in rydbergs in terms of an effective quantum number $v_{n_i\ell_i}$ leading to the $n_i\ell_i$ ionization threshold of the atomic ion, *i.e.*,

$$E^{\Lambda}_{n_i\ell_i n_o\ell_o}(N,Z) = -E^{I}_{n_i\ell_i}(N-1,Z) - \frac{Z^2_{\text{eff}}}{v^2_{n_i\ell_i}}, \qquad (5.31)$$

where $Z_{\text{eff}} = Z - N + 1$ is the effective nuclear charge and $E^{I}_{n_i\ell_i}(N-1,Z)$ is the ionization energy required to remove the $n_i\ell_i$ electron from the corresponding atomic ion of $N-1$ electrons.

The theoretical oscillator strengths in the dipole-length and dipole-velocity approximation, for an atomic transition from an initial state $| a \rangle$ to a final state $| b \rangle$, are given by [218]

$$f^{\ell}_{ba} = \frac{1}{3g_a}\Delta E_{ba} \sum_{\text{all } M} \left| \langle \Phi^{\Lambda_a}_a | \sum_{\alpha} \mathbf{r}_{\alpha} | \Phi^{\Lambda_b}_b \rangle \right|^2, \qquad (5.32)$$

and

$$f^{v}_{ba} = \frac{4}{3g_a}\Delta E^{-1}_{ba} \sum_{\text{all } M} \left| \langle \Phi^{\Lambda_a}_a | \sum_{\alpha} \nabla_{\alpha} | \Phi^{\Lambda_b}_b \rangle \right|^2, \qquad (5.33)$$

respectively. The transition energy

$$\Delta E_{ba} = E^{\Lambda_b}_b - E^{\Lambda_a}_a \qquad (5.34)$$

is given in rydbergs, and $g_a = (2S_a + 1)(2L_a + 1)$ is the degeneracy of the initial state. The oscillator strength f_{ba} is positive for an *absorption* when ΔE_{ba} is positive, and it is negative for an *emission* when ΔE_{ba} is negative.

For transitions limited to the two outershell electrons, the sum over the square of the dipole matrix in Eqs. (5.32) and (5.33) can be evaluated with a straightforward application of the angular momentum algebra, and the oscillator strengths in the dipole-length and dipole-velocity approximation are given by

$$f^{\ell}_{ba} = \delta_{S_b S_a}\frac{2L_b + 1}{3}\Delta E_{ba} \left| F^{\ell}_{ba} \right|^2 \qquad (5.35)$$

and

$$f^{v}_{ba} = \delta_{S_b S_a}\frac{4(2L_b + 1)}{3}\Delta E^{-1}_{ba} |F^{v}_{ba}|^2, \qquad (5.36)$$

respectively. The oscillator strength f^{emi}_{ab} for an *emission* from an upper state $| b \rangle$ to a lower state $| a \rangle$ is related to the oscillator strength f^{abs}_{ba} for an *absorption* from a lower state $| a \rangle$ to an upper state $| b \rangle$ by

$$f^{\text{emi}}_{ab} = -\left(\frac{2L_a + 1}{2L_b + 1}\right) f^{\text{abs}}_{ba}. \qquad (5.37)$$

The transition amplitude F_{ba} is given by

$$F_{ba} = \sum_{j,i} C_b^{\Lambda_b}(n'_j\ell'_j, n_j\ell_j) \, C_a^{\Lambda_a}(n'_i\ell'_i, n_i\ell_i) \, \mathbf{D}_{ba}^{ji}, \qquad (5.38)$$

where

$$\begin{aligned}
\mathbf{D}_{ba}^{ji} &= d_{ba}(j'j, i'i) + d_{ba}(jj', ii') + (-1)^{S_a} d_{ba}(j'j, ii') \\
&\quad + (-1)^{S_b} d_{ba}(jj', i'i) \qquad (5.39)
\end{aligned}$$

is the dipole transition matrix between configurations $(n'_j\ell'_j, n_j\ell_j)$ and $(n'_i\ell'_i, n_i\ell_i)$. For a configuration representing two equivalent electrons, a factor of $2^{-1/2}$ should be added. The matrix element d_{ba} is the product of the angular coefficient ρ and the one-particle radial integrals, *i.e.*,

$$d_{ba}(j'j, i'i) = \rho(\ell'_j\ell_j\ell'_i\ell_i; \Lambda_b\Lambda_a)\langle P_{n_j\ell_j} \mid P_{n_i\ell_i}\rangle\langle P_{n'_j\ell'_j} \mid t \mid P_{n'_i\ell'_i}\rangle, \qquad (5.40)$$

where $\langle P_{n'_j\ell'_j} \mid t \mid P_{n'_i\ell'_i}\rangle$ is the one-particle radial dipole matrix element and t is the radial part of the position and gradient operators in the length and velocity approximation, respectively. The overlap integral $\langle P_{n_j\ell_j} \mid P_{n_i\ell_i}\rangle = \delta_{n_jn_i}\delta_{\ell_j\ell_i}$ in the FCHF approximation. The angular factor ρ is given by

$$\begin{aligned}
\rho(\ell_1\ell_2\ell_3\ell_4; \Lambda_b\Lambda_a) &= (-1)^{\ell_1}\delta_{\ell_2\ell_4}[(2\ell_1 + 1)(2\ell_3 + 1)]^{1/2} \\
&\quad \times \begin{pmatrix} \ell_1 & 1 & \ell_3 \\ 0 & 0 & 0 \end{pmatrix}\begin{Bmatrix} L_a & 1 & L_b \\ \ell_1 & \ell_4 & \ell_3 \end{Bmatrix}, \qquad (5.41)
\end{aligned}$$

where Λ_α represents the quantum numbers S, L, M_S, and M associated with the state $\mid \alpha\rangle$.

5.3 Configuration-interaction method for continuum spectrum

In this section, the B-spline-based configuration-interaction procedure outlined in the previous section will be extended to the continuum spectrum of a two-electron or a divalent atom with an $ns^2\,{}^1S$ ground state. The spectra of two-electron or divalent atoms, at energies above the first ionization threshold, are often dominated by a series of doubly excited autoionization resonances embedded in a single continuum open-channel $ns\ell$. The state wave function Φ_E^Λ at an energy E can be expressed as the sum of two terms, *i.e.*,

$$\Phi_E^\Lambda = \Xi_{E,ns\ell}^\Lambda + \sum_{\mu_0\ell_0\ell'} \Xi_{E,\mu_0\ell_0\ell'}^\Lambda, \qquad (5.42)$$

where the first term represents the $ns\ell$ ionization channel and the second term denotes the combined contribution of doubly excited configurations from all closed channels. The kinetic energy ε and the momentum k of the ionized electron are given by $\varepsilon = k^2/2 = E + E_I$, where E_I is the ionization energy of the remaining ns electron after the removal of the first ns electron.

In a direct scattering calculation, the correct wave function of an outgoing ℓ electron with momentum k is given by the expression

$$\left(\frac{2}{\pi k}\right)^{1/2} \sin\left[kr + \frac{q}{k}\ln(2kr) - \frac{\ell\pi}{2} + \delta_\ell^C + \delta_\ell\right]; \tag{5.43}$$

as $r \to \infty$, where q is the effective nuclear charge experienced by the outgoing electron and δ_ℓ is the scattering phase shift due to the short range interaction. In the present calculation, the scattering phase shift is determined by comparing the oscillating part of the configuration series function $\Xi_{E,ns\ell}^\Lambda$ for the $ns\ell$ open-channel with the asymptotic expression given by Eq. (5.43). First, we express the configuration series function $\Xi_{E,ns\ell}^\Lambda$ in a form identical to the Slater determinant function $\Psi_{ns\xi_{\varepsilon\ell}}^\Lambda$, where the second radial function is replaced by a one-particle radial function

$$\xi_{\varepsilon\ell}(r) = \sum_v C_E^\Lambda(ns, v\ell)P_{v\ell}(r). \tag{5.44}$$

Second, the numerical function $\xi_{\varepsilon\ell}(r)$ is matched at a finite r against an asymptotic expression [219]

$$\xi_{\varepsilon\ell}(r) \longrightarrow A\left[\frac{k}{\zeta(r)}\right]^{\frac{1}{2}} \sin[\phi(r) + \delta_\ell] \tag{5.45}$$

as $r \to R$, where ζ and ϕ are functions of r. As $r \to \infty$, $\zeta \to k$ and

$$\phi \longrightarrow \left(kr + \frac{q}{k}\ln(2kr) - \frac{\ell\pi}{2} + \delta_\ell^C\right). \tag{5.46}$$

An accurate representation of the continuum by a discretized finite basis set depends critically on the matching between the calculated radial function and the correct asymptotic expression over extended r with a constant amplitude A. Although the oscillating function $\xi_{\varepsilon\ell}(r)$ is a sum of one-particle radial functions which are subject to a nuclear charge of Z, asymptotically, the effective nuclear charge $q = Z - 1$ for a two-electron or divalent atom. Fig. 5.2 represents two calculated $\xi_{\varepsilon\ell}(r)$ at $k = 0.10949$ and 0.40495, respectively, for electron-hydrogen scattering below the $n = 2$ threshold. The matching between the calculated $\xi_{\varepsilon\ell}(r)$ and the expression given by Eq. (5.45) is nearly perfect. As a result, the scattering phase shift δ_ℓ can be determined without help from any elaborate fitting procedure such as the one proposed by Martin and Salin. [204]

At energy near the doubly excited resonances, the phase shift increases rapidly by a total of π. In our calculation, a set of closely populated energy eigenvalues is needed to describe the detailed energy variation of

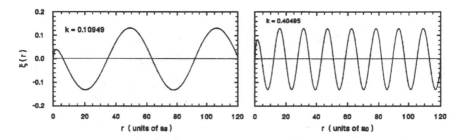

Fig. 5.2. The calculated one-particle functions $\xi(r)$ representing the H^- $1ss$ configuration series at momenta $k = 0.10949$ and 0.40495. Data are taken from [207].

the scattering phase shift across the resonance. This is carried out by repeating the calculation at slightly varying values of R. The energy E_r and the width Γ of the resonance are determined by a least squares fit of the phase shifts to the usual expression [220, 221]

$$\delta_\ell(E) = \sum_{i=0}^{2} a_i E^i + \tan^{-1} \frac{\Gamma/2}{E_r - E}. \tag{5.47}$$

With the orbital function of the outgoing electron normalized asymptotically by Eq. (5.43), the photoionization cross section (in unit of a_0^2) from an initial state $|I\rangle$ at photon energy E_γ (in atomic units) is given by

$$\sigma = \frac{8}{3}\pi^2\alpha \, g(E_\gamma) \, |\mathbf{D}_{EI}|^2 \, , \tag{5.48}$$

where α is the fine structure constant and $g(E) = E$ and E^{-1} for the dipole length and velocity approximations, respectively. The dipole matrix between the initial state $|I\rangle$ and the final state $|E\rangle$ is given by

$$\mathbf{D}_{EI} = \langle \Phi_E^\Lambda | \hat{\mathbf{D}}(1,2) | \Phi_I^\Lambda\rangle, \tag{5.49}$$

where

$$\hat{\mathbf{D}}(1,2) = \mathbf{D}(\mathbf{r}_1) + \mathbf{D}(\mathbf{r}_2), \tag{5.50}$$

and $\mathbf{D}$ represents the position and gradient operators for the length and velocity approximations, respectively.

With the final state wave function Φ_E^Λ expressed in terms of a discretized finite basis set, the normalization constant $(2/\pi k)^{\frac{1}{2}}$ in Eq. (5.43) should be replaced by the amplitude A given in Eq. (5.45), i.e., a constant

$$N_k = \frac{2/(\pi k)}{A^2} \tag{5.51}$$

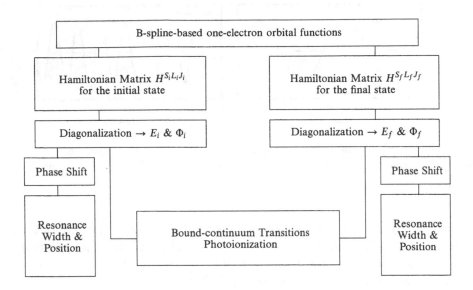

Fig. 5.3.   Computational procedure for the CIC calculation.

is added to Eq. (5.48) and the dipole matrix $\mathbf{D}_{EI}$ is replaced by $\mathbf{D}_{EI}^{d}$, or

$$\sigma = \frac{8}{3}\pi^2\alpha \; g(E_\gamma) \; N_k \; \left|7\mathbf{D}_{EI}^{d}\right|^2 . \tag{5.52}$$

The new dipole matrix element $\mathbf{D}_{EI}^{d}$, similarly defined by Eq. (5.49), is evaluated using the state wave functions Φ_E^Λ and Φ_I^Λ calculated with the discretized finite basis set. As a result, the photoionization cross section can be expressed in terms of the oscillator strength f_{EI} for absorption by the simple relation

$$\sigma = \frac{4\pi\alpha}{kA^2}f_{EI}, \tag{5.53}$$

where f_{EI} is calculated numerically following the same procedure for the bound-bound transitions given in Section 5.2.

The computational procedure employed in the CIC method is summarized in the flow chart shown in Fig. 5.3. In our numerical calculations, which typically involve a few excited states in each of the doubly excited resonant series, a radius of $R = 120$–$140a_0$ is used. The B-spline set is generally limited to a modest size of $n = 80$–100 with $K = 9$–11. Depending upon the energy variation of the phase shift, approximately fifteen to twenty-five R values are included in our calculation for a complete description of a series of doubly excited resonant states.

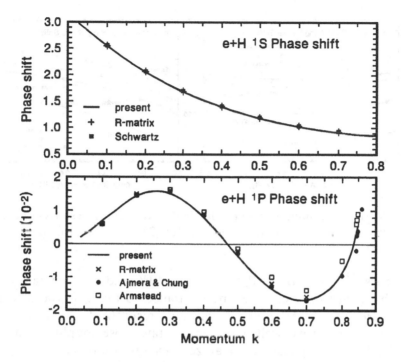

Fig. 5.4. The calculated S- and P-wave phase shifts for electron-hydrogen scattering. Solid line, present: Ref. [207]; $+$ and $\times$, R-matrix: Ref. [222]; boxes (solid, Schwartz, and open, Armstead): Ref. [223]; solid circles, Ajmera and Chung: Ref. [224].

5.4 Applications

The accuracy of the CIC method is first demonstrated by the excellent agreement shown in Fig. 5.4 between the calculated phase shifts for ^1S and ^1P electron-hydrogen scattering and a few selected earlier and recent theoretical calculations. [222, 223, 224] The contributions from different configuration series to the scattering phase shifts can also be examined in detail when we vary the size of the basis set of the CIC calculation. One such example is shown in Fig. 5.5 in terms of the variation of the nonresonant $e-He^+$ $^{1,3}P^o$ scattering phase shifts as we expand the basis set from a single $1sp$ configuration series to a total of 89 series. The phase shifts from the single ionization channel calculation, represented by the solid curves labeled as "$1sp$ only", differ significantly from our final results. The phase shift corrections due to the addition of $2sp$, $2ps$, and $2pd$ series to the $1sp$ series, account for about half of the phase shift difference. The calculated phase shifts, with contribution limited to that from the $1sp$, $2sp$, $2ps$, and $2pd$ series, agree well with the $1s2s2p$ close-coupling results

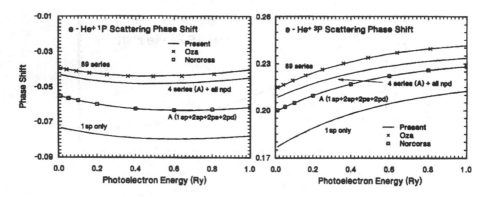

Fig. 5.5. Variation of nonresonant $e-\text{He}^{+\ 1,3}P$ scattering phase shifts as functions of photoelectron energy. Solid line, present: Ref. [237]; ×, Oza: Ref. [226]; boxes, Norcross: Ref. [225].

of Norcross [225] as shown. Between 30%-40% and approximately 25% of the phase shift difference can be attributed to the *npd*-type of configuration series for the ^1P and ^3P continua, respectively. Our converged phase shifts agree very well with the 20-state close-coupling results by Oza. [226] Also, the phase shifts at zero energy approach the expected values of $\pi\mu$ (where μ is the quantum defect), which equal -0.038 and 0.214 for the ^1P and ^3P series, respectively. All two-electron configuration functions included in the nonresonant photoionization calculations are of the bound-bound or bound-continuum type, *i.e.*, at least one negative energy single-particle orbital function is included in each of the two-electron basis functions. We have tested and found that the contribution from the continuum-continuum type of two-electron functions to the scattering phase shift in the nonresonant region is negligible.

Fig. 5.6 presents the nonresonant H^- and He photoionization cross sections. Our calculated photodetachment cross sections from the H^- ground state agree very well with most of the more accurate earlier theoretical calculations. [203, 227, 228, 229] A more recent calculation using hyperspherical coordinates [230] has led to cross sections in the length form which are slightly higher near the peak region and slightly lower at higher energy. In addition, a simple model calculation by Crance and Aymar [231] (not shown) has led to cross sections which are significantly smaller at low energy but larger at high energy than most of the more elaborate theories. Only the dipole-length results are shown in Fig. 5.6. Our dipole-velocity and dipole-length results agree to about 1%-2% or better for the entire energy range. Our calculated photoionization cross sections for transitions from the He ground state also agree very well with the experimental data compiled recently by Samson, [232] which are accurate to 1%-2%.

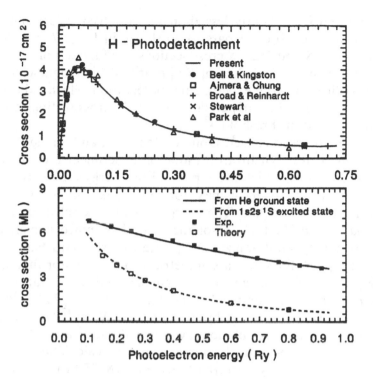

Fig. 5.6. The nonresonant photoionization cross sections for H$^-$ and He. Solid line, present: Ref. [207]; solid circles, Bell and Kingston: Ref. [227]; rectangles, Ajmera and Chung: Ref. [228]; +, Broad and Reinhardt: Ref. [203]; ×, Stewart: Ref. [229]; open triangles, Park et al.: Ref. [230]; solid rectangles, experiment: Ref. [232]; open rectangles, theory: Ref. [233].

Also shown in Fig. 5.6 is the close agreement between the present calculation and the earlier theory by Jacobs [233] for the nonresonant photoionization cross sections from the excited $1s2s\ {}^1S$ state of He. In fact, for the $1s2s\ {}^1S^e \to {}^1P^o$ photoionization, most of the theoretical cross sections, including the present results and those by Burgess and Seaton, [234] Jacobs, [233] and Dalgarno and co-workers, [235] are in close agreement with each other and they also agree well with the observed data near the ionization threshold. [236] Theoretical photoionization cross sections at selected photoelectron energies are compared in Table 5.1. Near the threshold, our theoretical results differ noticeably from the close-coupling calculation by Norcross, [225] which employed an essentially uncorrelated initial state wave function. For the $1s2s\ {}^3S^e \to {}^3P^o$ photoionization, our theoretical results also agree reasonably well with other earlier theoretical results. [233, 234, 235] Similarly to the earlier results of Burgess and Seaton, [234] our near-threshold $1s2s\ {}^3S^e \to {}^3P^o$ cross sections shown in

Fig. 5.7 appear to be slightly less than the observed values but clearly within the experimental error bars. Also shown in Fig. 5.7 is the ratio R between the $1s2s$ ^1S and $1s2s$ ^3S cross sections as a function of wavelength. At shorter wavelength, the values of R appear to deviate significantly from unity. This is different from experimental observation, which indicates that the $1s2s$ ^1S cross section nearly equals the $1s2s$ ^3S cross section at energies close to the ionization threshold. [236]

As we have remarked earlier, Domke, Remmers, and Kaindl [199] have recently resolved in a synchrotron radiation experiment the He $(sp, 2n^-)$ and $(2pnd)$ $^1P^o$ series, which are separated by energy differences ranging from a maximum of 16.4 ± 0.4 meV for the $4^-/3d$ pair to an estimated 4 ± 2 meV for the $7^-/6d$ pair. The energy separations between the neighboring $2pnd$ and $sp, 2(n+1)^-$ resonances are determined numerically by fitting the measured photoionization spectra to a standard Fano formula for the $sp, 2(n+1)^-$ lines and a symmetric monochromator function corresponding to a weighted sum of a Gaussian and a Lorentzian profile for the $2pnd$ lines. Theoretically, the $(2pnd)$ and $(sp, 2n^-)$ $^1P^o$ series were first studied quantitatively by Burke and McVicar [220] using a four-channel close-coupling calculation over two decades ago. Other early studies include a nine-channel coupled-equation calculation within the framework of many-body perturbation theory (MBPT) by Salomonson, Carter, and Kelly, [238] an L^2-based R-matrix method by Gersbacher and Broad, [239] and a STO-based L^2-basis method by Sánchez and Martin. [205] These two narrow $^1P^o$ series have also been studied qualitatively in other recent theoretical works, including, among others, the L^2-basis calculation by Moccia and Spizzo, [204] the variational R-matrix calculation by Hamacher and Hinze, [241] and the spline-based multi-configuration Hartree-Fock (MCHF) calculation by Froese Fischer and Idrees. [206]

Our calculated photoionization profiles in the vicinity of the $(sp, 23^-)$, $(2p3d)$, and $(sp, 24^-)$ $^1P^o$ resonances, using the B-spline-based CIC method, are shown in Fig. 5.8. We have shifted the theoretical energies to the lower energy side by 3 meV for all three resonances to facilitate a direct comparison with the observed spectra. Since, experimentally, only the relative data are given, to normalize the observed spectra we have to match two values of the *photoionization yield* to the calculated cross sections. For a specific photoionization yield ϕ, we start by measuring the energy difference Δ between corresponding photon energies on opposite sides of the observed resonance. To determine the photoionization cross section σ corresponding to the photoionization yield ϕ, we then match the experimental energy difference Δ to the theoretical energy difference between the two photon energies, which correspons to σ. Following this procedure, the photoionization yields corresponding

Table 5.1. The nonresonant photoionization cross sections (in Mb) from He $1s2s$ $^{1,3}S$ metastable states. The photoelectron energy ε is given in rydbergs. Only length results are listed. (Values taken from Ref. [237])

ε (Ry)	$1s2s$ ^1S			$1s2s$ ^3S		
	Change & Zen[a]	Jacobs[b]	Norcross[c]	Change & Zen[a]	Jacobs[b]	Norcross[c]
0.01	8.798	8.753	9.350	5.345		4.751
0.05	7.258	7.128	7.083	4.804		4.332
0.10	5.803	5.671	5.413	4.188		4.042
0.15	4.722	4.473		3.654	3.537	
0.20	3.902	3.821	3.904	3.199	3.157	3.509
0.25	3.268	3.233		2.812	2.796	
0.30	2.770	2.768		2.485	2.480	
0.40	2.047	2.068	2.122	1.968	1.970	1.993
0.60	1.225	1.233	1.258	1.302	1.311	1.268
0.80	0.795	0.794	0.858	0.913	0.926	0.918
1.00	0.546	0.543	0.490	0.671	0.685	0.629
1.20	0.390	0.388	0.319	0.512	0.527	0.500
1.40	0.285	0.284	0.277	0.404	0.418	0.449
1.60	0.212	0.211	0.219	0.328	0.342	0.372

[a] Ref. [237]
[b] Jacobs [233]
[c] Norcross [225]

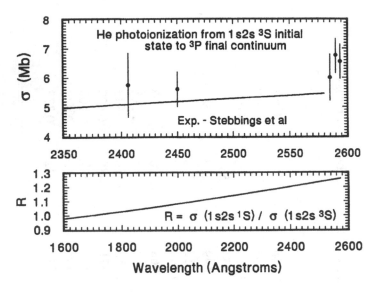

Fig.   5.7.   Near-threshold   photoionization   cross   sections   from   He $1s2s\,^3S$
metastable state and the ratio R between the $1s2s\,^1S$ and $1s2s\,^3S$ photoionization cross sections. Experiment, Stebbings *et al.*: Ref. [236].

to $\Delta = 5$ meV and 7.5 meV for the $sp, 23^-$ resonance are normalized
to theoretical cross sections of 1.37 Mb and 1.32 Mb, respectively. We
have also applied the same normalization procedure to the observed $2p3d$
and $sp, 24^-$ spectra shown in Fig. 5.8. The calculated and the observed
structure profiles appear to agree well. Our calculation clearly suggests
a rapid drop to a nearly zero cross section on the high energy side of
the $sp, 2n^-$ resonances. The absence of an expected sharp dip in the
experimental spectra (with a "width" less than 2 meV) next to the $2p3d$
resonance can be attributed to the lack of experimental energy resolution. Similarly, the top portions of the $2pnd$ resonance structures are
also absent in the observed profiles when their widths become less than
2 meV.

The calculations by Salomonson, Carter, and Kelly [238] and Sánchez
and Martin [205] have also suggested an asymmetric resonance profile for
the $2pnd$ series with a significantly larger negative q value than the one
for the $sp, 2n^-$ series. As a result, in contrast to the observed spectra,
we conclude that a higher peak photoionization cross section σ_{max} is expected for the $2pnd$ resonances than that for the $sp, 2n^-$ resonances. The
numerical resonance widths and the peak photoionization cross sections
σ_{max} for selected $(sp, 2n^+)$, $(sp, 2n^-)$, and $(2pnd)^1P^o$ series below the He$^+$
$N = 2$ threshold are given in Ref. [208].

In contrast to the strongly asymmetric spectrum commonly found in

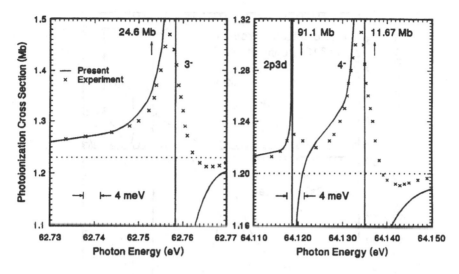

Fig. 5.8. Comparison of the calculated and the observed photoionization cross sections near the doubly excited $(sp, 23^-)$, $(2p3d)$, and $(sp, 24^-)$ $^1P^o$ resonances of the He atom below the $N = 2$ threshold. Solid line, present: Ref. [208]; ×, experiment: Ref. [199].

photoionization from the ground state, the resonance structures of the spectra from the bound excited states are generally symmetric with their peak cross sections several orders of magnitude greater than the cross sections from the ground state. For the $1s2s$ $^1S \rightarrow sp, 22^+$ ^1P transition, our theoretical $\sigma_{max} = 519.3$ Mb is in good agreement with the theoretical value of 541 Mb by Dalgarno and co-workers. [235] It differs noticeably from the close-coupling results of 436 Mb and 384 Mb by Norcross [225] and Jacobs, [233] respectively. This disagreement may be partially attributed to the approximately 15% overestimation in width from the close-coupling calculation [237] since σ_{max} is inversely proportional to Γ (see, e.g., Eq. (46) of Ref. [243]). For the $1s2s$ $^3S \rightarrow sp, 22^+$ ^3P transition, our theoretical $\sigma_{max} = 2554$ Mb is greater than the results from all three previous calculations. [225, 233, 235] This discrepancy can be attributed *entirely* to the difference in the calculated resonant widths (see, e.g., Table IV of Ref. [237]). In fact, σ_{max} would range from 2400 to 2500 Mb for these three calculations if adjustment due to the difference in resonance width is taken into account.

A more recent calculation, based on a comparison between the photoionizations from the $2s^2$ ^1S ground state and the $2s3s$ ^1S bound excited state of Be, shown in Fig. 5.9, has further demonstrated that both the resonant energy E_r and the resonant width Γ of a doubly excited autoionization state of a two-electron or a divalent atom (*e.g.*, alkaline-

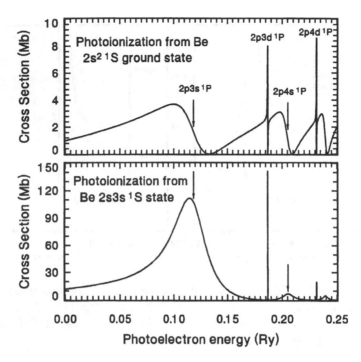

Fig. 5.9. Photoionization cross sections from the $2s^2{}^1S$ ground state and the $2s3s^1S$ bound excited state of the Be atom. Data are taken from [244].

earth atom) can be determined more directly from photoionization orig-
inating from a bound excited state. [244] Experimentally, the absorption
spectrum from the ground state of Be was first examined in detail by
Mehlman-Balloffet and Esteva [245] and Esteva *et al.* [246] Our theoretical spectrum from the ground state is in good agreement with most of
the earlier calculations [204, 247, 248, 249] as well as the observed absorption
spectrum. [246, 245] The ground state spectrum shown in Fig. 5.9 clearly exhibits a substantial overlap between the broad $2p(n+1)s^1P$ and the narrow
$2pnd^1P$ series in spite of a weak interaction strength between these two
series. [250] A qualitative estimate, based entirely on the observed absorption spectrum from the ground state, has led to a resonant width at least
two to three times greater than the earlier theoretical estimates [251, 252]
for the broad $2pns^1P$ resonances. Contrary to the ground state spectrum,
our calculation has shown that the broad and the narrow series are completely separated in the photoelectron spectrum from the bound excited
$2s3s$ ^1S state. As a result, the estimated resonant widths are reduced
substantially and are in closer agreement with the earlier close-coupling
results.

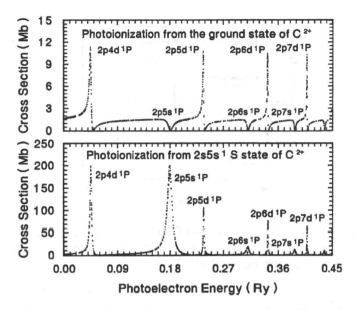

Fig. 5.10. Photoionization cross sections from the $2s^2\,^1S$ ground state and the $2s3s^1S$ bound excited state of Be-like C^{2+} ion. Data are taken from [244].

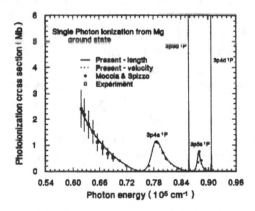

Fig. 5.11. Photoionization cross section for transitions to the ^1P continuum from the ground state of Mg. Solid and dotted lines, length and velocity: Ref. [209]; solid circles, Moccia and Spizzo: Ref. [201]; rectangles, experiment: Ref. [254].

The large energy separation between the E_{peak} and E_r (indicated by the arrows in Fig. 5.9) in the Be ground state spectrum represents a resonant profile corresponding to small q values. This is consistent with the $q \sim 0$ resonant profile from the ground state of C^{+2} shown in Fig. 5.10, as the nuclear charge Z increases along the Be isoelectronic sequence. Our C^{+2} calculation has shown that the transition amplitude

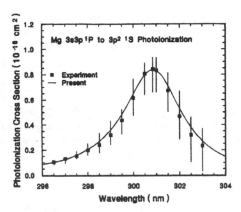

Fig. 5.12. Photoionization cross section from the singly excited $3s3p$ ^1P state of Mg to the doubly excited $3p^2$ ^1S resonance in the continuum. Solid line, present: Ref. [209]; rectangles, experiment: Ref. [256].

for the $2s^2$ $^1S \rightarrow 2pns^1P$ photoionization is completely dominated by the one-electron $2s^2 \rightarrow 2s\varepsilon p$ *bound-to-continuum* transition. The zero cross section results directly from the sign change of the dipole matrix at E_r due to the sign change of the effective radial function when the scattering phase shift increases rapidly by a total of π across the resonance. In contrast, the $2pns$ ^1P resonant profile in the photoelectron spectrum from the $2s5s$ ^1S bound excited state represents a completely different type of transition, *i.e.*, the one that is strongly dominated by the excitation of a *single bound* inner electron, *i.e.*, $2s \rightarrow 2p$, followed by a "shake-up" of the outer ns electron. This is similar to the process involving a two-electron excitation of core electron studied elsewhere. [253] Since the contribution from the direct $2s5s \rightarrow 2s\varepsilon p$ *bound-to-continuum* transition is small, the $2pns^1P$ autoionization state is represented by a nearly symmetric resonant profile corresponding to a large q value.

The CIC procedure has also been applied to the photoionization of neutral Mg from its ground state. [209] In Fig. 5.11, our calculation is compared with the recent theoretical results by Moccia and Spizzo [204] and the most recent *absolute* photoabsorption cross sections measured by Yih *et al.* [254] Our results are also consistent with the normalized experimental data by Preses *et al.* [255] In addition, our results are in close agreement with other earlier theoretical results, including the velocity results by Bates and Altick [221] and the MCHF results by Froese Fischer and Saha. [206] At energies near the ionization threshold, the cross sections from all the above mentioned theoretical calculations are noticeably less than the results from the earlier version of the eigenchannel R-matrix method by O'Mahony and Greene. [250] Fig. 5.12 compares our theoretical cross sections for the photoionization from the $3s3p\,^1P$ bound excited state in Mg at energies close to the $3p^2\,^1S$ autoionization state with the observed data compiled from an earlier two-color two-step experiment. [256]

The experimental results are normalized against the theoretical cross sections at the peak value, and they are in excellent agreement with the theoretical results. Our calculated length and velocity results agree to better than 5%. Only the length results are shown.

Based on the recent applications of the CIC method to multiphoton processes in Mg, [209] the energy spectrum of a *one-color* multiphoton process does not necessarily reveal atomic structure effects that are dominated by the final state multi-electron interactions in the continuum, in spite of the possibility of reaching directly the higher-L states of both odd and even parity as the number of photons increases. Perhaps, a better approach to studying the atomic structure effect in the continuum, instead, is to examine photoionization from a highly correlated bound excited state to a strongly correlated autoionization state in a *multi-color multi-step* process.

6

Direct double photoionization in atoms

Z. Liu

6.1   Introduction

In a double photoionization process one photon is absorbed and two electrons are ejected. It is an interesting process since it cannot occur without electron correlations, although it can be roughly approximated by shake-off or relaxation effects. Another important aspect is that the two escaped electrons and the residual ion form a three-body Coulomb system. The angular distribution of the photoelectrons and the energy dependence of the cross section near threshold is of great interest. With increasing use of synchrotron radiation in investigation of atomic and molecular systems, this process has attracted considerable attention, [257]-[261] and provides valuable information on dynamic effects of electron correlations.

Double photoionization of an atom can proceed in several ways. The simplest process is direct double photoionization where the photon absorption leads to the simultaneous ejection of two electrons, i.e.

$$\hbar\omega + A \longrightarrow A^{++} + 2e^-. \tag{6.1}$$

Double photoionization can also occur via Auger processes. In the case of two-step double ionization, photoionization of an inner-shell electron takes place in the first step, then after a certain lifetime the inner-shell hole is filled through an Auger transition, i.e.

$$\hbar\omega + A \longrightarrow A^{*+} + e^-, \tag{6.2}$$

$$A^{*+} \longrightarrow A^{++} + e^-. \tag{6.3}$$

In the second step, the excited ion state A^{*+} decays by changing either the electronic configuration of the excited core or its multiplet coupling in a spin-flip process. [261] Another class of indirect or sequential double photoionization is the occurrence of two-electron emission during a Auger decay where, in the first step,

$$\hbar\omega + A \longrightarrow A^*, \tag{6.4}$$

and then, either an Auger shake-off occurs

$$A^* \longrightarrow A^{++} + 2e^-, \tag{6.5}$$

or a two-step autoionization occurs

$$A^* \longrightarrow A^+ + e^-, \quad A^+ \longrightarrow A^{++} + e^-. \tag{6.6}$$

The net production of doubly charged ions comes from a delicate balance of competing processes due to direct or indirect mechanisms.

Both direct and indirect processes give significant contributions to the double charged ion production. A recent analysis shows that in the sudden limit up to 30%-40% of the double-ionization rate in neon and argon is due to Auger decay initiated by an excited electron. [261] By using photoelectron spectroscopy and coincidence techniques, [258] however, the direct and indirect double photoionization processes are distinguishable. Their contributions can be recognized by their energy partition pattern and the angular distribution of the fragments. In the direct double photoionization process, the excess energy beyond the double ionization threshold I is shared continuously between the kinetic energies of the two escaped electrons : $E_1 + E_2 = \hbar\omega - I$, and the measurement should be made through a triply differential cross section $d^3\sigma/dE_1 d\Omega_1 d\Omega_2$, where Ω_1 and Ω_2 are the solid angles of electrons 1 and 2, respectively. By contrast, the excess energy in the two-step double ionization process appears as $E_1 = \hbar\omega - E_{A^+}$ for the photoelectron and $E_2 = E_{A^+} - E_{A^{++}}$ for the Auger electron, where E_{A^+} and $E_{A^{++}}$ denote the binding energies of the singly and doubly charged ionic states. With these fixed kinetic energies, only a two-fold differential cross section $d^2\sigma/d\Omega_1 d\Omega_2$ is needed to establish the complete fragmentation pattern.

In this review, we focus only on the area of direct double photoionization where many-body perturbation theory (MBPT) has been successfully applied in several cases. [262]-[270] For previous review articles about this topic readers are referred to Starace, [164] Amusia [271] and Kelly. [272] Chang, Ishihara, and Poe carried out the first MBPT study for the double photoionization of neon. [262] Carter and Kelly performed the first MBPT calculation of double photoionization for an open-shell atom of carbon, [264] and developed a very useful method of choosing appropriate potentials to approximately take higher-order correlation effects into account. [267] Kelly and his coworkers have also applied MBPT to argon [263, 265, 268] and to other systems. [270] Recent measurements of the ratio of the double- to single-ionization rates at a photon energy of 2.8

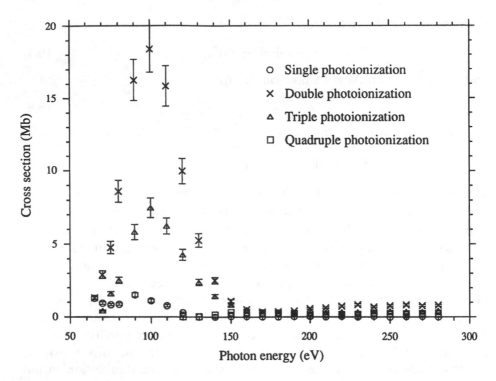

Fig. 6.1. Partial cross sections for photoionizations of Xe measured by Holland *et al.* [257].

keV for helium [273] have renewed interest in the asymptotic behavior of photoionization cross sections. [269, 274]

Here one point should be made. Up to now, except for helium, all works thus far have used MBPT to calculate the direct double photoionization process for many-electron atoms, since MBPT can give reasonable results by using a small number of partial wave pairs to describe the final-state wave function with two continuum electrons. MBPT also offers an advantage that through diagrams one can assess the relative importance of the different processes that are involved in the calculation.

It is known that the double ionization process can contribute a substantial percentage of the total ionization cross section. The ratio of double- to single-electron ionization probabilities varies from about 0.04 in helium to 0.1 in neon, 0.2 in argon, 1 in krypton and larger than 1 in xenon. [257, 275] Figure 6.1 shows cross sections for singly to quadruply charged ions of xenon as measured by Holland *et al.* [257] Van der Wiel and Chang [276] pointed out that the 4d shell polarized due to photon interaction may play an essential role to the production of Xe^{++} ions. However, there are still no attempts to calculate σ^{3+} or σ^{4+} to interpret the available data. The double ionization rate is also quite large for barium, and the absorption spectrum associated with 5p excitation is complex. [277, 278] Most of the previous measurements studied the cases when both photoelectrons are ejected from the outer shell with the same principal quantum number. Recently Farnoux *et al.* reported a double photoionization measurement of copper involving the 3d and 4s subshells over an energy range of 40-125 eV, [259] and Wuilleumier *et al.* measured double ionization of sodium from 3s and 2p electrons between 52 eV and 135 eV. [260] Analysis of these experimental data is certainly desirable to enhance our understanding of electron correlation effects.

In Section 6.2 we derive a basic formalism for MBPT calculation of direct double photoionization. Section 6.3 and 6.4 review how to include higher-order corrections to the electron correlation and dipole interactions, respectively. Finally, in Section 6.5, we list some topics for future research.

6.2 Basic formalism

The photoionization cross section $\sigma(\omega)$ is related to the imaginary part of the frequency-dependent dipole polarizability $\alpha_{pol}(\omega)$ by [169]

$$\sigma(\omega) = \frac{4\pi\omega}{c} \text{Im} \left[\alpha_{pol}(\omega)\right] , \qquad (6.7)$$

where ω is the photon energy and c is the speed of light. Atomic units are employed throughout except where noted otherwise.

The contribution to $\alpha_{pol}(\omega)$ from the doubly ionized final state $|\Psi_{pq}^{k_1k_2}\rangle$ is given by

$$\alpha_{pol}(\omega) = -\sum_{k_1k_2} |Z(pq \rightarrow k_1k_2)|^2 \left(\frac{1}{E_0 - E_f + \omega} + \frac{1}{E_0 - E_f - \omega}\right) , \quad (6.8)$$

where the dipole matrix elements can be expressed either in length-form as

$$Z(pq \rightarrow k_1k_2) = \langle\Psi_{pq}^{k_1k_2}| \sum_{i=1}^{N} z_i |\Psi_0\rangle \qquad (6.9)$$

or in velocity-form as

$$Z(pq \rightarrow k_1k_2) = \frac{1}{\omega}\langle\Psi_{pq}^{k_1k_2}| \sum_{i=1}^{N} \frac{\partial}{\partial z_i} |\Psi_0\rangle , \qquad (6.10)$$

E_0 and E_f are the energy eigenvalues corresponding to the ground state Ψ_0 and the final state $\Psi_{pq}^{k_1k_2}$. Both Ψ_0 and the final state $\Psi_{pq}^{k_1k_2}$ are eigenstates of the full atomic Hamiltonian in the absence of an external electric field. The sum over the continuum states k_1 and k_2 in Eq. (6.8) is converted to integration. The first denominator in Eq. (6.8) may vanish and is treated as

$$\lim_{\eta \to 0}(D + i\eta)^{-1} = PD^{-1} - i\pi\delta(D),\tag{6.11}$$

where P represents a principal-value integration. By the help of the δ-function in Eq. (6.11), one of the integrations over k_1 and k_2 in Eq. (6.8) can be carried out. With the excited-state single-particle continuum orbitals normalized in the k scale according to

$$\lim_{r \to \infty} R_k(r) \sim \frac{1}{r} \cos \left[kr + \frac{Q}{k} \ln(2kr) - \frac{\pi}{2}(\ell + 1) + \delta_\ell^C + \delta_\ell \right],\tag{6.12}$$

where the single-particle potential $V(r) \to -Q/r$ as $r \to \infty$ and Q is the asymptotic charge of the potential, the double photoionization cross section becomes

$$\sigma^{++}(\omega) = \frac{16\omega}{c} \int_0^{k_{max}} dk_1 \frac{|Z(pq \to k_1 k_2)|^2}{k_2},\tag{6.13}$$

where k_2 is related to k_1 by

$$k_2 = [2(\omega - k_1^2/2 - I)]^{1/2}\tag{6.14}$$

and the upper limit of integration over k_1 is restricted by

$$k_{max} = [2(\omega - I)]^{1/2}.\tag{6.15}$$

Numerically one can calculate the two-dimensional matrix $\mathcal{D}(k_1, k_2) = |Z(pq \to k_1 k_2)|^2$ for a set of preselected continuum k values, and then use an interpolation procedure to obtain the required values of $\mathcal{D}(k_1, k_2)$ on a finer k mesh for the integration in Eq. (6.13). In order to avoid numerical difficulty as $k_2 \to 0$ when $k_1 \to k_{max}$, one can use the relation $dk_1/k_2 = -dk_2/k_1$, and obtains

$$\sigma^{++}(\omega) = \frac{16\omega}{c} \int_0^{k_{max}/\sqrt{2}} dk_1 \frac{\mathcal{D}(k_1, k_2) + \mathcal{D}(k_2, k_1)}{k_2}.\tag{6.16}$$

This procedure is straightforward, but it is difficult to get reliable interpolation results when ω goes to a higher energy range. Alternatively, one can directly calculate the matrix elements in Eq. (6.16) for two sets of k_1 and k_2, related by Eq. (6.14), for a given ω value without invoking interpolation. Of course, the calculations will be more demanding. A third way is as follows. The values of k_1 and k_2 satisfy

$$k_1^2 + k_2^2 = k_{max}^2,\tag{6.17}$$

so they lie on a photon-energy arc. Let

$$k_1 = k_{max} \cos \theta, \quad k_2 = k_{max} \sin \theta \quad \text{with } 0 < \theta < \pi/2,\tag{6.18}$$

it is easy to show that

$$\sigma^{++}(\omega) = \frac{16\omega}{c} \int_0^{\pi/2} d\theta \, \mathscr{D}(k_{max}\cos\theta, k_{max}\sin\theta) \,. \qquad (6.19)$$

Neglecting the spin-orbit interaction and other relativistic effects, the perturbation expansion for $Z(pq \to k_1 k_2)$ is developed [196] with the atomic Hamiltonian

$$H = H_0 + H_c \,, \qquad (6.20)$$

where

$$H_0 = \sum_{i=1}^{N} -\frac{\nabla_i^2}{2} - \frac{Z}{r_i} + U(r_i) \qquad (6.21)$$

and

$$H_c = \sum_{i<j=1}^{N} v_{ij} - \sum_{i=1}^{N} U(r_i) \,. \qquad (6.22)$$

The term v_{ij} represents the Coulomb interaction between electron pairs, and the single-particle potential $U(r_i)$ is chosen to account for the average interaction of the ith electron with the other electrons. The unperturbed wave function is represented in the Russell-Saunders LSM_LM_S coupling scheme as a linear combination of determinants, each containing N different single particle states ϕ_n which are solutions of

$$[-\frac{1}{2}\nabla^2 - \frac{Z}{r} + U(r)]\phi_n = \varepsilon_n \phi_n \,. \qquad (6.23)$$

Usually U is taken as the Hartree-Fock potential V_{HF}^N for the N occupied orbitals. However, the full potential is not unique as is seen by considering the following general potential

$$U = V_{HF}^N + (1 - \mathbf{P})\hat{\Theta}(1 - \mathbf{P}) \,, \qquad (6.24)$$

where

$$\mathbf{P} = \sum_{i=1}^{N} |n_i\rangle\langle n_i| \qquad (6.25)$$

ensures orthogonality of excited- and ground-state orbitals. Note that $\hat{\Theta}$ is an arbitrary Hermitian operator which is chosen to represent a desired physical situation. [279] If one takes $V_{HF}^K = \langle \Psi_K | \sum_{ij} v_{ij} | \Psi_K \rangle$, where K represents N, $N-1$, or $N-2$ electron LS-coupled state, then the choice that $\hat{\Theta} = V_{HF}^{N-1} - V_{HF}^N$ corresponds to the widely used V^{N-1} potential. Carter and Kelly [267] showed that the choice that $\hat{\Theta} = V_{HF}^{N-2} - V_{HF}^N$, i.e. a V^{N-2} potential, is useful in the calculations for double photoionization which will be discussed in the next section.

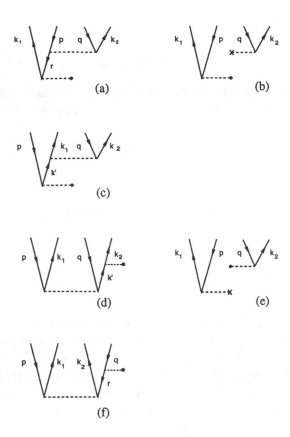

Fig. 6.2. Diagrams contributing to the matrix element $Z(pq \rightarrow k_1 k_2)$. Full circles indicate dipole interaction; dashed lines represent Coulomb interactions; crosses denote single-particle potential interactions. Exchange diagrams are also to be included.

Individual terms in the perturbation expansion for $Z(pq \rightarrow k_1 k_2)$ can be expressed by open diagrams. The lowest-order double photoionization diagrams contain one dipole interaction and one interaction with the electron-correlation perturbation H_c of Eq. (6.22), as shown in Figs. 6.2(a-f). [262, 265] The time ordering of the interactions proceeds graphically from bottom to top, with diagram 6.2(a-c) representing final-state correlations (FSC) and diagrams 6.2(d-f) representing ground-state correlations (GSC). Exchange diagrams are always understood to be included. In the following diagrams with single-particle potential terms such as Fig. 6.2(b) (Fig. 6.2(e)) are always considered together with terms such as Fig. 6.2(a) (Fig. 6.2(d)).

Figure 6.2(a), when $p = r$ and q and k_2 have the same angular momentum quantum number, is closely related to a monopole transition, and represents core rearrangement effects. [262] When an electron is ionized by dipole interaction, the rearrangement of the remaining core may cause the removal of another electron. This mechanism is often referred to as a shake-off process. Figure 6.2(a) with $p \neq r$ represents a virtual Auger transition. One inner-shell electron is the primary photoelectron excited to the continuum orbital through the dipole interaction. An outer-shell electron then fills the empty inner-shell state with simultaneous ejection of another electron. Figure 6.2(c) is an inelastic scattering diagram where an outgoing electron exchanges energy and momentum with another electron leading to the ionization of a second electron. Figures 6.2(d-f) represent contributions due to ground-state correlations which are also important in double photoionization. Figures 6.2(d) and 6.2(f) involve respectively two and three occupied orbitals in the ground state, respectively.

The general form of the energy denominators occurring in FSC diagrams is

$$D = \sum (\varepsilon_{\text{in}} - \varepsilon_{\text{out}}) + \omega \qquad (6.26)$$

and for GSC diagrams,

$$D = \sum (\varepsilon_{\text{in}} - \varepsilon_{\text{out}}), \qquad (6.27)$$

where ε_{in} (ε_{out}) is the single-particle energy associated with the hole-(particle-) orbital line, cut by a line immediately above the interaction line. In Fig. 6.2(c) the denominator Eq. (6.26) may vanish and should be treated according to Eq. (6.11)

Certain classes of diagrams may be taken [196] as a geometric sum to produce shifts in the single-particle energies corresponding to electron correlations. They can be included semiempirically by replacing the sum over single-particle energies ε_{in} of Eqs. (6.26), (6.27) with minus the experimental removal energy.

In Fig. 6.2(a) with $p \neq r$, when $\varepsilon_p + \varepsilon_q > \varepsilon_r$, the energy denominator can vanish, and the singularity should be removed by summing over certain Auger-transition diagrams to all orders. A detailed discussion will be given in Section 6.4.

The individual diagrams of Fig. 6.2 are normally evaluated in the usual manner by a sum over intermediate bound states and integral over the continuum. Diagram 6.2(d), however, should be evaluated by the differential-equation technique, [262, 280] because it involves calculations of the dipole matrix elements between pairs of continuum orbitals. In radial form (quantities with subscript $\mathscr{R}$ refer to radial parts only) diagram 6.2(d) becomes

$$\mathscr{D}_{\mathscr{R}} = \sum_{k'} \frac{\langle k_2 | r | k' \rangle_{\mathscr{R}} \, \langle k_1 k' | v | pq \rangle_{\mathscr{R}}}{\varepsilon_p + \varepsilon_q - \varepsilon_{k_1} - \varepsilon_{k'}}, \qquad (6.28)$$

where v represents the radial part of the Coulomb interaction and the

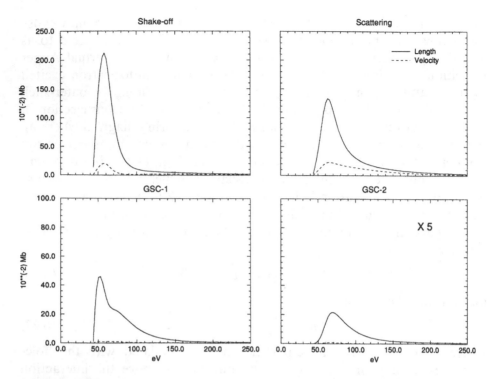

Fig. 6.3.   Cross sections from individual diagrams of Fig. 6.2 for the Ar $3p^2 \rightarrow$ *kpkd* transition. Solid lines, length form; broken line, velocity form.

sum over k' is limited to excited single-particle states. If one defines $|\Psi_{k_1}\rangle_{\mathcal{R}}$ such that

$$|\Psi_{k_1}\rangle_{\mathcal{R}} = \sum_{k'} \frac{|k'\rangle_{\mathcal{R}} \langle k_1 k'|v|pq\rangle_{\mathcal{R}}}{\varepsilon_p + \varepsilon_q - \varepsilon_{k_1} - \varepsilon_{k'}} , \qquad (6.29)$$

then $\mathcal{D}_{\mathcal{R}}$ is evaluated according to

$$\mathcal{D}_{\mathcal{R}} = \langle k_2|r|\Psi_{k_1}\rangle_{\mathcal{R}} . \qquad (6.30)$$

Letting $E = \varepsilon_p + \varepsilon_q - \varepsilon_{k_1}$, Eq. (6.29) takes on the form

$$(E - H_0)|\Psi_{k_1}\rangle_{\mathcal{R}} = \sum_{k'} |k'\rangle_{\mathcal{R}} \langle k_1 k'|v|pq\rangle_{\mathcal{R}} . \qquad (6.31)$$

Since $E < 0$, the solution $|\Psi_{k_1}\rangle_{\mathcal{R}}$ has the nature of a bound function.

For closed-shell atoms such as argon, a $3p^2$ pair may be ejected leaving the ion in one of the states $3p^4(^3P,^1D,^1S)$. Alternatively, a $3s3p$ pair can be ejected leaving the ionic state $3s3p^5(^{3,1}P)$ in *LS*-coupling. For the outgoing pair, there is an infinite number of possible partial waves

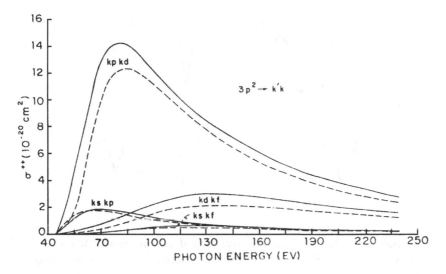

Fig. 6.4. Partial-wave cross sections for the four $3p^2 \rightarrow k_1 k_2$ channels [265]. Results are plotted for both dipole-length (solid curves) and dipole-velocity (dashed curves) matrix elements.

with the restriction that the total angular momentum of the ion to give a total ^1P state. Although the $3p^2$ pair contributions are the larger, the $3s3p$ pairs do contribute approximately 20% of the total double photoionization cross section at 89 eV photon energy. [265] For the $3p^2$ pairs, the largest contribution comes from the $kpkd$ partial wave with decreasing contributions from $kdkf$, $kskp$, and $kskf$ partial waves. It is reasonable to expect that including other higher-ℓ partial waves is not necessary. Figure 6.3 gives cross sections from contributions of the individual diagrams of Fig. 6.2 for the argon $3p^2 \rightarrow kpkd$ transition. Figure 6.4 gives partial cross sections for $3p^2 \rightarrow kpkd$, $kskp$, $kdkf$ and $kskf$ channels.

Figures 6.3 and 6.4 show that there are substantial cancellations between different diagrams. The matrix elements $Z(pq \rightarrow k_1 k_2)$ from Figs. 6.2(a) usually have the opposite sign to those from Figs. 6.2(c) and 6.2(d). At the maximum of the cross section the shake-off diagram of Fig. 6.2(a) is the largest. At higher energies the scattering diagram of Fig. 6.2(c) is relatively more important. Near the threshold the GSC diagram of Fig. 6.2(d) gives significant contributions.

The resultant cross section comes from a delicate interference of the different processes, and it is sensitive to the choice of multiple-basis sets and higher-order corrections.

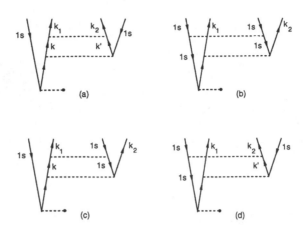

Fig. 6.5. Second-order final-state correlation diagrams associated with ladder, hole-hole, and hole-particle interactions. Exchange diagrams are also to be included.

6.3   Correction to electron correlation interaction

One important finding of the work of Carter and Kelly [267] is that, by using a mixture of V^{N-1} and V^{N-2} potentials to obtain multiple-basis sets, higher-order corrections to electron-electron correlation can be taken into account in the lowest-order MBPT calculation.

Second-order electron-correlation terms include hole-hole, particle- particle (ladder) and hole-particle interactions. Figures 6.5(a–d) show the four second-order correction diagrams arising from the scattering digram of Fig. 6.2(c). Each of the four diagrams can be individually large, but they tend to cancel among themselves. This is because each of these has about the same order of magnitude and diagrams 6.5(a),(b) have the opposite algebraic sign from diagrams 6.5(c),(d). As mentioned before, in calculations of the lowest-order diagrams, higher-order energy correlation sifts can be included by appropriate insertion in the denominator expressions of Eqs. (6.26), (6.27). This procedure correctly accounts for the hole-hole interactions to all orders, and reproduces the correct double ionization threshold. Now it is necessary to include both ladder and hole-particle interactions, otherwise an imbalance will occur.

Carter and Kelly [267] in their calculation for helium suggested that the ladder and hole-particle corrections in the final state can be significantly reduced by simply calculating one set of continuum orbitals k_1 in a V_{HF}^{N-1} potential and the other continuum orbitals k_2 in a V_{HF}^{N-2} potential. With this method, diagram 6.5(c) remains the same and corrects for the V^{N-1} potential used for electron k_1. However, diagram 6.5(d) now vanishes since electron k_2 has been calculated as if both $1s$ core electrons were missing. Ladder diagram 6.5(a) is then to a large degree canceled by the remaining diagram 6.5(c), and the cancellation continues order by order.

The ladder and hole-particle interaction corrections to the ground state, as shown in Figs. 6.6(a–d), must be treated somewhat differently. When a set of V^{N-2} orbitals is used, both diagrams 6.6(c) and 6.6(d) vanish. Diagram 6.6(b) and its higher orders can always be summed out to yield an energy shifted denominator in diagram 6.2(d). For the remaining diagram 6.6(a), one can use the following finding in MBPT. It is known from previous correlation-energy calculations [22] that the ratio R of a diagram in a given class in the nth order of perturbation expansion to the diagram in the same class at $(n-1)$th order is approximately constant. One can thus determine R by explicitly calculating the third-order ladder and second-order correlation energy diagrams and taking their ratio. Relative to diagram 6.2(d), the effects of ladder diagram 6.6(a) and all high orders can then be approximated by the geometric factor $(1-R)^{-1}$ which multiplies diagram 6.2(d). Generally the factor $(1-R)^{-1}$ will reduce the lowest-order amplitude.

Carter and Kelly [267] calculated σ^{++} of helium for the two partial-wave channels $1s^2 \rightarrow kskp\ ^1P$ and $1s^2 \rightarrow kpkd\ ^1P$. Additional channels of the form $1s^2 \rightarrow klkl+1\ ^1P$ with $\ell > 1$ are expected to be quite small and were omitted. With the use of combination V^{N-1} and V^{N-2} continuum orbitals, the results shown in Fig. 6.7 were in good agreement with experimental measurement. [281] Earlier calculations by Byron and Joachain [282] and by Brown [283] were based on a fully correlated initial-state wave function, but neglected correlations in the final state by representing the final state as a symmetrized product of uncorrelated Coulomb wave functions, each seeing a charge of $Z = 2$. The MBPT method, however, shows that ground- and final-state correlations are equally important, with final-state correlation terms actually being the larger of the two.

The final state of double photoionization can mix with channels which consist of single photoionization with excitations. In first-order perturbation theory, these final states are expressed as

$$|F\rangle = |\Psi_{pq}^{k_1k_2}\rangle + \sum_{k,n} |\Psi_{pq}^{kn}\rangle \frac{\langle \Psi_{pq}^{kn}|H_c|\Psi_{pq}^{k_1k_2}\rangle}{\varepsilon_{k_1} + \varepsilon_{k_2} - \varepsilon_k - \varepsilon_n} . \qquad (6.32)$$

Then the mixed terms lead to higher-order corrections. This coupling between the final-state channels can be included to all orders by solving a coupled system of integral equations. [183] The coupled integral equations

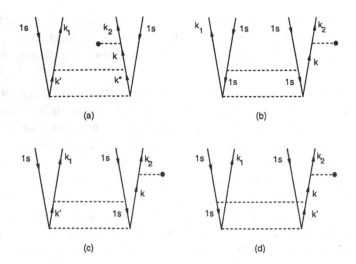

Fig. 6.6. Second-order ground-state correlation diagrams associated with ladder, hole-hole, and hole-particle interactions. Exchange diagrams are also to be included.

are written out explicitly as*

$$\bar{Z}(pq \to k_1 k_2) = Z(pq \to k_1 k_2) + \sum_{k,n} \frac{\langle k_1 k_2 | H_c | kn \rangle \bar{Z}(pq \to kn)}{\varepsilon_p + \varepsilon_q - \varepsilon_k - \varepsilon_n + \omega} , \quad (6.33)$$

$$\bar{Z}(pq \to kn) = Z(pq \to kn) + \sum_{k_1,k_2} \frac{\langle kn | H_c | k_1 k_2 \rangle \bar{Z}(pq \to k_1 k_2)}{\varepsilon_p + \varepsilon_q - \varepsilon_{k_1} - \varepsilon_{k_2} + \omega} , \quad (6.34)$$

where Z is an uncorrelated matrix element evaluated as in Fig. 6.2 and $\bar{Z}$ is a correlated matrix element to be determined. Generally, the mixing of double-photoionization channels and single-photoionization-with-excitation channels will reduce the double-photoionization cross section, because part of the incident photon flux is absorbed by satellite channels.

6.4   Correction to dipole interaction

Calculations of the dipole vertices of diagrams 6.2(a)-(c) can be improved by including correlation corrections, as shown in Fig. 6.8. Addition of these diagrams yields a substantial correction, particularly in the region near the maximum of the cross section, and also gives better length- and velocity-form agreement. [265]

---

* See Section .4.3.1.

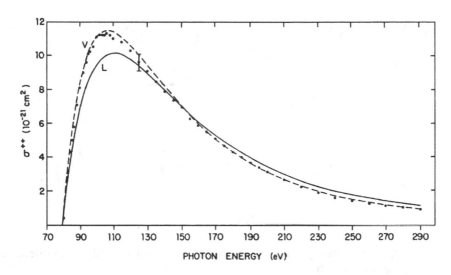

Fig. 6.7. Theoretical calculations of the total σ^{++} for He [267]. Curves for the correlated length (L) and velocity (V) cross sections include contributions from both *kskp* and *kpkd* channels. Experimental data are from Bizau *et al.* [281].

Another important higher-order correction is illustrated in Fig. 6.9.

Diagram 6.9(a) represents $r \to k_1$ photoionization followed by an Auger process in which the hole state r is filled by an outer shell electron p with a transition of another outer shell electron q to k_2. The unperturbed energy denominator for this Auger process is given by

$$D = \varepsilon_r - \varepsilon_{k_1} + \omega \ . \tag{6.35}$$

For a given ω, the denominator of Eq. (6.35) may vanish at a particular value of k_1 when r is an inner shell electron with respect to p and q. In this case the energy of the doubly excited final state $|\Psi_{pq}^{k_1 k_2}\rangle$ is degenerate with that of the singly excited intermediate state $|\Psi_r^{k_1}\rangle$. The pole in Eq. (6.35) can be removed by including higher-order diagrams such as shown in Fig. 6.9(b). The latter forms a geometric series with ratio

$$R = \left[-i\frac{1}{2}\Gamma(\omega) + \Delta(\omega) \right] / D \ , \tag{6.36}$$

where

$$\frac{1}{2}\Gamma(\omega) = \frac{2}{k} \, |\langle\Psi_r^{k_1}|v|\Psi_{pq}^{k_1 k_2}\rangle|^2 \tag{6.37}$$

and

$$\Delta(\omega) = \frac{2}{\pi} \mathrm{P} \int dk \frac{|\langle\Psi_r^{k_1}|v|\Psi_{pq}^{k_1 k}\rangle|^2}{\varepsilon_p + \varepsilon_q - \varepsilon_{k_1} - \varepsilon_k + \omega} \ . \tag{6.38}$$

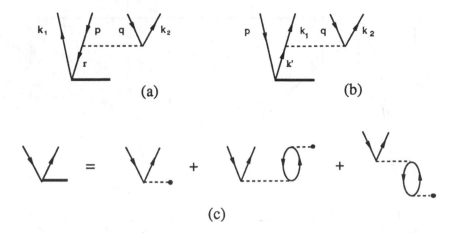

Fig. 6.8.   (a), (b) Diagrams from Fig. 6.2 involving the correlated dipole operator (c) for diagram (a) and (b).

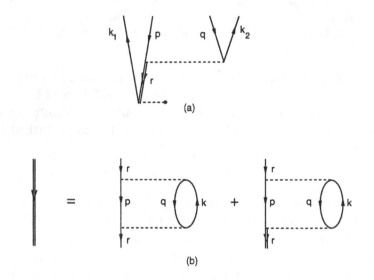

Fig. 6.9.   Virtual-Auger diagram (a) with its insertion segment (b).

The term Δ is interpreted as a contribution to the correlation energy of the $\Psi_r^{k_1}$ state. The term $\frac{1}{2}\Gamma$ is the lowest-order half width associated with the $\Psi_r^{k_1}$ state and determines the Auger rate for that state. When there is more than one channel, the geometric series can be similarly constructed by first summing over these channels in the diagram segment, then Δ and Γ involve a sum over the ionization channels.

Holland *et al.* [257] found that the argon double photoionization cross section, which decreases as the photon energy increases from 100 eV, starts

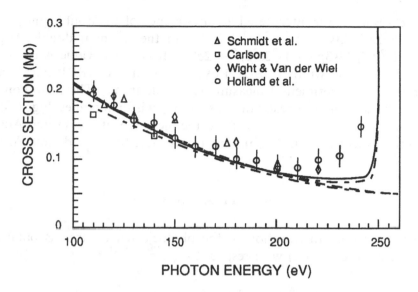

PHOTON ENERGY (eV)

Fig. 6.10. Double photoionization cross section for Ar leading to the $2p$ threshold. Theoretical calculations [268]: solid line, including relaxation effects and diagram 6.9(a); dashed and dotted line, including diagram 6.9(a) but without relaxation effects; dashed and double-dotted line, without diagram 6.9(a); experimental measurement by Schmidt *et al.* [284], Carlson [275], Wight and Van der Wiel [285] and Holland *et al.* [257].

to rise at photon energies near 220 eV which is well below the excitation energy of the $2p \rightarrow nd, ns$ resonance series.

Pan and Kelly [268] investigated this phenomenon by including the diagram 6.9 with $r = 2p$. The denominator for this diagram was shifted by insertions on the $2p$ hole line of the Auger process in which one $3p$ electron fills the $2p$ hole and a second $3p$ electron is ejected to the continuum. This diagram becomes dominant as the $2p$ threshold is approached. In principle, if this process is included there is no $2p$ single photoionization because the single $2p$ vacancy final state always decays by the Auger process and the correct final state has two electron hole states. Theoretical results for the argon σ^{++} cross section are given in Fig. 6.10 compared with experiment. It is seen that the calculated results show a sharp rise as the $2p$ threshold is approached. However, the calculated results, whether in the relaxed or unrelaxed basis, do not rise from 220 eV as measured by Holland *et al.* [257] This discrepancy needs further study, and it would be interesting to have measurements of σ^{++} leading up to inner shell threshold in other atoms.

Recently the Wuilleumier group [286] reported an observation of a Fano profile in the direct double photoionization process in sodium. The

observed resonance is interpreted as interference of the double photoionization $1s^22s^22p^63s \rightarrow 1s^22s^22p^5 + 2e^-$ with the resonant double Auger
process $1s^22s2p^63s3p \rightarrow 1s^22s^22p^5 + 2e^-$. Measured q value is equal to
$-1.40(15)$, and the width Γ $0.72(3)$ eV. Performing a MBPT calculation
to confirm the experimental measurement is desirable. The main contribution comes from the resonance $2s \rightarrow 3p$, which modifies Fig. 6.8(c).
A secondary contribution comes from the inclusion of the Virtual-Auger
insertion segment of Fig. 6.9(b), and is associated with a double Auger
rate. [287]

6.5　Future work

The kinetic energy distribution of the photoejected electron is obtained
by differentiating $\sigma^{++}(\omega)$ with respect to ε_{k_1}. [262, 264] Since $\varepsilon_{k_1} = \frac{1}{2}k_1^2$, we
have

$$\frac{d\sigma^{++}(\omega)}{d\varepsilon_{k_1}} = \frac{16\omega}{c} \frac{|Z(pq \rightarrow k_1k_2)|^2}{k_1k_2}. \tag{6.39}$$

The angular distribution, or the asymmetry parameter β, for the double
photoionization process, can be written [288, 289] as

$$\frac{d^2\sigma^{++}}{d\Omega_1 d\varepsilon_{k_1}} = \frac{1}{4\pi} \frac{d\sigma^{++}}{d\varepsilon_{k_1}} \left[1 + \beta(E) P_2(\hat{\mathbf{e}} \cdot \hat{\mathbf{k}}_1) \right], \tag{6.40}$$

where $\hat{\mathbf{e}}$ is a photon polarization vector, $\varepsilon_{k_1} + \varepsilon_{k_2} = E$, and P_2 is a
second-order Legendre polynomial. Eq. (6.40) is similar to the case of
single photoionization, and it implies that there is a "magic" angle where
the intensity is independent of the angular distribution. Certainly this
conclusion needs experimental evidence. In a recent experiment Wehlitz
et al. [290] studied electron-energy and -angular distributions in the double
photoionization of helium. Their results are different from theoretical
calculations based on Wannier's law. [288] It will be of interest to apply
MBPT to this problem to take into account electron correlations in both
the ground state and final state.

　　Another suggested area of future work is to calculate σ^{++} for open
shell atoms such as sodium [260] and copper. [259] The choice of the most
suitable potentials for basis sets and for interactions between electrons
in different shells is still an open question. The measured cross section
σ^{++} for sodium is surprisingly low. A sodium atom has a neon core plus
a loosely bonded $3s$ valence electron. However, the experimental values
of the σ^{++} cross section for $2p3s$ double photoionization of sodium are
only about half of the total double photoionization cross section of neon.
It will be very interesting to find out what mechanism can explain the
suppressed σ^{++} for sodium correctly.

7

Photoelectron angular distributions

Steven T. Manson

7.1 Introduction

Collision experiments provide a considerable portion of our information on the interactions between particles, as well as on the internal structure of the particles themselves, in all of the subfields of modern physics. When the interactions between particles are large, information is obtained primarily on the collision process itself; when the interactions are small, the collision probes principally the structure of the interacting particles. Since the coupling between the photon field and the electron is small, photoionization is an example of the latter type of process. Furthermore, since the photon is not thought to possess any internal structure, photoionization is clearly a probe of the target atom.

In photoionization, as in any collision process, measurements of total cross sections provide information on the absolute squares of matrix elements. Differential cross sections, angular distributions, provide information on the matrix elements themselves, including their relative phases. Thus, angular distribution studies allow us to peel back one more layer of the "onion." Furthermore, in many cases, photoelectron angular distributions are much more sensitive to many-body effects, *i.e.*, the effects of correlation, than are total cross sections; in fact, there are many instances in which many-body effects dominate the structure of the differential cross section.

The purpose of this chapter, then, is to provide a theoretical framework for the understanding of the physics of atomic photoelectron angular distributions. Within this framework, the separation between angular momentum geometrical factors and dynamical, wave-function-dependent, factors in the determination of the angular distribution will be highlighted. In addition, examples of processes displaying the range of photoelectron angular distribution phenomenology will be presented, along with examples which illustrate the physics which can learned from photoelectron angular distributions; these examples are intended to be illustrative rather than exhaustive. Before proceeding, however, it is useful to note that, although this chapter is concerned with atomic photoelectron angular distributions, many of the underlying concepts are independent of the nature of the target. Accordingly, much of this chapter is applicable to molecular and condensed matter targets as well.

7.2   Theory

7.2.1   General considerations

As a starting point, we shall consider only relatively low energy photoionization where the electric dipole approximation is excellent, *i.e.*, the photons can all be taken to have one unit of angular momentum (in units of $\hbar$). For electric dipole photoionization, it has been shown [291] from very general principles, conservation of angular momentum and parity, that for linearly polarized light impinging upon an unpolarized target atom, the angular distribution of the photoelectrons can be expressed as a linear combination of the Legendre polynomials $P_0(\cos(\theta)) = 1$ and $P_2(\cos(\theta)) = (3\cos^2(\theta) - 1)/2$ where θ is the angle between the photoelectron propagation and the photon polarization directions. Accordingly, the differential cross section for photoionization of the ith subshell of an unpolarized target by linearly polarized photons can be written as [292]

$$\frac{d\sigma_i}{d\Omega} = \frac{\sigma_i}{4\pi} \left[1 + \beta_i P_2(\cos(\theta))\right] , \qquad (7.1)$$

where σ_i is the total subshell cross section and β_i, which contains all of the dynamical angular distribution information, is known as the asymmetry parameter. From Eq. (7.1) it is clear that σ_i determines the overall intensity of the process. Equation (7.1) also shows that, since $-1/2 \leq P_2(\cos(\theta)) \leq 1$, then β_i is limited to the range $-1 \leq \beta_i \leq 2$ for all cases, since the differential cross section cannot be negative for any value of θ.

But linearly polarized incident light is not the only possibility. It is to be emphasized, however, that no additional dynamical information is obtained by the use of alternative polarizations. For example, unpolarized light may be considered as equivalent to a linear superposition of incoherent equal intensity linearly polarized beams oscillating along perpendicular x- and y- axes [293] and the differential cross section may be written, using Eq. (7.1), as

$$\frac{d\sigma_i}{d\Omega} = \frac{1}{2}\left(\frac{\sigma_i}{4\pi}\right)\left[1 + \beta_i P_2(\cos(\theta_x))\right] + \frac{1}{2}\left(\frac{\sigma_i}{4\pi}\right)\left[1 + \beta_i P_2(\cos(\theta_y))\right] . \quad (7.2)$$

Then, taking the photon beam to be along the z-axis, and using the geometrical relation

$$\cos^2(\theta_x) + \cos^2(\theta_y) + \cos^2(\theta_z) = 1 \,, \tag{7.3}$$

it is found that

$$\frac{d\sigma_i}{d\Omega} = \frac{\sigma_i}{4\pi} [1 - \beta_i P_2(\cos(\theta_z))] \tag{7.4}$$

for unpolarized photons. This same result is true for circularly polarized light which can be considered as a linear combination of *coherent* linearly polarized beams of equal intensity at right angles to each other. Similarly, elliptically polarized light, which can be considered as a superposition of circularly polarized and linearly polarized light, and partially polarized light, which is equivalent to a linear combination of unpolarized and either circularly polarized or linearly polarized light, [294] result in angular distributions whose dynamics depend only upon σ_i and β_i.

Note that, in both Eq. (7.1) and Eq. (7.4), the dependence of the differential cross section on the asymmetry parameter β_i vanishes at the angle θ at which $P_2(\cos(\theta))$ vanishes, *i.e.*, where $\cos^2(\theta) = 1/3$. At this angle, known as the "magic" angle (roughly 54°), the differential cross section of Eq. (7.1) or Eq. (7.4) is given by $\sigma_i/4\pi$. Thus, measurements at this "magic" angle provide a method of measuring total cross sections using photoelectron spectroscopy without collecting the electrons with 4π geometry.

Prior to launching into a detailed study of the asymmetry parameter, β_i, which determines the angular distribution, discussion of some general considerations is in order. Both the subshell cross section, σ_i, and the asymmetry parameter, β_i are generally energy- dependent. The cross section, σ_i, which is proportional to the sum of the absolute squares of the dipole matrix elements for transitions from the initial state to the various allowed final states, derives its energy dependence from the dependence of the dipole matrix elements on energy. The asymmetry parameter, β_i, is expressed as a ratio and is more complicated; the energy dependence comes about from interference of the matrix elements of the alternative possible photoionization channels so that the details of the relative magnitudes and phases of the various matrix elements are crucial. An obvious corollary is that for a photoionization process consisting of only a single possible channel, β_i *must* be energy-independent and determined only by geometrical factors: angular momentum geometry, of course.

7.2.2 *Angular momentum transfer analysis*

To calculate the differential cross section, thus, knowledge of both σ_i and β_i is required. A number of equivalent theoretical formulations exist. [164]

In this review the angular momentum transfer formulation of Dill and Fano [295, 296, 297] will be presented since the essential physics and the separation between angular momentum geometry and dynamics emerge most clearly therein. Consider the ejection of a photoelectron from an unpolarized atom A,

$$A(L_0 S_0 J_0 \Pi_0) + \hbar\omega(j_\gamma = 1, \Pi_\gamma = -1)$$
$$\rightarrow A^+(L_c S_c J_c \Pi_c) + e(\ell s j, \Pi_e = (-1)^\ell), \quad (7.5)$$

with Π the parity and L, S, J the orbital, spin and total angular momenta of the various particles. The angular momentum transferred between atom and ion is defined as

$$\mathbf{j}_t = \mathbf{j}_\gamma - \mathbf{l} = \mathbf{J}_c + \mathbf{s} - \mathbf{J}_0 = \mathbf{J}_{cs} - \mathbf{J}_0 \qquad (7.6)$$

The utility of this definition of j_t lies in the fact that the total subshell cross section is simply an incoherent sum of the cross sections for the alternative possible values of j_t, $\sigma_i(j_t)$, i.e.,

$$\sigma_i = \sum_{j_t} \sigma_i(j_t), \qquad (7.7)$$

and the asymmetry parameter is a weighted average of the $\beta_i(j_t)$, i.e.,

$$\beta_i = \sum_{j_t} \frac{\beta_i(j_t)\sigma_i(j_t)}{\sigma_i}, \qquad (7.8)$$

The possible values of j_t are obtained from Eq. (7.6), subject to the constraints of conservation of total angular momentum,

$$\mathbf{J}_0 + \mathbf{j}_\gamma = \mathbf{J}_c + \mathbf{j}, \qquad (7.9)$$

and total parity

$$\Pi_0 \Pi_\gamma = \Pi_c \Pi_e. \qquad (7.10)$$

The allowed values of j_t for a particular transition are classified as to whether $\Pi_0 \Pi_c = \pm(-1)^{j_t}$; those corresponding to the positive sign are known as parity-favored, while those corresponding to the negative sign are called parity-unfavored. Now these definitions are of considerable importance because the "favored-ness" or "unfavored-ness" of a transition really matters. The essential point is that while the asymmetry parameter for parity-favored transitions is, generally, energy-dependent, for a parity-unfavored transition, $\beta_i(j_t) = -1$, independent of energy. Using this fact, Eq. (7.8) can then be simplified and written as

$$\beta_i = \frac{\left[\sum_{j_t} \beta_i(j_t)_{\text{fav}}\sigma_i(j_t)_{\text{fav}} - \sigma_i(j_t)_{\text{unfav}}\right]}{\sigma_i}. \qquad (7.11)$$

Thus, from Eq. (7.1), the parity-unfavored transitions have $\sin^2(\theta)$ angular distributions, which peak in a direction perpendicular to the electric vector of the incident light (and vanish parallel to this direction), which is a very counter-intuitive result.

To obtain β_i we must obtain the other quantities in Eq. (7.11). They are computed in terms of the reduced scattering amplitudes, $S_\ell(j_t)$, to be discussed below, where ℓ is the photoelectron angular momentum. Then, for parity-favored transitions, [295, 296, 297]

$$\sigma_i(j_t) = \left(\frac{4}{3}\right) \pi^2 \alpha \hbar \omega \frac{2j_t + 1}{2J_0 + 1} \left[|S_+(j_t)|^2 + |S_-(j_t)|^2\right] , \qquad (7.12)$$

$$\beta_i(j_t) = \frac{1}{(2j_t + 1)\left[|S_+(j_t)|^2 + |S_-(j_t)|^2\right]} \Big((j_t + 2)|S_+(j_t)|^2 +$$
$$(j_t - 1)|S_-(j_t)|^2 - 6\left[j_t(j_t + 1)\right]^{\frac{1}{2}} \operatorname{Re}\left[S_+(j_t)S_-(j_t)^*\right]\Big) , \qquad (7.13)$$

and for parity-unfavored transitions

$$\sigma_i(j_t) = \left(\frac{4}{3}\right) \pi^2 \alpha \hbar \omega \frac{2j_t + 1}{2J_0 + 1} |S_0(j_t)|^2 , \qquad (7.14)$$

$$\beta_i(j_t) = -1 . \qquad (7.15)$$

In the above equations α is the fine-structure constant, $\hbar\omega$ is the photon energy, Re denotes real part and the subscripts $\pm$, and 0 refer to $\ell = j_t \pm 1$ and $\ell = j_t$ respectively. The reduced scattering amplitudes are given in terms of reduced matrix elements as [298]

$$S_\ell(j_t) = n(\lambda) \sum_J (-1)^{J_0 - J - 1} (2J + 1)^{\frac{1}{2}} \left\{ \begin{matrix} J_{cs} & \ell & J \\ 1 & J_0 & j_t \end{matrix} \right\}$$
$$\times \langle \gamma_0 L_0 S_0 J_0 \| \sum_{i=1}^{N} r_i^{(1)} \| (J_c s) J_{cs} \ell (J - 1), \qquad (7.16)$$

with λ the photon wavelength, n a proportionality factor dependent only on λ, γ_0 the set of quantum numbers needed to describe the initial state completely, the term in brackets the Wigner $6j$ symbol, and the minus sign denoting *incoming wave* boundary conditions of the final state. The reduced matrix elements are related to the ordinary dipole matrix elements via the Wigner-Eckhart theorem. [299] These scattering amplitudes are simply the linear combinations of the dipole matrix elements which recouple the angular momenta of the system to the various possible values of j_t, the angular momentum transfer. Further specification of the reduced matrix elements depends upon the particular process, Eq. (7.5), being studied.

7.2.3   The effects of approximations

In the above development, the *only* approximations that have been made are the use of first order perturbation theory for the photon-electron interaction, and the subsequent electric dipole approximation; aside from these, the treatment presented is completely general. It is to be emphasized, however, that these approximations are quite excellent for low energy photoionization with ordinary atomic physics photon fluxes. It is of interest, however, to investigate the effects of additional approximations with an eye towards the physics associated with each approximation. Since *LS* coupling is adequate in a variety of low-*Z* situations, we start there.

Within the framework of *LS* coupling, spin angular momentum is conserved since the photon field does not couple to the spin so that

$$\mathbf{S}_0 = \mathbf{S}_c + \mathbf{s} \,. \tag{7.17}$$

Under these conditions, then, the expression for j_t becomes

$$\mathbf{j}_t = \mathbf{L}_c - \mathbf{L}_0 \,, \tag{7.18}$$

and the expression for the reduced scattering amplitudes becomes [164]

$$S_\ell(j_t) = n(\lambda) i^\ell e^{-i\delta_\ell^C} \sum_L (2L+1)^{\frac{1}{2}} \left\{ \begin{array}{ccc} L_c & \ell & L \\ 1 & L_0 & j_t \end{array} \right\}$$

$$\times \langle \gamma_0 L_0 S_0 \| \sum_{i=1}^{N} r_i^{(1)} \| (L_c S_c) \ell (L-1) \,, \tag{7.19}$$

with δ_ℓ^C the Coulomb phase shift and the rest of the notation as in Eq. (7.16). Thus, *LS* coupling restricts the possible values of j_t as well as changing the expression for the reduced scattering amplitudes.

If we further restrict the wave functions of the initial and final states to be single configuration wave functions, *i.e.*, multiconfiguration and multichannel interactions are ignored, one has the Hartree-Fock (HF) approximation and we have an $n\ell \rightarrow \varepsilon\ell$ transition, ε being the photoelectron kinetic energy, with the possibility of core relaxation. In this case, the possible values remain as determined by Eq. (7.18) and the reduced scattering amplitudes are considerably simplified, [294]

$$S_\ell(j_t) \propto i^\ell e^{-i\delta_\ell^C} (-1)^{\ell_0 - \ell_>} \ell_>^{\frac{1}{2}} \sum_L (2L+1) \exp\left(-i\delta_{\varepsilon\ell}^{L_c S_c L}\right) R_{\varepsilon\ell}^{L_c S_c L}$$

$$\times \left\{ \begin{array}{ccc} L_c & \ell & L \\ 1 & L_0 & j_t \end{array} \right\} \left\{ \begin{array}{ccc} L_c & \ell & L \\ 1 & L_0 & \ell_0 \end{array} \right\} \,, \tag{7.20}$$

with $\ell_>$ the greater of ℓ and ℓ_0, and the $\delta_{\varepsilon\ell}^{L_c S_c L}$ and $R_{\varepsilon\ell}^{L_c S_c L}$ are the non-Coulomb phase shifts and radial dipole integrals of the indicated channels, the latter given by

$$R_{\varepsilon\ell}^{L_cS_cL} = \int_0^\infty P_{n\ell_0}(r)rP_{\varepsilon\ell}^{L_cS_cL}(r)dr , \qquad (7.21)$$

where the P's are r times the initial and final orbitals of the photoelectron. Note that, although discrete state correlation and interchannel coupling are omitted at the HF level, multiplet structure is still included. The wave function of the photoelectron in the final state generally depends upon the quantum numbers L_c and S_c, as well the total angular momenta of the final state, L and S_0. This dependence is the result of the exchange interaction between the emerging photoelectron and the ion core, an interaction that is explicitly non-central or anisotropic.

The next level of simplification involves approximating the anisotropic interactions by a central field. In this single-electron approximation the phase shifts and dipole matrix elements of Eq. (7.20) no longer depend upon L_c, S_c or L, but only on ε and ℓ. With anisotropic interactions omitted the phases and dipole matrix elements can be factored out of the summation of Eq. (7.20) and the sum over the $6j$ symbols may be performed analytically to obtain [294]

$$S_\ell(j_t) \propto \delta(\ell_0, j_t)i^\ell e^{-i\delta_\ell^C}(-1)^{\ell_0-\ell}>\ell_>^{\frac{1}{2}}(2\ell_0+1)^{-1}e^{-i\delta_{\varepsilon\ell}}R_{\varepsilon\ell} . \qquad (7.22)$$

Substituting this expression for the reduced scattering amplitude into Eq. (7.13), and employing Eq. (7.11), the expression for the asymmetry parameter becomes [300]

$$\begin{aligned}
\beta_i(j_t) = \frac{1}{(2\ell_0+1)\left[\ell_0R_-^2 + (\ell_0+1)R_+^2\right]} &\big((\ell_0+1)(\ell_0+2)R_+^2 \\
&+ \ell_0(\ell_0-1)R_-^2 - 6\ell_0(\ell_0+1)R_-R_+\cos(\delta_+^C + \delta_+ - \delta_-^C - \delta_-)\big) ,
\end{aligned} \qquad (7.23)$$

where, for simplicity, the subscripts $\pm$ refer to $\ell = \ell_0 \pm 1$ and the explicit dependence of the phase shifts and dipole matrix elements on ε has been omitted. This central-field result is known as the Cooper-Zare (CZ) formula.

Several points concerning the CZ formula are noteworthy. First, from Eq. (7.22), it is seen that only a single value of j_t is allowed, $j_t = \ell_0$. This is just the value to be expected in a single-particle picture since conservation of angular momentum becomes

$$\mathbf{l}_0 + \mathbf{j}_\gamma = \mathbf{l} , \qquad (7.24)$$

so that from the definition of j_t, Eq. (7.6), the only possible value is ℓ_0. Furthermore, it is easy to see that only parity-favored transitions are possible. This result suggests a useful physical picture of the process. In the central field model, the photon is absorbed by the atom, transferring its energy and angular momentum to the photoelectron. But this is only the first stage of the general process, which has a second stage, namely the emerging photoelectron interacts with the residual ion and angular momentum exchanges can take place. This allows alternative possibilities for j_t. From a physical point of view, angular momentum can only be changed by a torque; thus, the residual ion and the photoelectron must exert torques on each other in order for these angular momentum exchanges, and thereby the alternative possibilities for j_t, to take place. Torques can only result from non-central, *i.e.*, anisotropic, forces, exchange interactions, for example. This shows why the omission of exchange removes the possibility of the second stage of the photoionization process.

A second point of interest is the that the forms of the CZ result for the β parameter, Eq. (7.23), and the general result for parity-favored transitions, Eq. (7.13), are so similar. This implies that much of the physics of the photoionization process is retained in the simple CZ result. In fact, it is straightforward to show that for photoionization of closed-shell atoms, or of atoms with a single electron outside closed subshells, the general result, Eq. (7.13) reduces to the same *form* as the CZ formula, Eq. (7.23), within the framework of *LS* coupling, with the proviso that the reduced dipole amplitudes of Eq. (7.20) still take multiconfiguration and multichannel interactions into account. For the closed-shell case, for example, $L_0 = 0$ which results in only a single possible term for the reduced scattering amplitude in *LS* coupling, Eq. (7.20), and, due to the properties of the 6-*j* symbols, $j_t = \ell_0$, just as in the central-field CZ case.

7.2.4   *Application of the angular momentum transfer analysis*

As a first example of the angular momentum transfer analysis, consider the ground state of atomic hydrogen, since the result is known from other considerations. A linearly polarized electric-dipole photon induces oscillations in the spherically symmetric $1s$ charge distribution in the direction of the electric vector, inducing a Y_{10} symmetry relative to the photon polarization. Thus, the differential cross section $d\sigma/d\Omega$ [Eq. (7.1)] goes as $|Y_{10}|^2$ which is proportional to $\cos^2(\theta)$, meaning $\beta = 2$.

Consider now this process from the point of view of the angular momentum transfer analysis. In the form of Eq. (7.5), this process can be characterized as

$$H(1s\,{}^2S_{1/2}) + \hbar\omega \rightarrow H^+({}^1S_0) + e^- . \tag{7.25}$$

Thus,

$$J_0 = s = \tfrac{1}{2}, \quad j_\gamma = 1, \quad J_c = 0$$
$$\Pi_0 = 1, \quad \Pi_\gamma = -1, \quad \Pi_c = 1. \tag{7.26}$$

Angular momentum conservation,

$$\mathbf{J} = \mathbf{J}_0 + \mathbf{j}_\gamma = \mathbf{J}_c + \mathbf{s} + \mathbf{l}, \tag{7.27}$$

implies that $\ell \le 2$, but from parity conservation

$$\Pi_0\Pi_\gamma = \Pi_c\Pi_e, \tag{7.28}$$

$\Pi_e = -1 \ [= (-1)^\ell$ from Eq. (7.5)] so that ℓ must be odd; this restricts its value to $\ell = 1$. From Eq. (7.6), then, $j_t = 0$ or 1. Then, since $\Pi_0\Pi_c = 1$, $j_t = 0$ is parity-favored and, from Eq. (7.13), $\beta_i(j_t = 0) = 2$ while $j_t = 1$ is parity-unfavored so that $\beta_i(j_t = 1) = -1$. The expression for β_i, from Eq. (7.11), is then

$$\beta_i = \frac{2\sigma_i(j_t = 0) - \sigma_i(j_t = 1)}{\sigma_i(j_t = 0) + \sigma_i(j_t = 1)}. \tag{7.29}$$

The cross sections, given for the parity-favored and parity-unfavored cases by Eqs. (7.12) and (7.14) repectively, are [301]

$$\sigma_i(j_t = 0) \propto |S_1(0)|^2 = \frac{2}{3}\left[R_{1/2}^2 + 4R_{3/2}^2 + 4R_{1/2}R_{3/2}\cos(\delta_{3/2} - \delta_{1/2})\right], \tag{7.30}$$

$$\sigma_i(j_t = 1) \propto 3\,|S_1(1)|^2 = \frac{4}{3}\left[R_{1/2}^2 + R_{3/2}^2 - 2R_{1/2}R_{3/2}\cos(\delta_{3/2} - \delta_{1/2})\right], \tag{7.31}$$

where the R_j are the magnitudes of the radial dipole matrix elements for the $J = 1/2$, $3/2$ final states and the δ_j, their phases. But for the hydrogen atom, R_j and δ_j are independent of J, to an excellent approximation, with the result that $\sigma_i(j_t = 1)$ reduces to zero, and, from Eq. (7.29), $\beta_i = 2$.

Clearly this case constitutes an example of theoretical overkill. Several points are illustrated by this example, however, notably that where there is only one possible final state, β_i must be independent of energy. From a physical point of view, the variation of β_i with energy requires at least two outgoing channels with differing energy dependences to interfere with one another. In addition, $\beta_i = 2$ for an $s \rightarrow p$ transition. Although there are actually two possible transitions, corresponding to $J = 1/2$ and $3/2$ in the final state, the fact that the spin-orbit interaction in the hydrogen p-continuum is so small means that effectively there is only a single transition.

When the spin-orbit interaction is comparatively larger, however, it is possible that the matrix elements for the two channels can be different. The angular distribution can be dramatically altered in that situation. Such a case arises for the photoionization of the outer s-electron of the alkali atoms, Na through Cs. The angular momentum geometry is exactly the same as for hydrogen, discussed above. The principal difference is that the $s \rightarrow p$ dipole matrix elements undergo a change of sign, known as a

Cooper minimum, [169, 302, 303] in the continuum. At energies close to the zeros, then, the small spin-orbit effect can make a major difference and the dipole matrix elements $R_{1/2}$ and $R_{3/2}$ of Eq. (7.31) are no longer equal; they have zeros at slightly different energies and, thus, they can even, in fact, have opposite signs. To an excellent approximation, the phases are still the same, however. Hence, at an energy such that $R_{1/2} = -2R_{3/2}$, it is seen from Eq. (7.30) that $\sigma_i(j_t = 0) = 0$; it then follows that $\beta_i = -1$. The situation for these alkali atoms around the Cooper minimum has been calculated at the Dirac-Fock level, [301] which amounts to application of the general theory with the reduction of the reduced scattering amplitudes, Eq. (7.16), to the form obtained with single configuration relativistic wave functions; the results are shown in Figs. 7.1 where the transition from $\beta_i = 2$, away from the minimum, to $\beta_i = -1$ at the minimum, is seen. No experimental verification for Na exists, but it does for K, [304] Rb [305] and Cs. [294] In any case it is clear that the existence of two differing continuum channels for the photoionization process gives rise to an energy dependent β_i even for an $s \rightarrow p$ transition.

From a physical point of view, the orientation of the photoelectron spin was changed which allowed for the possibility of $j_t = 1$; this, of course required a torque, and the torque was supplied by the spin-orbit force, which is a non-central force. This emphasizes that the existence of a non-central, or anisotropic, interaction between the photoelectron and the ion core is required to give more than a single possibility for j_t. As discussed earlier, however, electrostatic forces, notably exchange, can also provide the non-central interaction.

Consider the photoionization of the $3s$ subshell of atomic Cl, for example

$$Cl(3s^2 3p^5\ ^2P) + \hbar\omega \rightarrow Cl^+(3s3p^5\ ^{1,3}P) + e^- . \tag{7.32}$$

Spin-orbit forces shall be ignored so that LS coupling is relevant. The angular momenta, in this case are

$$L_0 = 1, \quad j_\gamma = 1, \quad L_c = 1$$
$$\Pi_0 = -1, \quad \Pi_\gamma = -1, \quad \Pi_c = -1, \tag{7.33}$$

and, invoking parity and angular momentum conservation, it is found that $\Pi_e = -1$ and $\ell = 1$. The allowed values of j_t, then, are $j_t = 0, 1$, and 2 and the form of β_i becomes

$$\beta_i = \frac{2\sigma_i(j_t = 0) - \sigma_i(j_t = 1) + \frac{1}{5}\sigma_i(j_t = 2)}{\sigma_i(j_t = 0) + \sigma_i(j_t = 1) + \sigma_i(j_t = 2)} . \tag{7.34}$$

The three alternative values of j_t reflect the recoupling of the three possible final channels, corresponding to total orbital angular momentum $L = 0, 1, 2$ final states. The $S_1(j_t)$ are linear combinations of the dipole

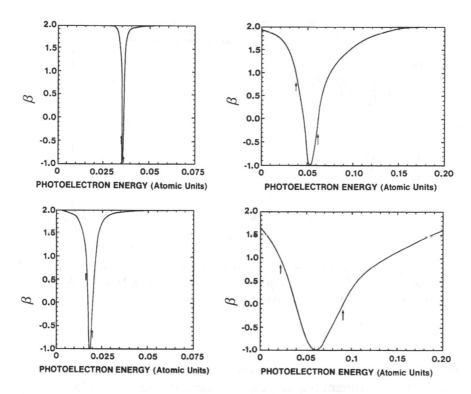

Fig. 7.1. Asymmetry parameter for photoionization. Top left Na 3s; top right, Rb 5s; bottom left, K 4s; bottom right, Cs 6s. Calculations were made using the Dirac-Fock approximation; from Ref. [301]. The arrows indicate the energies of the zeros in the dipole matrix element for final $J = 1/2$ (lower energy arrow) and $J = 3/2$ (higher energy arrow) states.

matrix elements for each of the final states, as given in Eq. (7.19). From a physical point of view, the ^3P final ionic state is of greatest interest, since the three dipole matrix elements have Cooper minima in this case, thus giving great importance to $S_1(1)$ and $S_1(2)$ in the vicinity of the minima. [306] Consequently, β_i varies rather significantly from the naive view for $s \rightarrow p$ transitions of $\beta_i = 2$; the result of a Hartree-Fock calculation [306] is shown in Fig. 7.2 where the deviation from the value of 2 is quite evident. A fuller discussion of photoelectron angular distributions of s-electrons is given in Ref. [294].

Looking now at $\ell_0 \neq 0$, a particularly interesting case is the process

$$N(p^3\,^2D) + \hbar\omega \rightarrow N^+(p^2\,^1S) + e^- . \qquad (7.35)$$

Within the framework of *LS* coupling

$$
\begin{aligned}
L_0 = 2, \quad j_\gamma = 1, \quad L_c = 0 \\
\Pi_0 = -1, \quad \Pi_\gamma = -1, \quad Pi_c = 1,
\end{aligned}
\qquad (7.36)
$$

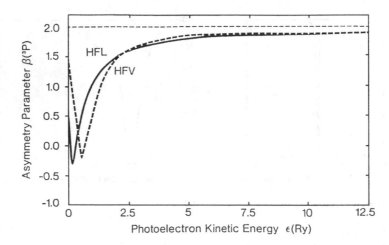

Fig. 7.2. Asymmetry parameter for Cl $3s$ photoionization leaving Cl$^+$ in the $3p$ state, Cl $3s^2 3p^5(^2P) + \hbar\omega \rightarrow$ Cl$^+ 3s3p^5(^3P) + e^-$, calculated in Hartree-Fock length (HFL) and velocity (HFV) formulations from Ref. [306].

so that $\Pi_e = 1$; Thus, $\ell = 2$ is the only possibility for the photoelectron's orbital angular momentum, *i.e.*, due to conservation of angular momentum, the $p \rightarrow s$ transition is not possible. Then, since spin is conserved in *LS* coupling [Eq. (7.17)], Eq. (7.6) shows that only $j_t = 2$ is possible. But, since $\Pi_0 \Pi_c = -1$, and $(-1)^{j_t} = 1$, this is a parity-unfavored transition, and $\beta_i = -1$, independent of energy; [307] this result has not been confirmed by experiment as yet. However, a recent experiment [308] has investigated satellite transitions in argon which are purely parity-unfavored, *e.g.*,

$$\text{Ar}(3p^6\ ^1S) + \hbar\omega \rightarrow \text{Ar}^+((3p^4\ ^1D)3d\ ^2P) + e^- , \qquad (7.37)$$

and found that $\beta_i = -1$ over the entire energy range studied.

The applications presented highlight the interactions which can take place when the photoelectron can only have a single orbital angular momentum, ℓ; the purpose of emphasizing such cases is threefold: to illustrate the use of the angular momentum tranfer technology in relatively uncomplicated situations; to highlight the separation between angular momentum geometry and dynamics; and to focus on the fact that, although ℓ is restricted to a single value, angular momentum exchanges with the ionic core, induced by spin-orbit or exchange forces, can allow multiple final state channels to interfere with one another and produce an energy-dependent β_i.

7.3   Illustrative examples

As a first example, consider a case where ℓ can have more than one value, In particular, we look at the photoionization of the $4d$ subshell of the closed shell atom Xe where $4d \to \varepsilon p$ and $4d \to \varepsilon f$ transitions are both allowed. As discussed in Sec. 7.2.3, the form of the β parameter is just the CZ expression, Eq. (7.23), but the dipole matrix elements and phase shifts are obtained from a many- body calculation. Thus, the variation of β_i with energy is determined only by the interference between the $4d \to \varepsilon p$ and $4d \to \varepsilon f$ channels. A great deal of both experimental and theoretical work has been done on this angular distribution; [153] some of the more recent experimental results are shown in Fig. 7.3 along with the results of some many-body calculations. The excellent agreement between theory and experiment is clear evidence that the inclusion of many-electron interactions is of importance; a single configuration HF calculation gives qualitatively correct results but the details differ from experiments in some respects. [309] The outstanding feature of β_i for this case is the significant variation with energy, measured experimentally and reproduced theoretically. In the energy region shown, β_i, exhibits four turning points. The oscillations of β_i give information about the dipole matrix elements and their phases. From threshold up to about 150 eV, the variation is caused primarily by the variation of the cosine of the phase difference between f- and p-waves; this arises from the last term in the numerator of Eq. (7.23). The rapid variation near threshold arises from the rapid variation in the Coulomb phase difference, while just above threshold, the variation is caused by the shape resonance in the f-channel, a rapid increase in the phase shift by roughly π over a small energy range. [169] Above 150 eV, the rapid decrease and subsequent recovery of β_i, results from a Cooper minimum in the f-channel matrix element; at the Cooper minimum the asymmetry parameter must be 0.2 for a d-subshell. [310] Thus, investigation of the β parameter provides a very sensitive method for the determination of the locations of Cooper minima.

Another case of interest is the photoionization of the $2p$ subshell of Ar, another closed shell system. Here a measurement of the β parameter [311] over a significant energy range gave qualitative, but not quantitative, agreement with the results of a HF calculation. [309] This was surprising because HF does so well for Ne $2p$, an outer subshell, and it seemed that a single configuration approximation should be even better for a tightly bound inner subshell like Ar $2p$. A calculation using the many-body relativistic random phase approximation (RRPA) [157] showed quite good agreement with experiment; the results are shown in Fig. 7.4. The

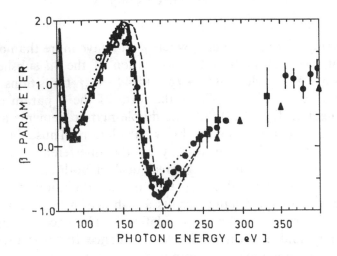

Fig. 7.3.   Asymmetry parameter for Xe 4*d* from Ref. [153]. The various points are experimental values and the curves are calculated values, which include many-body correlations.

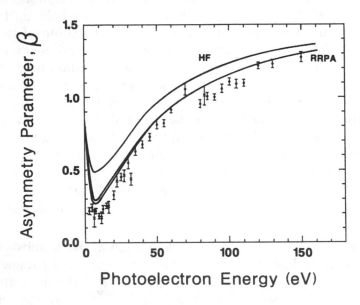

Fig. 7.4.   Asymmetry parameter for photoionization of Ar 2*p*. The experimental points are from Ref. [311]. The theoretical HF results are from Ref. [309] and the 5- and 14- channel RRPA results (upper and lower curves, respectively) from Ref. [311].

fundamental cause of the improvement turned out to be interchannel coupling between the $2p \rightarrow \varepsilon s$ and $2p \rightarrow \varepsilon d$ channels. The s-channel is roughly one-fifth the size of the d-channel near threshold and the interchannel interaction between the two channels causes a significant change in the smaller one. Since the asymmetry parameter is a ratio, Eq. (7.23), the alteration of just one of the matrix elements causes a significant change in β. Since the s-channel is so small a part of the total cross section, this change is not very evident in σ. The reason this does not occur in Ne is that the s- and d-channels are about the same magnitude so that there is no weaker channel to be modified by interacting with the stronger one.

Thus, from this example it is clear that many-body effects, interchannel coupling in this case, are crucial to a correct theoretical description of the process. In addition, when an outer shell becomes with increasing atomic number an inner shell, the effects of many-body interactions do not necessarily diminish. This also demonstrates that the photoelectron angular distribution can be far more sensitive to certain effects than the total subshell cross section.

The two previous examples presented cases where many-body correlations caused significant modifications to the photoelectron angular distribution asymmetry parameter. As a final example, a case where many-body correlations *dominate* the behavior of the β parameter is scrutinized. We consider $7s$ photoionization in the closed subshell Ra atom. In the absence of relativistic (spin-orbit) interactions, β_i must be equal to 2 and independent of energy since only a single $7s \rightarrow \varepsilon p$ channel is possible. [294] With the inclusion of relativistic interactions, however, $7s \rightarrow \varepsilon p_{3/2}$ and $7s \rightarrow \varepsilon p_{1/2}$ channels are allowed. The variation of β_i with energy is, therefore, determined by the interference between the two allowed channels; the expression for β_i, in this case, is given by [294]

$$\beta_i = \frac{2R_{3/2}^2 + 4R_{1/2}R_{3/2}\cos(\delta_{1/2} - \delta_{3/2})}{R_{1/2}^2 + 2R_{3/2}^2} , \qquad (7.38)$$

where the R_j are the radial dipole matrix elements to the εp_j continua and the δ_j are the phase shifts. Even for so heavy an element, $Z = 88$, the phase shifts of the two continua, split only by the spin-orbit interaction, are essentially equal. Then, if the two matrix elements are approximately equal, it is seen from, Eq. (7.38), $\beta_i \approx 2$. However, this transition does exhibit Cooper minima and the asymmetry parameter, calculated with the framework of the many-body RRPA, is shown in Fig. 7.5 where a very significant amount of structure is seen. [312] There are three major dips in β_i, each corresponding to a Cooper minimum. What makes this case particularly interesting is that the two low-energy dips, and associated

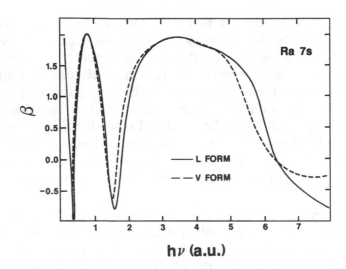

hν (a.u.)

Fig. 7.5. Asymmetry parameter for Ra 7*s* photoionization calculated in 20-channel RRPA dipole-length (L) and dipole-velocity (V) formulations from Ref. [312].

Cooper minima, are *entirely* due to the effects of interchannel coupling of the strong 6*p* and 5*d* channels; omitting consideration of these many-body interactions removes these two low energy dips and makes β_i constant and equal to 2 in this energy region. The high-energy dip is also caused by a Cooper minimum, but one that exists even without the presence of many-body interactions. The position and shape of β_i at the higher energies is, however, significantly affected by many-body correlations.

7.4   Final remarks

The theory of photoelectron angular distributions has been reviewed and the separation between angular momentum geometry and dynamics, *i.e.*, wave-function-dependent effects, has been highlighted. In addition, the effects of varying levels of approximation on the calculation of the β parameter have been discussed. Detailed examples to illustrate the application of the angular momentum transfer analysis to a number of situations, along with examples which portray the importance of many-body correlations, have been presented. Finally, it is important to reiterate that the study of photoelectron angular distributions provides information which cannot be obtained from integrated cross sections.

Part 3

ATOMIC SCATTERING:
A. General considerations

8

The many-body approach
to electron-atom collisions

M. Ya Amusia

8.1 Introduction

Many-body theory has been very successful in the *ab initio* calculation
of electron–atom scattering cross sections. Its application has made it
possible to take into account the target atom polarization in the collision
process without the introduction of a semi-empirical potential. The many-
body approach and the diagrammatic technique associated with it have
also permitted the illumination of hidden difficulties in the description of
the electron-atom scattering process.

The first calculations of electron–hydrogen elastic scattering cross sections using the diagrammatic technique of many-body theory were performed by H. P. Kelly about thirty years ago. [313] In that calculation,
as well as in subsequent investigations involving helium, argon and
xenon, [314] the polarization of the target atom by slow and medium
energy electrons was taken into account. The effects of the exchange
between the projectile electrons and target electrons were also accounted
for. Although, at first glance, the interaction between the incoming electrons and the atomic particles is of second order, the polarization of the
target in these works included a number of higher order corrections. This
inclusion has proven to be very important.

The subsequent application of many-body theory to the scattering of
electrons from highly polarizable atoms led to the development of methods
which permitted the incorporation of the corresponding corrections non-
perturbatively. [315] A connection between elastic scattering and negative
ion formation has also allowed one to use many-body techniques for the
calculation of electron affinities. [316]

The flexibility of the many-body technique is essential in applying it to
the scattering of non-electron particles on atoms, *e.g.* positrons and μ-
mesons. As an example, there is the $e^+ + He$ scattering process considered

in Ref. [317]. The many-body technique has permitted one to treat not only elastic, but also inelastic scattering. As targets one can investigate atoms as well as their ions.

This chapter is intended to present qualitatively the main features of the many-body technique, along with some concrete applications.

8.2   The diagrammatic method in the scattering problem

The main advantage of many-body theory is that it does not explicitly use the wavefunction of the multiparticle system. On the contrary, it uses only those functions that describe single particle motion, and which are directly engaged and altered in the process under consideration. For the remainder of this chapter ψ_E will be used to denote a correlated projectile wavefunction and ϕ_i will be used to denote an approximate initial wavefunction. Atomic units are used throughout this chapter.

The many-body theory, and the diagrammatic language which makes the approach transparent, is described at length in a number of books (*e.g.* Ref. [318]). Here we will review some of the essential features of this approach, and we will apply it to the scattering problem.

The principal element of the diagrammatic technique is a line, which we shall take as directed to the *right* (*left*) describing an electron in an *excited state* (*a vacancy in a closed shell*). The continuous spectrum is considered as an excited state. An atom with only closed subshells is considered as a vacuum. The Coulomb interaction between electrons is denoted by a vertically oriented zigzag line. The action of the projectile electron, or any of the atomic electrons, upon the target electrons leads to the creation of electron–vacancy pairs. To draw diagrams the following rules are applied, which represents a physical process that develops in time from the left to the right.

- The initial and final states of the process under consideration must be drawn. These states will be represented by horizontal arrows denoting the total number of electron and vacancy levels.

- The states must be connected using the electron–vacancy horizontally oriented lines, and the vertical zigzag lines denoting the interelectron interaction.

- New electron–vacancy lines can only start or end with a Coulomb interaction, a photon emission, or a photon absorption.

The propagation of an electron in the field of the nucleus, in the absence of interaction with another electron is represented by a single line with a

dot at the end

$$\longrightarrow\!\!\bullet \qquad\qquad (8.1)$$

The first order interaction of this electron with target atom electrons is described by the two diagrams

$$(8.2)$$

(a)   (b)

where the single Coulomb interaction line (zigzag line) connects the projectile and the target electrons.

The loop in diagram (8.2a) is the target atom electron (a vacuum electron). The entire diagram (8.2a) describes the interaction between the projectile electron and the target electron density. The diagram (8.2b) represents the exchange between the projectile electron and target electrons. In the higher orders of the Coulomb interaction, two types of diagrams will appear: those that describe the projectile–target interaction (*e.g.* diagram (8.2) and more complicated diagrams that repeat the diagrammatic elements), and those that correct the projectile–target interaction itself. We will start with the first type. For ease of drawing let us present only diagram (8.2a) instead of both diagrams (8.2a,8.2b). Then the scattering process, or the propagation of an incoming particle, is described by the following sequence of diagrams

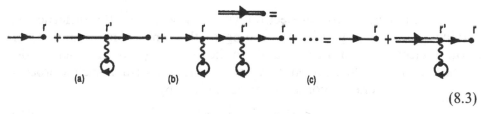

$$(8.3)$$

where the double line takes into account the infinite number of "projectile electron–target electron" interaction events. It is seen that the infinite series of diagrams is converted into a diagrammatic integral equation. Analytically, the corresponding equation is derived keeping in mind that the motion of the electron from point $\mathbf{r}'$ to point $\mathbf{r}$ is described by a one-particle Coulomb Green's function. The loops in diagram (8.3) represent the unperturbed density of noninteracting electrons in the Coulomb field of the atomic nucleus. If we correct these loops, using the definition of the wavefunction given by diagram (8.3), an equation is derived for both the projectile electrons and the target electrons

$$\longrightarrow \!\!\bullet = \longrightarrow \!\!\bullet + \longrightarrow \!\!\bullet + \longrightarrow \!\!\!\!\!\!\! \qquad (8.4)$$

(a)   (b)

Analytically, it is presented in the following way

$$\left(-\frac{\nabla^2}{2} - \frac{Z}{r} + \int \rho(\mathbf{r}') \frac{d\mathbf{r}}{|\,\mathbf{r}' - \mathbf{r}\,|} \right) \phi_E(\mathbf{r}) - $$

$$\sum_{i \leq N} \int \phi_i^*(\mathbf{r}') \frac{d\mathbf{r}'}{|\,\mathbf{r}' - \mathbf{r}\,|} \phi_E(\mathbf{r}') \phi_i(\mathbf{r}) = E \phi_E(\mathbf{r}) , \qquad (8.5)$$

where $\rho(r) = \sum_{i \leq N} |\,\phi_i(\mathbf{r})\,|^2$ is the atomic electron density, and the summation $i \leq N$ includes all N orbitals occupied in the target atom. Here N is the number of electrons in the target atom.

It is seen that Eq. (8.5) is the Hartree-Fock equation. This is a one-electron approach in the sense that each of the electrons moves independently in a static field potential. This potential is self-consistent, however, since it is determined by the common action of all of the atomic electrons. The simplest diagram beyond the Hartree-Fock approximation that describes the perturbation of the target atom by a projectile electron is given by

$$E \longrightarrow \!\!\bullet\; E' \atop \qquad \begin{matrix} \varepsilon n \\ i \end{matrix} \qquad (8.6)$$

This is the amplitude for an electron with energy E to scatter inelastically, leaving it with energy E', and leading to the ionization (excitation) of another electron from the orbital i into the orbital ε (n) or, in other words, to the creation of an electron–vacancy pair. If not virtual, such a process must satisfy the energy conservation law, given by

$$E = E' + \varepsilon - \varepsilon_i , \qquad (8.7)$$

ε_i being the energy of the single electron orbital i. As illustrated below, diagram (8.6) can be an element of a more complicated diagram. In this case Eq. (8.7) can be violated and the process of diagram (8.6) is virtual.

In most calculations which apply many-body theory to atoms, the Hartree-Fock equation (8.5) is considered as an initial or zero approximation. For the remainder of this chapter, it is implied that all of the diagrams in Eqs. (8.3) and (8.2) are included in the definition of each line, describing both the electrons and the vacancies.

Let us limit ourselves to elastic scattering only. The simplest correction to elastic scattering beyond the Hartree-Fock approximation is given by the following two diagrams (direct and exchange)

$$(8.8)$$

Notice that the elastic scattering diagrams end with the same number of lines as at their start, each of them having the same energy.

Diagram (8.8a) describes an intermediate state virtual (or real) excitation of the target atom by the projectile electron. The projectile electron polarizes the target, and then is in turn affected by this polarization. Diagram (8.8a) leads to the concept of the polarization potential when the energy dependence of diagram (8.8a) is neglected.

To disclose the origin of the polarization interaction, let us limit ourselves to second order in the Coulomb interaction. Analytically, diagram (8.8a) is described by the expression

$$\langle E \mid \widehat{\Sigma}_d^{(2)} \mid E \rangle = \sum_{i \leq N} \oint_{\substack{E' \\ \varepsilon}} \frac{\langle Ei \mid V \mid E'\varepsilon \rangle \langle E'\varepsilon \mid V \mid Ei \rangle}{E + \varepsilon_i - E' - \varepsilon + i\eta}, \qquad (8.9)$$

where the subscript "d" denotes that only the direct term in second order is included. In Eq. (8.9)

$$\langle Ei \mid V \mid E'\varepsilon \rangle = \int \phi_E^*(\mathbf{r}_1)\phi_i^*(\mathbf{r}_2)\frac{1}{\mid \mathbf{r}_1 - \mathbf{r}_2 \mid}\phi_{E'}(\mathbf{r}_1)\phi_\varepsilon(\mathbf{r}_2)d\mathbf{r}_1 d\mathbf{r}_2 \qquad (8.10)$$

is the Coulomb interaction matrix element. The symbol $\oint$ in Eq. (8.9) denotes a summation and integration over the discrete and continuous spectrum of vacant one-electron orbitals. The infinitesimally small imaginary quantity *in* shows how to treat the singularity in the denominator of Eq. (8.9).*

Equation (8.9) demonstrates the correspondence rules, which permit one to write an analytic equation corresponding to a particular diagram.

- Each zigzag line connecting pairs of solid arrows represents a matrix element in the numerator, and is given as indicated by Eq. (8.10).

- Each intermediate state leads to an energy denominator. This denominator is the sum of all the vacancy energies minus the sum of

---

* Cf. Eq. (4.45), $\lim_{\eta \to 0^+}(D + i\eta)^{-1} = P(D^{-1}) - i\pi\delta(D)$.

all the electron energies in the intermediate state plus the incoming energy.

- A summation (integration) is performed over all the intermediate occupied ($i \le N$) and vacant (ε) orbitals.

- The sign of the diagram is given by $(-1)^{h+l}$, l being the number of closed electron–vacancy loops and h being the number of internal vacancy lines.

Equation (8.9) shows that the dependence of $\widehat{\Sigma}_d^{(2)}(\mathbf{r},\mathbf{r}',E)$ upon the energy E originates in the energy denominator. The imaginary part of $\widehat{\Sigma}_d^{(2)}$ is given by

$$\langle E \mid \mathrm{Im}\widehat{\Sigma}_d^{(2)} \mid E \rangle = -\pi \sum_{i \le N} \oint_{\substack{E' \\ \varepsilon}} |\langle Ei \mid V \mid E'\varepsilon \rangle|^2 \, \delta \left(E + \varepsilon_i - \varepsilon - E' \right) ,$$

(8.11)

the absolute value of which is the probability of inelastic scattering of the electron in the orbital E calculated in the lowest order of perturbation theory and including only the amplitude from diagram (8.6). Notice that the inclusion of diagram (8.8b) leads to an exchange matrix element in Eqs. (8.9) and (8.11): $\langle E'\varepsilon \mid V \mid Ei \rangle$ is replaced by $-\langle E'\varepsilon \mid V \mid iE \rangle$ in the analytic expression related to diagram (8.8b).

The higher order corrections to the scattering amplitude of diagram (8.8) are generated by the inclusion of Coulomb interactions which either connect electrons and vacancies already existing in diagram (8.8), or create new electron–vacancy pairs. The first possibility is illustrated by

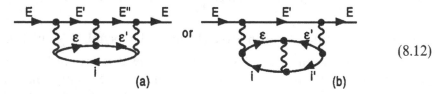

(a)   or   (b)

(8.12)

while the second is given by

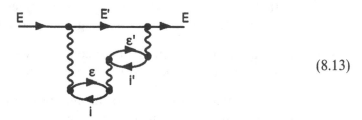

(8.13)

In a similar way, diagrams that describe higher order corrections may be constructed. To take into account all of them is impossible, being equivalent to a precise solution of the Schrödinger equation describing the projectile together with the target atom. Notice, however, that even to take into account the interaction between the two electrons E', ε in diagram (8.8) means solving a three-body problem: two interacting electrons moving in an atomic core potential. Thus, only the several lowest order terms can be taken into account entirely, which forms the basis of the many-body perturbation theory (MBPT), extensively developed and applied by H.P. Kelly and his group. [319] The other possibility is to sum an infinite series of diagrams, but in each order to account for only part of the terms. This approach forms the basis of the random phase approximation with exchange (RPAE). [168] In MBPT the number of diagrams increase with the order of perturbation theory, so that in practice fourth order terms are rarely included. On the other hand, the neglect of all but a chosen sequence of diagrams in the RPAE leads to inaccuracy. In practice, there will be a mixing of these two approaches: the technique referred to as MBPT will include a summation of selected infinite sequences of diagrams, while some elements of perturbation theory are very often included in RPAE calculations.

8.3 The Dyson equation and the optical model

In the higher orders of perturbation theory, two types of corrections to the wavefunction of the projectile are possible: the inclusion of new interactions in the intermediate state of diagram (8.8), and the repetitions of diagram (8.8) according to the procedure illustrated in diagram (8.3).

Suppose that all the corrections in the intermediate states of diagram (8.8) are taken into account, and that we denote the sum of all the corresponding diagrams by a circle containing a summation sign. Then an integral equation for the incoming electron wavefunction ψ_E can be derived

$$\tag{8.14}$$

In many-body theory the circle containing Σ is called the "self-energy part" and is denoted by $\widehat{\Sigma}$. It is evident from diagram (8.8) that $\widehat{\Sigma}$ describes a non-local interaction, being dependent upon the scattered particle. The equation for the projectile motion is given by

$$\left(-\frac{\nabla^2}{2} - \frac{Z}{r} + \widehat{V}_{HF}(\mathbf{r}) \right) \psi_E(\mathbf{r})$$

$$+ \int \widehat{\Sigma}(\mathbf{r}, \mathbf{r}', E)\, \psi_E(\mathbf{r}') d\mathbf{r}' = E\psi_E(\mathbf{r}), \tag{8.15}$$

where, according to Eq. (8.5), the operator of the Hartree-Fock self-consistent field is defined by[†]

$$\widehat{V}_{HF}(\mathbf{r})\psi_E(\mathbf{r}) \equiv \int \rho(\mathbf{r}')\frac{d\mathbf{r}'}{|\,\mathbf{r}' - \mathbf{r}\,|}\psi_E(\mathbf{r}) \qquad (8.16)$$

$$- \sum_{i \leq N}\int \phi_i^\star(\mathbf{r}')\frac{d\mathbf{r}'}{|\,\mathbf{r}' - \mathbf{r}\,|}\psi_E(\mathbf{r}')\phi_i(\mathbf{r})\ .$$

Equation (8.15) is similar to the usual Schrödinger equation, but with an energy dependent and non-local potential. These features, mainly the first one, make Eq. (8.15) very complex, leading to problems with the orthogonality of the wavefunction $\psi_E(\mathbf{r})$ and the completeness of the set of $\psi_E(\mathbf{r})$.

To illustrate this, let us consider the normalization condition. For the energy dependent interaction, which determines the incoming electron wavefunction $\psi_E(\mathbf{r})$, we rewrite Eq. (8.15) as

$$\left(H^{HF} + \widehat{\Sigma}(E)\right)\psi_E(\mathbf{r}) = E\psi_E(\mathbf{r})\ . \qquad (8.17)$$

For different energies the functions must be orthogonal, *i.e.* $\langle\psi_E \mid \psi_{E'}\rangle = 0$ if $E \neq E'$. The normalization to a δ-function requires that the relation $\langle\psi_E \mid \psi_{E'}\rangle \propto \delta\left(E - E'\right)$ be valid. Multiplying Eq. (8.17) by $\psi_{E'}^*(\mathbf{r})$ from the left side, one obtains

$$\langle\psi_{E'} \mid \psi_E\rangle \left(E - E' - \widehat{\Sigma}(E) + \widehat{\Sigma}(E')\right) = 0\ . \qquad (8.18)$$

Expanding $\widehat{\Sigma}(E')$ in a Taylor series

$$\widehat{\Sigma}(E') = \widehat{\Sigma}(E) + \frac{\partial\widehat{\Sigma}}{\partial E}(E' - E)\ ,$$

one obtains from Eq. (8.18)

$$(E' - E)\left(1 - \frac{\partial\widehat{\Sigma}}{\partial E}\right)\langle\psi_{E'} \mid \psi_E\rangle = 0\ , \qquad (8.19)$$

leading to the following normalization condition

$$\left(1 - \frac{\partial\widehat{\Sigma}}{\partial E}\right)\langle\psi_{E'} \mid \psi_E\rangle = \delta\left(E' - E\right)\ . \qquad (8.20)$$

[†] Cf. Eq. (2.15)

Thus, it is seen that, due to the energy dependence of the polarization interaction $\widehat{\Sigma}(E)$, the wavefunction becomes renormalized

$$\langle \psi_{E'} \mid \psi_E \rangle = Z\delta\left(E' - E\right),$$

$$Z = \frac{1}{1 - \dfrac{\partial\widehat{\Sigma}}{\partial E}}, \tag{8.21}$$

where the quantity Z is called the renormalization constant. It is possible to demonstrate using Eq. (8.9) that $\partial\widehat{\Sigma}^{(2)}/\partial E < 0$, which means that $Z < 1$. The general proof that $Z < 1$ was given by Migdal. [320]

We have shown, therefore, that the polarization interaction $\widehat{\Sigma}(\mathbf{r}, \mathbf{r}', E)$ differs from the ordinary polarization potential not only by its energy dependence and non-locality but also by the presence of an imaginary part (see Eq. (8.11)).

The introduction of $\widehat{\Sigma}$ gives a general description of scattering by a structured target, which may be virtually or really excited by the projectile in the collision process. Equation (8.15) is called the Dyson equation and justifies the optical model for elastic scattering in which the target's action upon the projectile is described by a phenomenological potential with real and imaginary parts, which is energy dependent in general.

8.4 Calculation of the elastic scattering cross section

The scattering cross section $\sigma(E)$ is determined [172] by the square modulus of the amplitude $f(E, \theta)$, θ being the scattering angle

$$\sigma(E) = \int |f(E, \theta)|^2 \, d\Omega. \tag{8.22}$$

The amplitude is determined by

$$f(E, \theta) = -\frac{1}{2\pi} \left\langle k\varphi_{\mathbf{k}}^{(0)} \left| \widehat{V}_{HF} + \widehat{\Sigma}(E) \right| \psi_{\mathbf{k}'}^{+} \right\rangle, \tag{8.23}$$

where $\varphi_{\mathbf{k}}^{(0)} = \exp(i\mathbf{k} \cdot \mathbf{r})$, and $\psi_{\mathbf{k}'}(r)$ is the solution of Eq. (8.15) normalized as $\varphi_{\mathbf{k}}^{(0)}$. Here $\mathbf{k}$ is the momentum of the projectile. The variable θ is the angle between the vectors $\mathbf{k}$ and $\mathbf{k}'$. Asymptotically, these functions will behave as[‡]

$$\lim_{r \to \infty} \varphi_{\mathbf{k}}^{(0)} \sim \sum_{\ell=0}^{\infty} \sum_{m_\ell} 4\pi i^\ell Y_{\ell m_\ell}(\hat{\mathbf{r}}) Y_{\ell m_\ell}^*(\hat{\mathbf{k}}) \frac{\cos\left(kr - \dfrac{\pi(\ell+1)}{2}\right)}{kr},$$

[‡] Notice that these asymptotic expansions are valid for *neutral* targets. Cf. Eqs. (4.9) and (5.43) when $q \neq 0$.

$$\lim_{r\to\infty} \psi_{\mathbf{k}'}^+ \sim \sum_{\ell=0}^{\infty}\sum_{m_\ell} 4\pi i^\ell e^{i\delta_\ell(E)} Y_{\ell m_\ell}(\hat{\mathbf{r}}) Y_{\ell m_\ell}^*(\hat{\mathbf{k}}')$$

$$\times \frac{\cos\left(kr - \frac{\pi(\ell+1)}{2} + \delta_\ell(E)\right)}{kr}. \tag{8.24}$$

The operator $\widehat{\Sigma}$ is defined as

$$\widehat{\Sigma}\psi_{\mathbf{k}} \equiv \int \widehat{\Sigma}\mathbf{r}, \mathbf{r}', E)\psi_{\mathbf{k}}(\mathbf{r}')d\mathbf{r}'.$$

In electron–atom scattering the target is spherically symmetric, so the total amplitude can be expanded as a sum over partial waves containing the scattering phase shifts δ_ℓ

$$\begin{aligned}
f(E,\theta) &= \sum_{\ell=0}^{\infty}(2\ell+1)f_\ell(E,\theta)P_\ell(\cos(\theta)) \\
&= \sum_{\ell=0}^{\infty}\frac{(2\ell+1)}{2ik}\left(e^{2i\delta_\ell(E)}-1\right)P_\ell(\cos(\theta)), \tag{8.25}
\end{aligned}$$

and

$$\begin{aligned}
\sigma(E) &= 4\pi\sum_{\ell=0}^{\infty}(2\ell+1)|f_\ell(E)|^2 \\
&= \frac{4\pi}{k^2}\sum_{\ell=0}^{\infty}(2\ell+1)\sin^2(\delta_\ell). \tag{8.26}
\end{aligned}$$

The total phase shift $\delta_\ell(E)$ is a sum of two terms: the Hartree-Fock contribution $\delta_\ell^{HF}(E)$ and an additional part $\Delta\delta_\ell(E)$ coming from $\widehat{\Sigma}$. The phase shift $\delta_\ell^{HF}(E)$ is given by

$$e^{i\delta_\ell^{HF}(E)}\sin(\delta_\ell^{HF}(E)) = -\frac{k}{2\pi}\langle\varphi_{\ell,E}^{(0)} \mid \widehat{V}_{HF} \mid \phi_{\ell,E}^{HF+}\rangle, \tag{8.27}$$

where the index ℓ denotes the ℓ-component of the free particle and Hartree-Fock wavefunctions,

$$\lim_{r\to\infty}\varphi_{\ell,E}^{(0)} \sim \sqrt{4\pi}i^\ell Y_{\ell m_\ell}(\hat{\mathbf{r}})\frac{\cos\left(kr - \frac{\pi(\ell+1)}{2}\right)}{kr},$$

$$\lim_{r\to\infty}\phi_{\ell,E}^{HF+} \sim \sqrt{4\pi}i^\ell \exp[i\delta_\ell^{HF}(E)]\, Y_{\ell m_\ell}(\hat{\mathbf{r}})$$

$$\times \frac{\cos\left(kr - \frac{\pi(\ell+1)}{2} + \delta_\ell^{HF}(E)\right)}{kr}. \tag{8.28}$$

The correction to $\delta_\ell^{HF}(E)$ is determined by the matrix element

$$\exp[i\Delta\delta_\ell(E)] \sin(\Delta\delta_\ell(E)) = -\frac{k}{2\pi} \exp[-2i\delta_\ell^{HF}(E)] \left\langle \phi_{\ell,E}^{HF-} \mid \hat{\Sigma} \mid \psi_{\ell,E}^+ \right\rangle , \tag{8.29}$$

with

$$\lim_{r\to\infty} \phi_{\ell,E}^{HF-} \sim \sqrt{4\pi} i^\ell \exp[-i\delta_\ell^{HF}(E)] Y_{\ell m_\ell}(\hat{\mathbf{r}}) ,$$

$$\times \frac{\cos\left(kr - \frac{\pi(\ell+1)}{2} + \delta_\ell^{HF}(E)\right)}{kr}$$

$$\lim_{r\to\infty} \psi_{\ell,E}^+ \sim \sqrt{4\pi} i^\ell \exp[i\delta_\ell(E)] Y_{\ell m_\ell}(\hat{\mathbf{r}})$$

$$\times \frac{\cos\left(kr - \frac{\pi(\ell+1)}{2} + \delta_\ell(E)\right)}{kr} . \tag{8.30}$$

Thus to determine the scattering cross section one needs to know the self-energy $\hat{\Sigma}$ and the projectile wavefunction ψ_E. Notice that if one rewrites Eq. (8.29) using wavefunctions normalized according to the energy scale,

$$\lim_{r\to\infty} \phi_{\ell,E}^{HF} \sim \left(\frac{2}{\pi k}\right)^{1/2} Y_{\ell m_\ell}(\hat{\mathbf{r}}) \frac{\cos\left(kr - \frac{\pi(\ell+1)}{2} + \delta_\ell^{HF}(E)\right)}{r} ,$$

$$\lim_{r\to\infty} \psi_{\ell,E} \sim \left(\frac{2}{\pi k}\right)^{1/2} Y_{\ell m_\ell}(\hat{\mathbf{r}}) \frac{\cos\left(kr - \frac{\pi(\ell+1)}{2} + \delta_\ell(E)\right)}{r} , \tag{8.31}$$

then Eq. (8.29) becomes

$$\exp[i\Delta\delta_\ell(E)] \sin(\Delta\delta_\ell(E)) = -\pi\langle\phi_{\ell,E}^{HF} \mid \hat{\Sigma} \mid \psi_{\ell,E}\rangle . \tag{8.32}$$

In the case when $\hat{\Sigma}$ is small, the expression for $\Delta\delta_\ell(E)$ becomes much simpler

$$\Delta\delta_\ell(E) \approx -\pi\langle\phi_{\ell,E}^{HF} \mid \hat{\Sigma} \mid \phi_{\ell,E}^{HF}\rangle . \tag{8.33}$$

Equations (8.27) and (8.33) are convenient for determining $\Delta\delta_\ell(E)$, since they only require knowledge of the wavefunction inside the atom.

Equations (8.22)-(8.33) are applicable to any projectile, not just electrons.

An equation can be derived directly for the matrix element of Eq. (8.32), escaping the need for explicit calculation of the projectile wavefunction $\psi_E(r)$. To do this, Eq. (8.15) must be transformed into the integral form, symbolically given by

$$\psi_{\ell,E}(r) = \phi_{\ell,E}^{HF}(r) + \frac{1}{E - \hat{H}_\ell^{HF} + i\eta} \hat{\Sigma}\psi_{\ell,E}(r) , \tag{8.34}$$

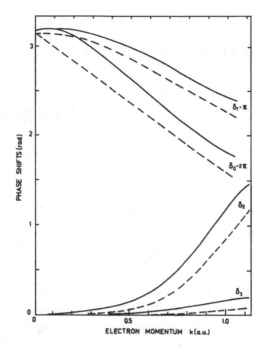

Fig. 8.1.   Electron-argon elastic scattering phase shifts for $\ell = 0, 1, 2, 3$. Solid lines, HF + polarization correction; dashed lines, HF. From Ref. [315].

then multipling by $\widehat{\Sigma}$ (the ℓ-component of $\widehat{\Sigma}$) and by $\phi_{\ell,E'}^{HF^{*}}(r)$ from the left side and integrating over r, yields

$$\langle E'\ell \mid \widehat{\Sigma} \mid \widetilde{E\ell}\rangle = \langle E'\ell \mid \widehat{\Sigma} \mid E\ell\rangle$$

$$+ \sum_{E''}\!\!\!\!\!\!\int \langle E'\ell \mid \widehat{\Sigma} \mid E''\ell\rangle \frac{1}{E - E'' + i\eta}\langle E''\ell \mid \widehat{\Sigma} \mid \widetilde{E\ell}\rangle . \qquad (8.35)$$

The $\sim$ sign above the energy and angular momentum of the orbital indicates that the matrix element is calculated with the wavefunction $\psi_{\ell,E}(r)$. Integration over the continuum and summation over the discrete excitations are performed for the intermediate projectile orbitals. It is seen that the diagonal part of the matrix given by Eq. (8.35) determines an additional phase shift coming from the inclusion of the polarization interaction $\widehat{\Sigma}$. Thus the entire problem of determining $\Delta\delta_{\ell}$ is separated into two steps, the first of which is the calculation of $\langle E'\ell \mid \widehat{\Sigma} \mid E\ell\rangle$ and the second is the solution of Eq. (8.35).

In the first step, the approximation for $\widehat{\Sigma}$ is important. In the calculations performed to date, diagram (8.8) and the atomic electron–vacancy interactions found in diagrams (8.12b) and (8.13) were taken into account, while the projectile–atomic electron interaction of diagram (8.12a) is neglected. It has been demonstrated that this approximation has the

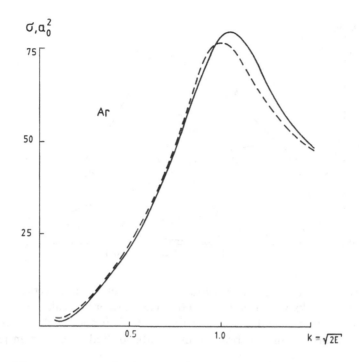

Fig. 8.2. Electron-argon elastic scattering cross section. Vertical axis, σ/a_0^2; horizontal axis, $k = \sqrt{2E}$. Solid line: calculations in RPAE; dashed line: experiment [321]

asymptotic form of a polarization potential $-\alpha_{\mathrm{pol}}/(2r^4)$, α_{pol} being the target atom dipole polarizability. If the phase shifts are determined in such a way, that as $E \to \infty, \delta_\ell(E) \to 0$, then at $E = 0$ the following equality is valid

$$\delta_\ell(0) = \pi(n_\ell^0 + n_\ell^{(v)}),\qquad(8.36)$$

where n_ℓ^0 is the number of occupied energy levels in the target atom and $n_\ell^{(v)}$ is the number of vacant bound orbitals in the "projectile + target atom" system with angular momentum ℓ.

Schematically, the behaviour of elastic scattering phase shifts $\delta_\ell(\ell = 0, 1, 2, 3)$ for electron–argon collisions is presented in Fig. 8.1. It is seen that the total phase shift δ_0 is equal to $n\pi$ not only at $E = 0$, but at some higher energies where δ_1 (and all other δ_ℓ with $\ell > 1$) are still small. According to Eq. (8.26), this leads to a minimum (called the Ramsauer minimum) in the total elastic scattering cross section.

The results of calculations in the approximation described above are illustrated by two examples: Ar [321] and Ca, [322] found in Figs. 8.2 and 8.3. In the latter case the polarization interaction is very strong, leading to

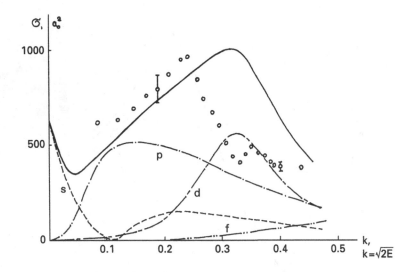

Fig. 8.3. Electron–calcium elastic scattering cross section. The vertical scale gives σ/a_0^2; the scale marks are at 500 and 1000. The horizontal scale gives $k = \sqrt{2E}$; the scale marks are at 0.1–0.5. Broken line, s; broken and single-dotted line, p; long-and-short broken line, d; broken and double-dotted line, f; from [322].

the formation of a stable negative ion. For Ca⁻ the extra electron is in an almost pure $4p$-orbital similar to that of the neighboring Sc atom. Good agreement with experiment is demonstrated for Ar, while the minimum in the $e^- + $ Ca elastic cross section is at such a low energy that it is difficult to observe.

The angular distributions of scattered electrons, which are much more sensitive to the details of the polarization interaction $\widehat{\Sigma}$, are also reproduced reasonably well. The inclusion of $\widehat{\Sigma}$ dramatically modifies the cross section, compared to the Hartree-Fock values, but for small E only. With an increase in energy the influence of $\widehat{\Sigma}$ vanishes rapidly.

Although the agreement with experimental data is satisfactory in general, it is of great interest to improve the theory by including also the interaction between the projectile and electron–vacancy target atom excitations: Eq. (8.12a) and similar higher order corrections. It requires, as was mentioned above, the solution of a three-body problem, accounting for the interaction of two continuous spectrum electrons E' and ε moving in the self-consistent atomic field and in the field of the vacancy i.

Let us briefly discuss positron–atom elastic scattering. Two specific features are essential in the consideration of $e^+ + A$ collisions: there is no exchange with the target electrons and the projectile can form a bound state called positronium. The first feature leads to the neglect of the Fock term in the one-electron approximation, which is achieved by discarding diagram (8.2b), as well as the last term from the left hand side of Eq. (8.5). The e^+–atomic electron interaction (diagram (8.12a) and similar diagrams) leads to virtual positronium formation in the intermediate state and may be taken into account effectively [317] by adding to the $\varepsilon_{e^+} + \varepsilon_{e^-}$ energy in the denominator of Eq. (8.9) the positronium binding energy $\varepsilon_{Ps} < 0$, thus decreasing the virtuality of the intermediate state and strengthening the polarization interaction.

8.5 Resonance scattering

The intermediate states E' and ε in the polarization interaction given by diagram (8.8a) and analytically by Eq. (8.9) can be both continuous and discrete. If the incoming electron energy is close to the energy of an intermediate state, $E_k + \varepsilon_n - \varepsilon_i$, the interaction $\widehat{\Sigma}^{(2)}$ becomes large, leading to a dramatic alteration of the cross section when compared to its Hartree-Fock value. For electron energies close to one of the discrete excitations of the "projectile + target" system, Eq. (8.9) may be parametrized in the following way

$$\widehat{\Sigma}^{(2)}_{EE'} = \frac{g_{EE'}}{E - E_R} + \overline{\Sigma}^{(2)}_{EE'} . \tag{8.37}$$

Here $E_R = E_k + \varepsilon_n - \varepsilon_i$ is the resonance energy, close to that of the projectile, $g_{EE'} = \langle E_i \mid V \mid kn \rangle \langle kn \mid V \mid E'i \rangle = C_E C^\dagger_{E'}$, and $\overline{\Sigma}^{(2)}$ is the non-singular part of $\widehat{\Sigma}^{(2)}$. For a neutral atom, the intermediate state "$|kn\rangle$" in $g_{EE'}$ is that of a negative ion formed by an electron in the field of an excited atom. Able to decay by emitting an electron with energy E, this negative ion is unstable.

Close to the pole in Eq. (8.37) $\widehat{\Sigma}$ must be calculated in a non-perturbative manner by solving

$$\langle E\ell \mid \widehat{\Sigma} \mid \widetilde{E'\ell} \rangle = \frac{C_E C^\dagger_{E'}}{E - E_R} + \frac{1}{E - E_R} \int_{E''} \frac{C_E C^\dagger_{E''} dE''}{E - E'' + i\eta} \langle E''\ell \mid \widehat{\Sigma} \mid \widetilde{E'\ell} \rangle ,$$
$$\tag{8.38}$$

which is obtained from Eq. (8.35) by substituting Eq. (8.37) and neglecting the $\overline{\Sigma}^{(2)}_{EE'}$ contribution. Writing $\langle E\ell \mid \widehat{\Sigma} \mid \widetilde{E'\ell} \rangle$ as the product $C_E C^\dagger_{E'} f(E)$, Eq. (8.38) transforms into an algebraic equation with the following solution

$$\langle E\ell \mid \widehat{\Sigma} \mid \widetilde{E'\ell} \rangle = \frac{g_{EE'}}{E - E_R - P \int_{E''} \frac{g_{E''E''} dE''}{E - E''} + i\pi g_{EE}} + \overline{\Sigma}^{(2)}_{EE'} , \tag{8.39}$$

where the non-singular term is taken into account in lowest order, assuming for simplicity that its contribution is small. The P preceding the integral denotes its principal value.

The discrete "two excited electrons–one vacancy electron" contribution and the one electron continuum orbitals lead to a shift in the resonance energy E_R given by

$$\tilde{E}_R = E_R + P \int_{E''} \frac{g_{E''E''} dE''}{E - E''} , \qquad (8.40)$$

and to a width determined by

$$\frac{\Gamma_E}{2} = \pi g_{EE} . \qquad (8.41)$$

The additional phase shift $\Delta \delta_\ell$ in the partial wave ℓ, where the singularity in $\hat{\Sigma}^{(2)}$ exists as described by Eq. (8.37), is determined in accordance with Eq. (8.32) by the following

$$e^{i\Delta\delta_\ell} \sin(\Delta\delta_\ell) = \frac{-\Gamma_E/2}{E - \tilde{E}_R + i\Gamma_E/2} . \qquad (8.42)$$

Near the singularity in Eq. (8.42) the dependence of g_{EE} upon E is neglected and $\Gamma_{\tilde{E}_R}/2 = \pi g_{\tilde{E}_R \tilde{E}_R}$. At $E = \tilde{E}_R$ the additional phase shift takes the value $\Delta \delta_\ell = \pi(n + 1/2)$, thus leading to a maximum in the elastic scattering cross section (when the HF phase shift δ_ℓ^{HF} is small) or to a resonance in the ℓ partial wave. Resonances which are discrete excitations embedded in a one particle continuum are called Feshbach resonances, since they were first considered by H. Feshbach in nuclear scattering theory. One should distinguish them from shape resonances, which are due to the specific radial dependence of a one particle potential, or in atomic physics, the Hartree-Fock potential.

The calculation of the resonance energy E_R, as well as the interaction matrix element connecting the continuum with a discrete excitation, can require the accounting of more complicated diagrams, *i.e.* those beyond the second order. For example, the interaction between the two excited electrons and the vacancy i shifts the energy E_R and alters the matrix element g_{EE}. The inclusion of the non-singular $\overline{\Sigma}_{EE'}$ may be also important.

The incorporation of the non-resonant part of $\hat{\Sigma}^{(2)}$ transforms Eq. (8.42) into

$$\exp[i\Delta\delta_\ell(E)] \sin(\Delta\delta_\ell(E)) = \frac{-\Gamma_E/2}{E - \tilde{E}_R + i\Gamma_E/2} + Q_\ell , \qquad (8.43)$$

where Q_ℓ can be considered energy independent in the vicinity of $\tilde{E}_R$.

Using Eq. (8.43), one can derive for the ℓ-wave partial cross section the contribution

$$\sigma_\ell(\epsilon) = \sigma_\ell^{HF} \left\{ \frac{[\epsilon_R - \epsilon + q]^2}{(\epsilon_R - \epsilon)^2 + 1} + (1 + qQ_\ell)^2 + Q_\ell^2 - 1 \right\} , \qquad (8.44)$$

where $\epsilon \equiv 2E/\Gamma$, $\epsilon_R \equiv 2\tilde{E}_R/\Gamma$, and $q \equiv \cot(\delta_\ell^{HF})$. Thus the cross section acquires a Fano-type resonance profile.

8.6   Formation of negative ions and the scattering problem

According to Eq. (8.36), the calculation of the elastic scattering phase shift at $E = 0$ gives information on the number of bound states in the "projectile + target" system. Thus the existence of stable negative ions can be demonstrated by a calculation of the phase shift from electron–atom scattering theory. For example, if δ_ℓ is calculated in the Hartree-Fock approximation and $n_\ell^{(v)} \neq 0$, it means that the negative ion already exists at this level of approximation.

It is essential to keep in mind, that the behavior of the phase shift δ_ℓ as a function of energy is quite different in the Hartree and Hartree-Fock approximations. In the Hartree approximation, Eq. (8.36) has the usual form of the Levinson theorem [172]

$$\delta_\ell^H(0) = \pi n_\ell^{(v)}, \tag{8.45}$$

while the inclusion of exchange leads to Eq. (8.36) instead of Eq. (8.45), which means qualitatively different behaviour at small energies. Of course, with an increase of E the role of exchange decreases rapidly.

The Fock term works like an attractive potential for electron–atom collisions. The polarization interaction $\tilde{\Sigma}$ is basically a second-order correction (see Eqs. (8.8) and (8.9)) and is negative for any charge on the projectile, thus increasing the attraction of the target. To directly disclose whether a negative ion exists for a given atom, Eq. (8.5) for Hartree-Fock binding or Eq. (8.15) for correlational binding must be solved, giving the wavefunction and the affinity simultaneously. In practice, such calculations require one to simplify the interaction $\hat{\Sigma}$. The first step is to restrict the types of diagrams included in $\hat{\Sigma}$: usually this restriction is to the direct (and exchange) interactions of an atomic electron and a vacancy, like diagram (8.13) and all similar higher order terms. The second step is to neglect the E-dependence of $\hat{\Sigma}$. This makes it similar to a non-local potential. The latter assumption considerably simplifies the problem. [323]

However, it is possible to derive information about the existence of a negative ion directly from the scattering data using Eq. (8.36). If $n_\ell^{(v)} > 0$, then the incoming electron can be bound in the state with angular momentum ℓ. This method is also useful when a given state, corresponding to a resonance in an electron–atom scattering cross section is transformed for neighboring atoms with higher nuclear charge Z into a bound state. In the resonance case the corresponding phase shift reaches a value of $\pi(n + \frac{1}{2})$ at some resonance energy E_R. Then for neighboring atoms with higher Z, δ_ℓ becomes greater than $\pi(n + \frac{1}{2})$, E_R becomes smaller, finally leading, if the atomic attraction acting upon the projectile starts to be strong enough, to $\delta_\ell = \pi(n_\ell + 1)$.

The affinity, obtained in the Hartree-Fock approximation, is usually much smaller than the experimental value. Thus one must include the polarization interaction. There are cases, however, where there are no negative ions in the Hartree-Fock approximation, but with the inclusion of $\widehat{\Sigma}$ (calculated in second order with higher order corrections, similar to diagram (8.13)) leads to stable negative ion states, with the affinity reasonably close to experimental data. For example, the calculated result for Ca^- is 0.58 eV, [316] while the measured value is 0.43 ± 7 eV. [323] Just recently, however, a new affinity was found experimentally for Ca^-, [324] which is about three times smaller. Thus affinity calculations for alkaline earth negative ions require further investigation.

In the structure calculations for negative ions performed to date, only those cases were considered where the extra electron is in an almost pure one electron state. More complicated is the situation when the negative ion is formed due to binding of an extra electron to an excited target atom state, the simplest case being the binding of the projectile to an "electron k"–"vacancy i" excitation of the target atom. In this case the diagram, equivalent to $\widehat{\Sigma}^{(2)}$, is given by

$$(8.46)$$

The vacancy i is in lowest order a spectator of the projectile–excited electron collision process, but in higher orders the interaction with i may be important. Because the excitation energy $(\varepsilon_{k'} - \varepsilon_k)$ is smaller than that in Eq. (8.9) $(\varepsilon_i - \varepsilon)$, the polarization interaction is stronger and the ability to bind an extra electron is larger. In order to obtain the affinity and wavefunction of such a state, an equation similar to Eq. (8.15) with $\widehat{\Sigma}$ given by diagram (8.46) must be solved.

8.7 Collisions with open shell and excited atoms

Many-body theory is most convenient when applied to collisions with closed shell targets, whose ground states are nondegenerate. However, without too much effort, open shell targets may also be considered.

Let us start with the case when one electron is outside a closed core. In this case the collision must be considered as really a three body problem: the outer atomic electron in the discrete orbital k (similar to diagram (8.46), but without a vacancy) colliding with a projectile of energy E. The zero approximation wavefunctions for the projectile and outer electron may be calculated in the Hartree-Fock field of the target core. This will automatically include the action of the self-consistent potential of the core, as well as the exchange with the core electrons. The interaction between the projectile and the outer electron is included by summing the following infinite series of diagrams

$$(8.47)$$

Analytically, the summation of this sequence of diagrams is equivalent to the solution of a Schrödinger equation describing the motion of two particles in the core field and interacting with each other, *i.e.* a three-body equation. It is essential to note that summation over E' and k' in the intermediate states of diagram (8.47) includes not only integration (summation) over vacant orbitals, but also a summation over discrete occupied levels. Diagrammatically, this corresponds to the inclusion of terms which must be added to diagram (8.47), for instance

$$(8.48)$$

If the core itself is easily polarizable, corrections must be included which take into account the polarization interaction of the projectile and the outer electron with the core. This is achieved by using solutions of Eq. (8.15) instead of Hartree-Fock wavefunctions, with $\hat{\Sigma}(\mathbf{r}, \mathbf{r}', E)$ determined by the core's virtual excitations in an expression similar to Eq. (8.9). For easily polarizable cores the inclusion of only the direct Coulomb interaction between the projectile and outer electron is insufficient, and the modification of this interaction due to core virtual excitations must be taken into account. If this correction is large, a perturbative account of it is inadequate and a so-called effective interaction $\hat{\Gamma}$ must be introduced.

Usually, if included at all, it is taken into account within the random phase approximation with exchange (RPAE), which is determined by the following series of diagrams

$$\hat{\Gamma} \quad = \quad \} + \quad + \quad + \cdots + \begin{array}{c} \text{Exchange} \\ \text{Terms} \end{array} \quad = \quad \} + \quad \hat{\Gamma} \qquad (8.49)$$

The heavy shaded line denotes the effective interaction $\hat{\Gamma}(\omega)$. Analytically, the effective interaction in the RPAE is determined by [319]

$$\langle kj \mid \hat{\Gamma}(\omega) \mid in \rangle = \langle kj \mid U \mid in \rangle$$

$$+\sum_{\substack{k' \\ j' \leq N}} \left[\frac{\langle kj' \mid U \mid ik' \rangle \langle k'j \mid \hat{\Gamma}(\omega) \mid j'n \rangle}{\varepsilon_{j'} - \varepsilon_{k'} + \omega + i\eta} \right.$$

$$\left. + \frac{\langle kk' \mid U \mid ij' \rangle \langle j'j \mid \hat{\Gamma}(\omega) \mid k'n \rangle}{\varepsilon_{j'} - \varepsilon_{k'} - \omega} \right] , \qquad (8.50)$$

where $\langle kj \mid U \mid in \rangle \equiv \langle kj \mid V \mid in \rangle - \langle kj \mid V \mid ni \rangle$ includes direct and exchange Coulomb matrix elements, and ω is the energy transferred via $\hat{\Gamma}(\omega)$.

It is also very simple to consider the scattering from atoms with a vacancy instead of an extra electron in the outer shell. The collision of an electron in such a case may be represented by an infinite sequence of electron–vacancy scattering, pair annihilation and pair creation processes. A simple sequence of diagrams describing these processes is given by

$$(8.51)$$

This sequence of diagrams determines the scattering amplitude for both the elastic ($E' = E$, $i' = i$) and inelastic ($E' \neq E$, $i' \neq i$) collisions. It is important to note that the scattering of an electron with an atom having a vacancy (*i.e.* an ion) is expressed via the neutral atom effective interaction

$\widehat{\Gamma}$, given by diagram (8.50), which is non-local and energy dependent. The addition to the Hartree-Fock phase shift is given by an expression similar to Eq. (8.32) [325]

$$\exp[i\Delta\delta_L]\,\sin(\Delta\delta_L) = -\pi\langle Ei\,|\,\widehat{\Gamma}_L(\omega)\,|\,Ei\rangle\,, \qquad (8.52)$$

where $\omega = E - \varepsilon_i$, L is the total angular momentum of the electron–vacancy pair, and $\widehat{\Gamma}_L$ is the L-component of $\widehat{\Gamma}(\omega)$. The imaginary part of Eq. (8.52) describes a number of inelastic processes, such as the spin-flip of the projectile or the floating up of the vacancy i if it is deep in the initial state, *i.e.* a transition $i \to i'$ which may be accompanied by a change in the energy, angular momentum, and spin of the projectile. If the projectile is not an electron, the exchange terms found in diagrams 8.51(b'),(c'),(c'') must be omitted.

Atoms with half-filled shells may be treated in the same way as those with closed shells, but consisting of two types of electrons, "up" and "down," depending on their spin projection. Indeed, according to Hund's rule all electrons in a half-filled shell have the same spin projection. Let it be the "up" direction. Due to exchange with other atomic electrons every level is split into two "up" and "down" sublevels. Because the Coulomb interaction is unable to act upon spins, the target reacts upon the projectile as a closed shell system with both "up" and "down" electrons. Then many-body theory can be applied after a rather straightforward generalization to systems consisting of two kinds of particles instead of one. The cross section becomes strongly dependent upon the spin of the projectile–whether it is "up" or "down," leading to almost completely different results for those two cases. [326]

The consideration of collisions with excited atoms with one electron–vacancy pair is similar to that of atoms with one outer electron. If the vacancy is passive, the problem is completely equivalent to that described by diagram (8.47). However, the presence of an excited orbital opens new possibilities. Both superelastic scattering and elastic scattering via an unexcited atomic state as given by

$$(8.53)$$

where "$n\,i$" is the electron–vacancy excitation in the initial state. More complicated diagrams include the polarization interaction of the projectile with the excited electron and the core, as well as the modification of the Coulomb interaction described by diagram (8.49).

8.8   The scattering of muonic hydrogen by electrons

The scattering of muonic hydrogen by electrons is an example of a process where only a few low-order perturbation theory corrections are needed. Each Hartree contribution to the scattering amplitude is much smaller than the muonic hydrogen H_μ radius a_μ

$$f_H = \beta_\mu a_\mu,$$

$$\beta_\mu = \frac{1}{m_\mu}, \tag{8.54}$$

where m_μ is the muon mass (in atomic units $m_\mu \approx 205$). The repetition of the Hartree term, as in diagram (8.3c), leads to corrections which are a factor of β_μ smaller. Thus it is necessary to calculate only the contribution of the polarization term of diagram (8.8a). To emphasize the difference between e^- and μ particles, let us redraw this diagram

$$\tag{8.55}$$

where the thick line denotes the μ-meson. It is not necessary to take into account the interaction of the electron in the initial, intermediate or final state with the μ-meson in the 1s level: each time such an interaction is included an extra power of β_μ will appear. It is possible to demonstrate that the interaction between the intermediate state electron and the excited μ-meson also leads to small corrections. Therefore, only diagram (8.55) with free electron wavefunctions must be taken into account, whose analytic expression is given by Eq. (8.9). At first glance the electron energy difference ($E' - E$) may be neglected as compared to the excitation energy of H_μ, which is larger than $1/\beta_\mu$. However, even crude estimates based on the uncertainty principle show that for an electron to be close to the muonic hydrogen atom, a large amount of kinetic energy is required, namely $E' \sim a_\mu^{-2} \sim \beta_\mu^{-2} \gg E_{1s}^{(\mu)} = \beta_\mu^{-1}$. Therefore, in fact the energy difference $E_{2p,s}^{(\mu)} - E_{1s}^{(\mu)}$ may be neglected as compared to $(E' - E)$. Thus, one has a picture resembling an electron scattering from a degenerate target state. This leads to a linear Stark effect, resulting in a $[-1/(r^2)]$ instead of a $[-\alpha_{\text{pol}}/(2r^4)]$ potential. Of course, asymptotically the potential is always $[-\alpha_{\text{pol}}/(2r^4)]$, but the contribution of this region to the scattering

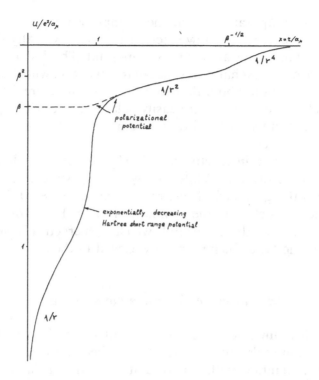

Fig. 8.4. Radial dependence of the potential of muonic hydrogen acting upon the incoming electron. The vertical scale gives μ in units of e^2/a_μ; the scale markings from top to bottom are β^2, β, 1. The horizontal scale gives r in units of $2/a_\mu$; the scale markings are, from left to right, 1 and $\beta^{-1/2}$. Near the r-axis the curve has the form $1/r^4$ and then flattens out to a $1/r^2$ dependence. On reaching the muonic hydrogen the potential seen by the electron rapidly deepens to take an exponentially decreasing Hartree short-range potential; this deepens again to a $1/r$ form near the vertical axis. The form of the polarization potential would continue along the broken line.

amplitude given by Eq. (8.23) for an $(e^- + H_\mu)$ collision is small. The radial dependence of the Hartree potential and the potential given by diagram (8.55) are depicted in Fig. 8.4.

The main contribution to the polarization amplitude and the phase shift of Eq. (8.33) comes from the intermediate distances $a_\mu < r < a_\mu \beta_\mu^{-1/2}$, where the radial dependence is $\sim -1/r^2$, leading to a term $\sim \beta_\mu^{3/2}$ in the scattering amplitude [327]

$$f = \beta_\mu + 3\beta_\mu^{3/2} . \tag{8.56}$$

The investigation of $e^- + H_\mu$ scattering yields a completely new approach to the theory of electron collisions with atoms. Indeed, it has been shown from numerical computations, that the asymptotically large distances are not important even for very low electron energies. Actually distances of order of the atomic radius are the most important. This is demonstrated by the importance of the exchange term of diagram (8.8b), whose contribution comes from distances of about the atomic radius. Therefore, it makes more sense, especially for collisions with easily polarizable atoms, to parametrize the correlation interaction by adding a term which decreases as $-1/r^2$ at intermediate distances.

Also interesting and instructive is $e^- + H_\mu^{(2s)}$ scattering, when the target is in an excited $2s$ orbital. While for the ground state the cross section is smaller than the geometrical one by a factor of $\beta_\mu^2 \approx 10^{-5}$, the excited state cross section is close to the geometrical value. This is due to the very large polarizability of the $2s$ orbital, which is energetically very close to the $2p$ orbital, the latter being virtually excited in the $e^- + H_\mu^{(2s)}$ collision process.

8.9   Inelastic electron-atom scattering

The simplest diagram describing inelastic scattering is diagram (8.6). In the final state the atomic electron can be either excited or ionized. For electrons as projectiles a diagram with a permutation of the E' and $\varepsilon(n)$ orbitals must be taken into account. Calculated with Hartree-Fock wavefunctions, such diagrams represent the distorted-wave first Born approximation. However, if the target atom is easily polarizable, higher-order corrections must be taken into account. These corrections are illustrated by

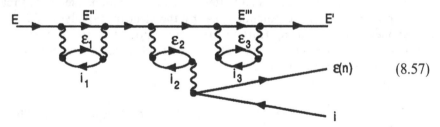

$$(8.57)$$

where the incoming and outgoing electron wavefunctions are modified by the polarization interaction in lowest order. The lowest order correction to the projectile–target electron interaction (via the $\varepsilon_2 i_2$ virtual excitation) is also included. Such a process, together with higher order corrections, is almost impossible to calculate. However, it is possible to include the polarization interaction by introducing a phenomenological polarization potential. Notice that, if in the final state E' and ε are discrete levels, diagrams (8.6) (and (8.57)) describe dielectronic capture.

Two-electron ionization (excitation) in electron–atom collisions is described in lowest order by diagrams representing a double interaction with the target or a single projectile–target interaction preceded or followed by an intra-target interaction

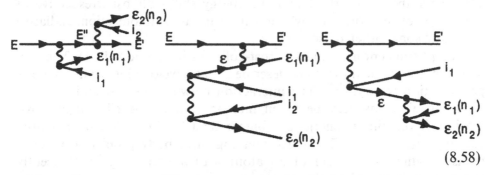

$$(8.58)$$

These diagrams are difficult to calculate, and with the inclusion of the corrections from diagram (8.57), they become even more complicated.

A special problem is found in inelastic scattering if the electrons in the final state are moving slowly. The interaction between them must be taken into account, requiring the solution of the three body problem in order to obtain the final state wavefunction. While the case of two very slow electrons has attracted a great deal of attention, starting with the famous Wannier paper, the case of three slow electrons in the final state has not been considered at all.

An inelastic collision close to the threshold of an autoionization resonance is quite interesting, and can be represented by

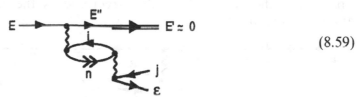

$$(8.59)$$

If the vacancy i is much deeper than j, the atomic electron ε leaves the target quickly. Therefore the field acting upon the outgoing electron instantly changes from that of an excited atom to that of an ion with a vacancy j. This alteration of the field modifies the shape of the energy distribution of the electron ε: instead of being a symmetrical profile with the width Γ determined by the decay rate of the "$n\ i$" resonance state, it becomes asymmetrical, much broader, and its maximum is shifted to energies higher than that of the "$n\ i$" excitation. This extra energy can be estimated as $\Delta E \sim -1/r_s$, where r_s is the distance traveled by the slow electron with the velocity $k_s = \sqrt{2E'}$ before the decay takes place, that is, during the time $\tau \sim 1/\Gamma$. Thus

$$\Delta E' \approx -\frac{\Gamma}{\sqrt{2E'}}.$$

$$(8.60)$$

If $E' + \Delta E' < 0$, the effect of the extra vacancy j is so strong that the slow electron E' will be excited to a discrete instead of a continuum level. The faster the decay, the larger the energy shift—almost irrespective of the origin of the width Γ: whether it is due to autoionization, radiative decay, or some other reason.

This phenomenon is called a post-collision interaction (PCI). The first quantum mechanical theory to describe it was developed using many-body perturbation theory. [328] The interaction between the fast and the slow outgoing electrons may be taken into account in the eikonal approximation, if the kinetic energy ε is bigger than the inter-electron potential energy, $\Gamma/(|\mathbf{k} - \mathbf{k}'|)$. The corresponding three body problem (two electrons moving in the field of an atom with a vacancy j) was recently incorporated in the PCI theory. [329]

How one treats the inelastic scattering problem is important, not only in electron–atom collisions, but also in photoionization. Indeed, on its way out of the atom, a photoelectron emitted from inside an atom will collide with the outer electrons, exciting and/or ionizing them. It appears that such a process is rather important, [330] leading to a modification of the energy and angular distribution of the photoelectrons. (See chapter 7 by S. Manson).

Another example of inelastic electron–atom scattering is the Bremsstrahlung process: the emission of photons due to the deceleration of the projectile. This process can proceed in two different ways: by emitting a photon in the static field of the target and by the polarization of the target by the projectile

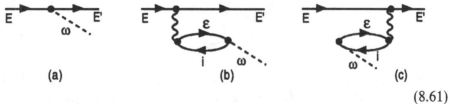

$$(8.61)$$

Diagram (8.61b) describes a process in which the photon is emitted not by the projectile, but by the excited target in a virtual (or real) manner. The analytic expression for this amplitude is given by

$$\langle E \mid \mathbf{d} \mid E' \rangle + \sum_{\substack{i \leq N \\ \varepsilon}} \Bigg[\left\langle E_i \left| (\mathbf{r} - \mathbf{r}')^{-1} \right| E'\varepsilon \right\rangle \frac{1}{\omega - \varepsilon + \varepsilon_i - i\eta} \langle \varepsilon \mid \mathbf{d} \mid i \rangle$$

$$- \left\langle E\varepsilon \left| (\mathbf{r} - \mathbf{r}') \right|^{-1} E'i \right\rangle \frac{1}{\omega + \varepsilon - \varepsilon_i} \langle i \mid \mathbf{d} \mid \varepsilon \rangle \Bigg] , \qquad (8.62)$$

where $\mathbf{d}$ is the operator describing the photon–electron interaction.

A number of calculations were performed recently, [331] demonstrating the important role played by the contributions from diagrams (8.61b,c).

If the target atom is easily polarizable, the Hartree-Fock approximation for the projectile is not sufficient and corrections of the type presented in diagram (8.57) must also be included. However, this has not yet been done.

The starting point for the investigation of inelastic scattering accompanied by photon radiation is given diagrammatically by

$$(8.63)$$

As an example, the analytic expression for diagram (8.63b) is given by

$$\oint_{E_1} \left[\langle E \mid \mathbf{d} \mid E_1 \rangle \frac{1}{E - \omega - E_1 + i\eta} \langle E_1 i \left| \frac{1}{\mathbf{r} - \mathbf{r'}} \right| E'\varepsilon \rangle \right.$$
$$\left. + \langle Ei \left| \frac{1}{\mathbf{r} - \mathbf{r'}} \right| E_1\varepsilon \rangle \frac{1}{E + \varepsilon_i - \varepsilon - E_1 + i\eta} \langle E_1 \mid \mathbf{d} \mid E' \rangle \right] . \qquad (8.64)$$

This process has not yet been calculated for continuum photons and ionized electrons.

The many-body approach is convenient for the investigation of inelastic scattering from excited atoms, where the excitation energy is transferred to the projectile. This so-called superelastic process in lowest order of perturbation theory is represented by diagram (8.53a). The collision with the metastable state can change the angular momentum or spin of the excited electron, thus transforming this state into a rapidly decaying one. This is illustrated by inelastic or quasielastic scattering with subsequent electron or photon emission

$$(8.65)$$

The polarization interaction may also affect the cross section for inelastic scattering by excited atoms.

8.10 Concluding remarks

In the previous sections we have demonstrated how many-body theory and the diagrammatic technique can be applied to a number of scattering problems. Of course, the application of this approach is much broader, allowing for the consideration of many problems in atomic physics. The approach is very convenient in formulating the initial approximation to many processes. In regards to obtaining calculational results of high accuracy, this depends not only upon the particular process under consideration, but also on the particular atom. Indeed, when the inter-electron interaction is not generally weak, higher-order corrections may be important and their calculation is a very difficult task. However, this difficulty is connected with the nature of atoms as multiparticle systems and has nothing to do with the theoretical apparatus. In fact, the advantage of diagrammatic many-body theory is that it reveals, in a very transparent way, the complexity of the problem and possible sources for solution.

This chapter has presented a many-body approach to scattering theory which, to a large extent, was developed by H. P. Kelly and his collaborators. Almost all of the branches of atomic many-body theory have received important contributions from him, and the high level of achievement is due to his tremendous efforts. Until the last day of his life, he actively participated in many-body research. For the numerous and unforgettable discussions on a variety of many-body problems, including the electron–atom and other particle collision processes, I am deeply grateful to him.

9

Theoretical aspects of electron impact ionization, (e, 2e), at high and intermediate energies

P. L. Altick

9.1 Introduction

Beginning with the pioneering experiments of Ehrhardt *et al.* [332] and Amaldi [333] more than 30 years ago, a field of considerable activity has developed in the area of measuring and theoretically predicting the angular and energy correlations of two electrons leaving an ion following ionization by electron impact, the so-called $(e, 2e)$ problem. This field is daunting both experimentally and theoretically; the experiments require coincidence measurements with typically small counting rates while the adequate description of two light charged particles with modest kinetic energy moving in the Coulomb potential of a heavy ion, one version of the Coulomb three body problem, has provided a continuing challenge to theorists.

The underlying goal of this research has been to achieve a better understanding of the ubiquitous process of ionization by studying the basic process of an incoming electron transferring a measured amount of momentum to a target atom with the result that a secondary electron is ejected, its momentum being measured also. Such experiments then measure all kinematic variables and provide the most detailed information possible about the event. The only observable averaged over is the spin. A practical application of this activity is that, if theories can be developed to get these differential cross sections right, then integration should produce reliable total ionization cross sections which are in constant demand in other areas, astrophysics and the controlled fusion program for example.

In its several decade existence the $(e, 2e)$ work has already spread into several subfields. Much interest has been focused on the threshold region where the emerging electrons' kinetic energy is dominated by the Coulomb potential energy for a substantial distance. This is the intriguing Wannier regime, but will not be a subject here because attempts to understand this

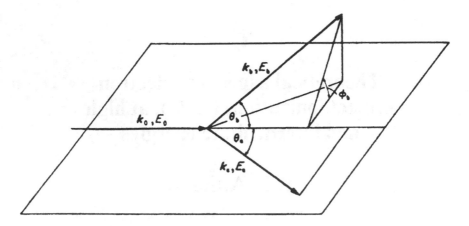

Fig. 9.1. The kinematic variables in an $(e, 2e)$ experiment. Because the ion can absorb momentum, one electron can leave the scattering plane.

energy range will be covered by Pan and Starace in chapter 11. Rather the present emphasis will be on the process as it manifests itself at high to intermediate incident energy. The terms "high" and "intermediate" need further clarification which will be provided below.

The kinematic variables in an $(e, 2e)$ experiment are depicted in Fig. 9.1. To completely specify the momenta of electrons a and b following ionization clearly takes six variables. Since the kinetic energy of the ion after the collision is changed negligibly from that before, there is an approximate conservation of energy which removes one free variable. Of the four angles required to specify directions, one, say ϕ_a, can be put equal to zero since the x-axis can be oriented arbitrarily. If the detection of the two outgoing electrons takes place in the scattering plane, coplanar geometry, then the other azimuthal angle is fixed at either $0°$ or $180°$. There are then three free variables remaining, and the experiments thus measure what is called a triply differential cross section (TDCS), but it is really a special case of a five-fold cross section. Although we will focus on the coplanar situation, out of plane measurements exist, as well. [334]

When the $(e, 2e)$ experiments became feasible, a good question was: what are the interesting values of the kinematical parameters, or, put another way, in the three dimensional parameter space, where are the regions that show interesting physics? Several types of measurements emerged from such considerations. The asymmetric coplanar geometry refers to detection of electron a, the "fast" electron at a typically small angle with respect to incidence and a small energy loss. Electron b, the "slow" electron is then ejected with low energy and is detected at different angles which extend from the forward to the backward direction. Since H.

Ehrhardt's group in Kaiserslautern, Germany first used and analyzed this geometry, it is frequently denoted as the "Ehrhardt Geometry." This is the principal geometry that we will be involved with. Early on, researchers in Adelaide, Australia, following the work of Amaldi *et al.* [333] saw that $(e, 2e)$ experiments could be used to extract information about the momentum distribution of target atomic and molecular orbitals. To accomplish this, the non-coplanar symmetric or the coplanar symmetric geometry is employed. The latter is of interest here. Using it, two electrons share energy equally after the collision and are detected at equal angles on opposite sides of the incident beam. The angle is the only free variable. This geometry illuminates other aspects of the $(e, 2e)$ problem as will be seen below.

Electron impact ionization has now grown and diversified beyond the range of any review of reasonable length. Accordingly the discussion below will be restricted to theoretical approaches in coplanar symmetric or asymmetric geometries. Particular emphasis will be given to the problem of representing the final state interaction of the two free electrons, *i.e.* selection of an asymptotic state following the collision. The theoretical models depend strongly on the value of the incident energy, so it is convenient to organize the models considered here into separate energy regimes.

9.2 The highest energies: the optical and impulsive limits

As the incident energy approaches infinity while the momentum transfer approaches zero, the impinging electron delivers an electric jolt to the target that resembles that of a radiation field. This is the optical limit and is a simple one, theoretically, as we now show.

The TDCS for electron impact ionization has been derived in several places; see for example Peterkop. [335] It is given by

$$\frac{d^3\sigma}{d\Omega_a d\Omega_b dE_b} = (2\pi)^4 \frac{k_a k_b}{k_0} |T_{fi}|^2,\qquad(9.1)$$

with

$$T_{fi} = \left\langle \Psi_f^- |V| \Psi_i \right\rangle.\qquad(9.2)$$

There are various approximations for the initial state, Ψ_i, and the final state, Ψ_f, discussed as we move along. They contain free electron wave functions, however, and, as written, these functions are normalized to a delta function in momentum. This is the prior form in which the incident wave and target are described by a channel function and the final state is, in principle, a complete scattering state, *i.e.* a full solution to the Schrödinger or Lippmann-Schwinger equations. The potential, V, contains, accordingly, all the interactions of the incoming electron with the target particles.

When the energy is sufficiently high, the simple form of the Born approximation (BA) is employed whereby the fast electron is described by a plane wave. Since we will be comparing theory to experiments on helium initially, we write the T-matrix for this case:

$$T_{fi} = \sqrt{2} \int d^3\mathbf{r}_a \int d^3\mathbf{r}_b \int d^3\mathbf{r}_c (2\pi)^{-3/2} e^{-i\mathbf{k}_0 \cdot \mathbf{r}_a} \Psi_{\text{He}}(\mathbf{r}_b, \mathbf{r}_c)$$

$$\times \left[-\frac{1}{|\mathbf{r}_a - \mathbf{r}_b|} - \frac{1}{|\mathbf{r}_a - \mathbf{r}_c|} \right] \Psi_{\text{He}^+}(\mathbf{r}_c)(2\pi)^{-3} e^{i\mathbf{k}_a \cdot \mathbf{r}_a} \Psi_{\mathbf{k}_b}^{-}(\mathbf{r}_b). \qquad (9.3)$$

The functions $\Psi_{\text{He}}(\mathbf{r}_b, \mathbf{r}_c)$ and $\Psi_{\text{He}^+}(\mathbf{r}_c)$ are the ground states of helium and He$^+$ respectively. The ejected electron wave function, $\Psi_{\mathbf{k}_b}^{-}(\mathbf{r}_b)$, assuming the asymmetric geometry, is represented by some sort of distorted wave using, for example, the static potential of the He$^+$ core. An additional simplification is made here in that only the incident electron-ejected electron interaction is kept. The other terms in V would drop out if the ejected electron wave function were orthogonal to the ground state. This orthogonality is frequently imposed as an additional condition, and we can assume that is the case here. The Bethe integral is carried out over $\mathbf{r}_a$ resulting in

$$T_{if} \propto \int d^3\mathbf{r}_b \int d^3\mathbf{r}_c \Psi_{\text{He}}(\mathbf{r}_b, \mathbf{r}_c) e^{-i\mathbf{K} \cdot \mathbf{r}_b} \Psi_{\text{He}^+}(\mathbf{r}_c) \Psi_{\mathbf{k}_b}^{-}(\mathbf{r}_b). \qquad (9.4)$$

If $\mathbf{K}$, the momentum transfer is small enough, the exponential is replaced by the dipole operator with a polarization direction along $\mathbf{K}$. In this limiting case, we have arrived at a definite prediction. With $\mathbf{K}$ held constant, sweeping $\mathbf{k}_b$ should produce a $\cos^2$ distribution, *i.e.* two peaks, centered about the momentum transfer direction with FWHM (full width at half maximum) of 90°. This two peak pattern is quite typical over a large range of asymmetric geometries. The peaks are usually labeled "binary" for the forward and "recoil" for the backward one. This terminology does not seem appropriate at this stage, but it will later.

Experiments have been performed at energies around 8 keV with helium and neon as targets. Lahmam-Bennani *et al.* [336] found strong evidence that the optical limit was being reached. They extrapolated the FWHM to zero $\mathbf{K}$ and observed the trend toward 90°, Fig. 9.2. The binary and recoil peaks also approached equal intensity.

Another limit can be reached by letting the momentum transfer become large with the additional momentum being given to the ejected electron which will now be represented by a plane wave. The $\mathbf{r}_b$ integral in Eq. (9.4) reduces to the Fourier transform of the target ground state wave

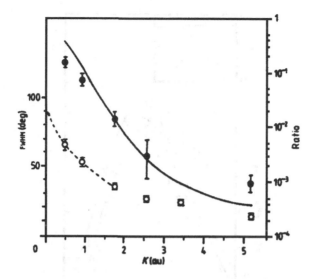

Fig. 9.2. Experimental measurement of the ratio of the binary peak to the recoil peak (solid line, scale on right), and the width, FWHM, of the binary peak (broken line, scale on left). The experimental points approach the optical limit, ratio = 1, FWHM = 90°, as the momentum transfer goes to zero. The target is helium, $E_a = 8000$ eV, $E_b = 100$ eV and 337 eV. The incident energy, $E_0 = 8000$ eV $+ E_b + 24.6$ eV. From Ref. [336].

function. The helium ground state must be represented in a separable approximation, the one electron wave function denoted by $\phi_{\mathrm{He}}(\mathbf{r})$.

$$T_{fi} \propto \int d^3\mathbf{r}_b \phi_{\mathrm{He}}(\mathbf{r}_b) e^{-i(\mathbf{K}-\mathbf{k}_b)\cdot\mathbf{r}_b}. \tag{9.5}$$

This is the impulse approximation. The cross section factors into a scattering part and a form factor. This type of approximation, although more sophisticated than that presented here, forms the basis for the large body of work done principally in Adelaide but also elsewhere wherein the form factors and hence the target wave functions are experimentally determined. Atoms, molecules, and even solid surfaces have been investigated. To describe this interesting aspect of $(e, 2e)$ would take us too far from the main theme here, and, anyway, thorough and up to date reviews exist. [337]

When the impulse approximation is compared with laboratory data in the same sequence of experiments mentioned above, it is found that it works best for a unique set of kinematical parameters called the "Bethe ridge." This set corresponds to a collision in which the residual ion absorbs no energy or momentum. An example of this situation is given in Fig. 9.3, where one observes that the impulse approximation crosses the experimental curve just on the Bethe ridge. In strong contrast to the

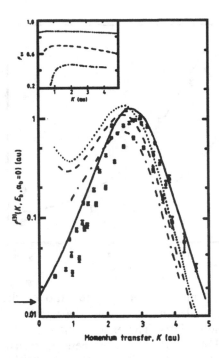

Fig. 9.3. A plot of the generalized oscillator strength, directly related to the TDCS, versus momentum transfer for three incident energies: $\square$, −8124.6 eV; $\triangle$, −1524.6 eV; $\bullet$, −524.6 eV. $E_b = 100$ eV throughout. To be noted are the dotted, broken, and chain curves which are plane wave impulse calculations at the three energies, resp. The Bethe ridge mentioned in the text occurs around $K \sim 2.75$ a.u., which is the area of greatest agreement between theory and experiment. From Ref. [337].

optical limit, the impulsive limit produces just one peak, now appropriately called a binary peak, *i.e.* it results from a binary collision with the ion having little influence.

Sweeping between these limits, even at this 8 keV incident energy, the plane wave BA does not provide a good description of the TDCS. The reason is that the ejected electron does not have a high energy, typically a few tens of electron=volts, and a plane wave description is not adequate. Lahmam-Bennani *et al.* [338] tried a Coulomb wave with unit charge, orthogonalized to the ground state orbital, for the ejected electron, the OCW (orthogonal Coulomb wave) variation of the BA. This does a much better job overall, but still has its failings. Fig. 9.4 shows that the binary peak is generally well fitted, but there are difficulties with the small recoil peak.

Data has been taken also on neon at 8 keV. [339] The most striking result is that, when the target orbital is $2p$ rather than $1s$, the TDCS has

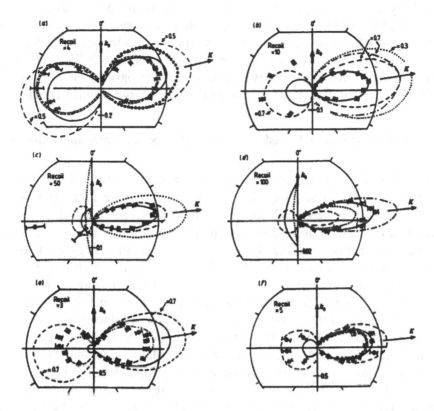

Fig. 9.4. Triply differential cross sections for electrons on helium with $E_a = 8000$ eV and held fixed. Spectra (a) to (d) have $E_b = 46$ eV and $\theta_a = 1°$, $2°$, $4°$, and $6°$. For spectra (e) and (f), $E_b = 20$ eV and $\theta = 1.25$ and $2°$. The theoretical model discussed in the text, OCW (orthogonal Coulomb wave), is presented as a solid line. From Ref. [337].

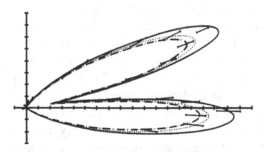

Fig. 9.5. Relative triply differential cross section for a Ne target with 8000 eV incident energy. $E_b = 200$ eV and $\theta_a = 9.44°$. The electron is ejected from a $2p$ orbital. See text regarding the theoretical curves (solid line). From Ref. [339].

an entirely different appearance, Fig. 9.5. This is not so surprising if one realizes that the optical limit is now complicated because $p \to d$ and $p \to s$ transitions are both possible, so the $\cos^2$ shape is no longer expected. The theory that has been applied is a modification of the OCW in that the charge on the Coulomb wave is considered a free parameter. Fair fits are obtained, but only at the price of unrealistically large pseudo-charges.

9.3   High energies: distorted waves and the second Born approximation

With the OCW approximation faltering at 8 keV, as the incident energy is lowered further, other models must be used. There is a general lack of data between 8 keV and 600 eV with the exception of the program to determine electron momentum densities, mentioned above. However, one such experiment on the $3p$ orbital of Ar at an energy of 1 keV is interesting here because a more refined theory is used, which is a distorted wave impulse approximation. [340] The scattering amplitude is written as in Eq. (9.2). Rather than improve just one side of the matrix element, the prior form is abandoned and the plane waves are replaced with distorted waves computed in a suitable atomic potential. Next a factorization is accomplished. This is an exact feature of the plane wave matrix element; it is an additional approximation here. The comparison with experiment reveals that the distorted waves are necessary for this target in this energy regime. The theory is still basically reliable only on the Bethe ridge, but the plane wave impulse approximation is not satisfactory even there. Fig. 9.6 shows the situation.

As lower energies are looked at, the pace of both experimental and theory pick up. There has been a lot of work done in the 250-600 eV range to which we now turn.

Since the first Born approximation is no longer adequate, even for small momentum transfer, one route toward a better theory would be to attempt to develop a second Born amplitude. This has been accomplished in a series of papers by Byron *et al.* [341] The second Born is too difficult for a complete evaluation, so these authors use an average excitation energy and employ the closure approximation. They use for the slow electron a Coulomb wave which is orthogonalized to the ground state. Within the framework of the first Born, the binary and recoil peaks are necessarily opposite each other and aligned symmetrically around the momentum transfer direction. At the energies now being considered, however, there is a noticeable misalignment in the experimental data with the two peaks bending away from each other. It is definitely a success of this simplified second Born theory that this bending is reproduced. In Fig. 9.7 some 600 eV data for helium is shown with the second Born calculations.

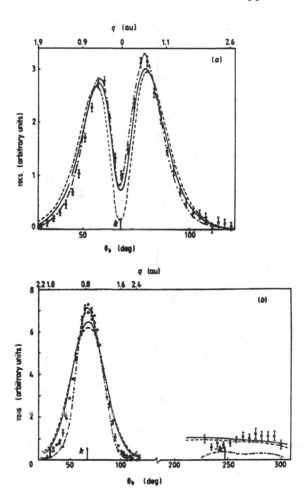

Fig. 9.6. Triply differential cross sections for electrons on an argon target, $3p$ orbital. $E_0 = 1015.8$ eV, $E_b = 120$ eV. In (a) $\theta_a = 20°$; in (b) $\theta_a = 14°$. The broken curve is the impulse approximation and the chain curve is the distorted wave result. In the recoil lobe in (b), the experimental data and the distorted wave curve are multiplied by 2 while the impulse approximation curve is multiplied by 300. From Ref. [340].

The expected optimal kinematics for this approach are small momentum transfer and, as a consequence, a large asymmetry in the energies of the two electrons. That is the situation shown in Fig. 9.7. Within this limited range of kinematics, the simplified second Born approximation represents a significant improvement over the first order theories.

An effort to include still more of the interactions in the scattering amplitude is contained in the work of Mota Furtado and O'Mahony. [342] These authors also use a simplified second Born as above, but, in addition,

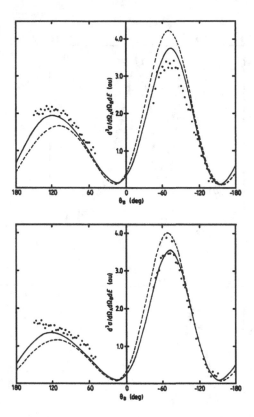

Fig. 9.7. Triply differential cross sections for a helium target, $E_0 = 600$ eV. The measurements are absolute. The solid curves givethe second Born calculation. The broken curves gives the first Born. Upper figure, $E_b = 2.5$ eV; lower figure, $E_b = 5$ eV. $\theta = 4°$. From Ref. [341].

perform a one channel R-matrix calculation to find the wave function of the slow electron. In Fig. 9.8, their results can be seen to follow the experiment well at 250 eV as long as the momentum transfer does not get too large. It is a bit of a mystery why the helium results at 600 eV are not better than they are, Fig. 9.9. They should be on a par with the Byron *et al.* [341] work discussed above. Perhaps the problem is in their representation of the ground state which is computed in the framework of the R-matrix theory also.

Up to this point, none of the theoretical approaches has used the correct asymptotic form for the final state. The long range interaction between the two exiting electrons has been accounted for only indirectly, as in a second Born calculation, or not at all. The analytic form of the wave function for three unbound charged particles when all mutual distances are large has been a matter of theoretical interest for a long time. Although a solution

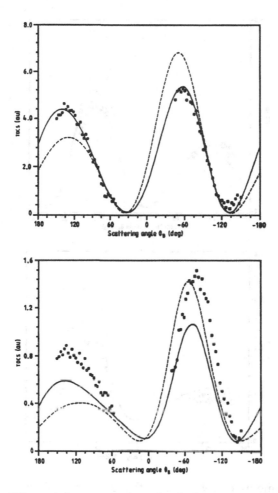

Fig. 9.8.   Triply differential cross sections for helium, $E_0 = 256$ eV, $E_b = 3$ eV. The solid curves give the second Born calculation described in the text, from [342]. The broken curves give the first Born approximation. Upper figure, $\theta_a = 4°$; lower figure, $\theta = 10°$.

to the problem was found by P. Redmond, it was not published until quoted by Rosenberg. [343] The quoted form with the Redmond phase is

$$\Psi \sim \exp(i\mathbf{k}_a \cdot \mathbf{r}_a + i\mathbf{k}_b \cdot \mathbf{r}_b + i\gamma), \tag{9.6}$$

with

$$\gamma = \frac{Z}{k_a} \ln(k_a r_a + \mathbf{k}_a \cdot \mathbf{r}_a) + \frac{Z}{k_b} \ln(k_b r_b + \mathbf{k}_b \cdot \mathbf{r}_b) - \frac{1}{k_{ab}} \ln(k_{ab} r_{ab} + \mathbf{k}_{ab} \cdot \mathbf{r}_{ab}). \tag{9.7}$$

This is a beautifully symmetric form. The Coulomb interaction between electrons manifests itself as an additional logarithmic phase containing the

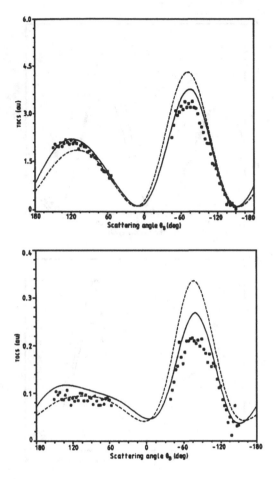

Fig. 9.9.   Same as Fig. 9.8, except that $E_0 = 600$ eV and $E_b = 2.5$ eV.

interelectron coordinates. This form was not used in any $(e, 2e)$ calculations until it was generalized by Brauner, Briggs and Klar in 1989. [344] In this important paper, the final state wave function was written as

$$
\begin{aligned}
\Psi_{BBK} \;=\; & C e^{i\mathbf{k}_a \cdot \mathbf{r}_a} e^{i\mathbf{k}_b \cdot \mathbf{r}_b} \\
& \times F\left(-\frac{i}{k_a},\, 1,\, -i\left(k_a r_a + \mathbf{k}_a \cdot \mathbf{r}_a\right)\right) \\
& \times F\left(-\frac{i}{k_b},\, 1,\, -i\left(k_b r_b + \mathbf{k}_b \cdot \mathbf{r}_b\right)\right) \\
& \times F\left(\frac{i}{k_{ab}},\, 1,\, -\frac{i}{2}\left(k_{ab} r_{ab} + \mathbf{k}_{ab} \cdot \mathbf{r}_{ab}\right)\right),
\end{aligned} \qquad (9.8)
$$

with

$$C = \frac{1}{(2\pi)^3} e^{\frac{\pi}{2k_a}} \Gamma\left(1 + \frac{i}{k_a}\right) e^{\frac{\pi}{2k_b}} \Gamma\left(1 + \frac{i}{k_b}\right) e^{-\frac{\pi}{2k_{ab}}} \Gamma\left(1 - \frac{i}{k_{ab}}\right), \quad (9.9)$$

i.e. using the full hypergeometric functions F rather than their asymptotic forms. It should be stressed that Eq. (9.8) is an asymptotic solution to the Schrödinger equation only when all radial distances are large. The hypergeometric function in the interelectron coordinates is a convenient continuation to $r_a \to 0$, but it is probably not the correct one, as we will see below.

Having written Eq. (9.8), the authors proceeded to calculate the prior form of the matrix element, Eq. (9.1) for the $(e, 2e)$ hydrogen problem. The theory remains first order in the interaction between the incident electron and the target, as in a first Born, but, in the asymptotic sense anyway, the final state interaction or post collision interaction is fully accounted for.

Further insight into the BBK (Brauner, Briggs, Klar) wave function is achieved when it is realized that this function becomes an exact solution to the Schrödinger equation everywhere but within range of the atomic field when the electrons are in the asymptotic field of a neutral atom, *e.g.* an electron detachment process. The Hamiltonian in question is just

$$H \sim \frac{\mathbf{p}_a^2}{2m} + \frac{\mathbf{p}_b^2}{2m} + \frac{1}{r_{ab}}. \quad (9.10)$$

This is easily seen by ignoring the hypergeometric functions which arise from the ion-electron interactions and then rearranging the kinetic energy into a center of mass term, which motion is now described by a plane wave, and a motion relative to the center of mass described exactly by a Coulomb wave function in the relative coordinates.

On the one hand this is an almost trivial observation, but it gets to the essence of the validity of the BBK function. When all three Coulomb potentials are present, the kinetic energy cannot be correctly partitioned as described in the last paragraph except in the limit of kinetic energy $\gg$ potential energy, which defines the asymptotic region. In terms of trajectories, the hyperbolic Coulomb curves must become almost straight before the BBK description is correct. An analytic study of the residual terms when the BBK function is employed with the full three body Hamiltionian shows that they arise from the kinetic energy operator and are rather complex in form. A discussion of these terms in parabolic coordinates has been given by Klar. [345]

Having digressed concerning the meaning of the structure of the BBK final state, how well does it work in practice? The answer is that, within its realm of validity, very well indeed. In the first applications, TDCS for hydrogen at 250 eV and 150 eV were found and compared

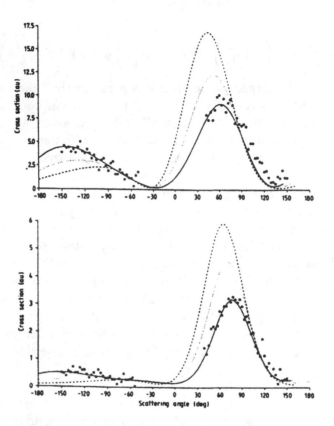

Fig. 9.10.   Triply differential cross sections for hydrogen at $E_0 = 250$ eV and E_b = 5 eV. Upper figure, $\theta_a = 3°$; lower figure, $\theta_a = 8°$. The solid curve is the BBK result discussed in the text description of other curves. From Ref. [344].

with absolute measurements from Kaiserslautern. [349] At 250 eV, the comparison between theory and experiment is just about perfect under the kinematical conditions of high energy asymmetry and small to modest momentum transfer, Fig. 9.10. For this situation the theory would appear to give better overall results than the modified second Born work reported by Baliyan and Srivastava. [346] Also in Ref. [344] are results for hydrogen at 150 eV. As to shape, these theoretical data also look extremely good, see Fig. 9.11. However arbitrary normalization factors between theory and experiment, close to one at 250 eV, are now growing, so that the theory can fall 30% below experiment for the largest momentum transfers. This is an indication of worse things to come as lower energies are looked at, but first, there is still more to consider in the present energy range.

A quite opposite theoretical approach to the hydrogen electron impact ionization problem has been developed in Belfast by Curran *et al.* [347, 348] These workers concentrate on the initial state rather than the final, *i.e.* the

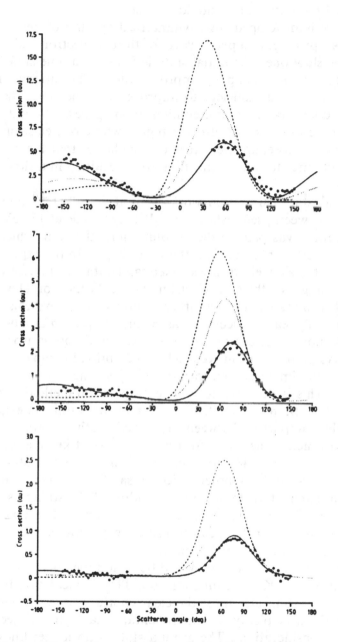

Fig. 9.11.   Triply differential cross sections for hydrogen at $E_0 = 150$ eV and E_b = 5 eV. Top figure, $\theta_a = 4°$; middle figure, $\theta_a = 10°$; in bottom figure, $\theta_a = 16°$. The solid curve is the BBK result. From Ref. [344].

"post" form of the matrix element in which the final state is a channel function and the initial state should be exact.

The theory is developed for asymmetrical sharing of energy, and the final state is a product of a plane wave for the fast electron and a Coulomb wave for the slow one. The initial state is found in a nine state, including six pseudostates, close coupling approximation. The rationale is that, in scattering theory, it is sufficient to improve only one state, the initial or the final, and here the region of configuration space where both electrons are close to the nucleus and also the region where one electron is distant and the other is close, are well treated. One hopes that these regions are the important ones for a an ionization event resulting in a slow and a fast electron.

This pseudostate calculation works well at 250 eV, [347] and a careful comparison between pseudostate and BBK was made at 150 eV. [348] Particular attention was paid to the absolute normalization, which, perhaps surprisingly, is harder to get right than the shapes. In the experiments, E_b, the energy of the slow electron was fixed, and data was taken by changing θ_b the polar angle of the slow electron. This collection of TDCS's are all normalized with respect to each other as no energies have changed in the detectors. Finally, each collection was separately put on an absolute scale by extrapolation to zero momentum transfer and comparison to optical data. [349] As pointed out in Ref. [348], the optimal comparison between theory and experiment is then to introduce just one normalization factor for each E_b. This is done because the extrapolation method is considered more uncertain than the relative measurements within one group. On this basis the comparison between theoretical methods produces no clear choice as to which is the most effective. Nine sets of kinematical variables are discussed in Ref. [348], for some of them the pseudostate theory is better than BBK and for others, vice versa. Unfortunately, no obvious trends are apparent to help our understanding of the strengths and weaknesses of each. As an illustration of the situation, in Fig. 9.12 two cases are shown representing the greatest discrepancy between the two theories and showing each one better for one case.

A natural next step for theory would be to add the final state interaction contained in the BBK wave function to a TDCS calculation for an atom other than hydrogen. Helium is the clear choice as it is the next simplest atom, and there is plenty of experimental data. This project involves several new considerations. The ground state is no longer known and so must be approximated. Correlation does not appear to be an important ingredient when the wave function is used in the first order matrix element, and so a description at the Hartree-Fock level is used. The slow electron is now moving in a residual He$^+$ core, and it is worthwhile to represent this continuum state by a sum of partial waves computed in the potential

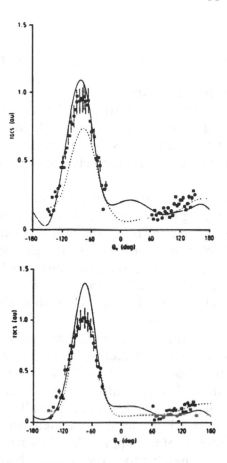

Fig. 9.12. Triply differential cross sections for hydrogen at $E_0 = 150\text{eV}$ and $\theta_a = 16°$. In the upper figure, $E_b = 3\text{eV}$; in the lower, $E_b = 5\text{eV}$. The solid curves are the pseudostate results and the dotted curve is the BBK theory. Experiment and theory have been normalized on the curve with the smallest momentum transfer (not shown). From Ref. [348].

of the ionic core. This is in the spirit of the Mota Furtado-O'Mahony work, [342] and this procedure has also been stressed by Sharma and Srivastava. [350]

Once partial waves are introduced rather than the known expressions for the Coulomb plane wave, the evaluation of the first order amplitude becomes more complicated. This is the case for both the slow and the fast electron, which is now described by the asymptotic form of a pure Coulomb wave. In the calculations done and discussed below, the geometry is kept highly asymmetric and the energy is kept fairly high. In this regime the hypergeometric function in the interelectron coordinates used by BBK, the "correlation factor", can safely be replaced by its asymptotic form, *i.e.*

$$F\left(\frac{i}{k_{ab}}, 1, -\frac{i}{2}\left(k_{ab}r_{ab} + \mathbf{k}_{ab} \cdot \mathbf{r}_{ab}\right)\right) \sim \left(k_{ab}r_{ab} + \mathbf{k}_{ab} \cdot \mathbf{r}_{ab}\right)^{-\frac{i}{k_{ab}}}, \qquad (9.11)$$

to within a normalization constant.

This is in turn also expanded into partial waves to return to single particle coordinates. After quite a bit of manipulation, the scattering amplitude is represented as a sum over six ℓ values with their accompanying m_ℓ sums. The final expressions and the details of the derivation are given in Ref. [351].

The results indicate that the inclusion of the correlation factor plays a significant role in bringing the theory into closer accord with experiment. At 600 eV and 400 eV, comparing to the absolute experimental data of Jung *et al.* [352] and Schlemmer *et al.*, [353] the agreement is very good. A sampling is given in Fig. 9.13. At 250 eV, discrepancies between theory and experiment become apparent, and at 150 eV, the theoretical description has become poor. This trend is quite understandable as high energy approximations were used throughout.

As with hydrogen, alternative theoretical approaches which avoid the difficulties of the final state interaction seem capable of reproducing the experimental data about as well as the previously discussed work. Srivastava and Sharma [354] have utilized a first Born amplitude with an improved treatment of the slow electron, essentially as described above, and have added a second Born term, computed with simpler functions using closure. One example of this approach is shown is Fig. 9.14. Again we are in the situation of being able to say very little about what interactions dominate the process because these two theories seem to sample different parts of configuration space.

A distinct measurement geometry, the coplanar symmetric, has also been studied in this energy range. Recall that in this geometry, both electrons come off with equal energy, so the slow-fast distinction and the theoretical advantages in modeling asymmetric energy sharing are lost. This is the impulse approximation geometry, but it breaks down already at higher energies, so a more sophisticated theory is necessary, and it appears that the Born approximation is viable. In a recent measurement, the TDCS in coplanar symmetric geometry were measured for helium and neon at 500 eV and 100 eV. [355] This cross section drops precipitously, but not necessarily monotonically, as the angle between the electrons increases, Fig. 9.15. A two order of magnitude decrease is typical. When the first measurements in this geometry were made, [356] the data did not show the sharp minimum visible in Fig. 9.15 at about 90° too clearly. Nevertheless second Born calculations of Bryon *et al.* [357] and Mota Furtado and O'Mahony [358] showed that the ef-

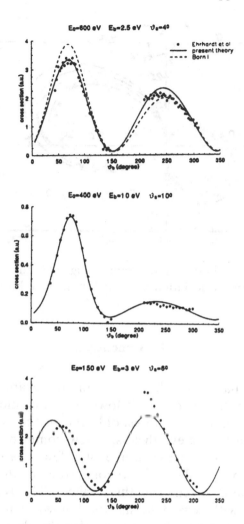

Fig. 9.13. Triply differential cross sections for helium for various sets of kinematical parameters. Top figure, $E_0 = 600$ eV, $E_b = 2.5$ eV, $\theta_a = 4°$; middle figure, $E_0 = 400$ eV, $E_b = 10$ eV, $\theta_a = 10°$; bottom figure, $E_0 = 150$ eV, $E_b = 3$ eV, $\theta_a = 6°$. Data points, Erhard *et al.* The experiments are absolute. The theory (solid lines) is first order with final state interaction included [351]; the broken lines give the Born result.

fect of higher order terms, *i.e.* the second Born term, was crucial. This term produced the minimum which is not present in the first Born treatment.

As can be seen in Fig. 9.15, the distorted wave theory does a good job at 500 eV and is beginning to fail at 100 eV. The theory is not as satisfactory for neon.

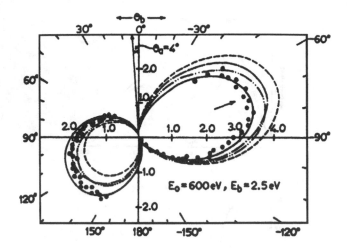

Fig. 9.14.   Same data as in the top diagram of Fig. 9.13. but on a polar plot. The solid curve is the hybrid calculation described in the text. From Ref. [354].

9.4   Conclusion

The logical continuation of the review in this chapter is to consider what happens when the energy E_0 is lowered to the interesting threshold region.   However, $(e, 2e)$ near threshold will be the subject of chapter 11 by Pan and Starace. Nevertheless, a few comments are in order to complete this story. The energy range of a few tens of electron=volts above threshold is very difficult for theory. None of the approaches discussed above remains effective in this regime. The BBK method does not do too badly with regard to shapes, but when normalized experiments were available, it became apparent that the absolute scale is badly off.

Distorted waves do better, but still are not capable of the kind of agreement we have seen at higher energies. So the development of a sensible theory here remains to be accomplished.

Returning to high energies, other types of interesting experiments are now in progress whose theoretical interpretation will undoubtedly be challenging. In Paris, the first $(e, 3e)$ experiments have been done [359] opening up the space of kinematical parameters to even more dimensions.

The process of simultaneous excitation and ionization has been measured and some theoretical treatments exist. [360, 361, 362] In this process three electrons change their state during the collision and different mechanisms are possible to accomplish this.

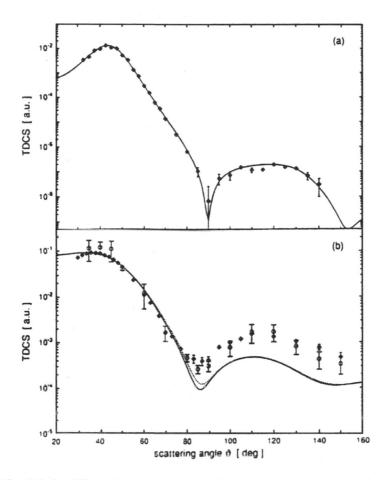

Fig. 9.15. Triply differential cross sections in the coplanar symmetric geometry for helium at (a) $E_0 = 500$ eV and (b) $E_0 = 100$ eV. The solid curve is a distorted wave theory. From Ref. [355].

The $(e, 2e)$ field, only partially presented here, is now maturing, but the combination of unsolved theoretical problems and new experimental directions maintain the sense of novelty and vitality that have characterized it from the beginning.

Part 3

ATOMIC SCATTERING:
B. Low-order applications

10

Perturbation series methods for calculating electron-atom differential cross sections

D. H. Madison

10.1 Introduction

The problem of electron–atom scattering is one of the oldest problems in atomic physics dating back to the early part of this century. This problem has often been studied and there were periods for which the problem was considered solved if one defined solved to mean that there was agreement between the existing experimental data and theoretical calculations. The early experimental measurements were typically total cross sections—that is cross sections summed over spins, summed over magnetic sub levels and integrated over scattering angles. For the case of atomic excitation, the measurements were normally performed for the allowed transitions. These cross sections were generally in reasonable agreement with the highly popular Born approximation which treats the projectile as a plane wave.

In the late 1960's and early 1970's, improved technology permitted the measurement of cross sections differential in the electron scattering angle following the collision. These more detailed cross sections revealed that the Born approximation was totally incorrect for large scattering angles. It was not possible to see this fact from the total cross sections which are dominated by the small angle differential cross sections (DCS). These DCS measurements revealed that an improved theoretical model was required. Most of the efforts in this direction since the early 1970's fall either into the category of a perturbation series approach or a non-perturbative close-coupling approach. In the perturbation series approach, the T-matrix is expanded in a power series for the interaction potential with the expectation that this interaction is small enough that the series will converge quickly. The primary advantage of the perturbation method is that the first term is relatively easy to calculate and the disadvantage is that each successive term is increasingly more difficult to evaluate.

In the close-coupling method, the wave function for the system is expanded in terms of a complete set. The advantage of this method lies in the fact that exact results would be obtained if a complete set could actually be used. The disadvantage lies in the fact that each complete set contains contributions from continuum wave functions and these wave functions cannot be directly included in the close-coupling expansion. Reviews of these methods may be found in Nakazaki, [363] Anderson *et al.*, [364] Itikawa, [365] Walters, [366] and Bransden and McDowell. [367, 368]

In terms of the perturbation series approach, the first order distorted wave approach (DWB1) or equivalent first order many body theory (FOMBT) has been very successful for understanding and interpreting the DCS measurements (a sample of these calculations may be found in references [369]–[384]). If one examines the physical effects contained in each of the terms of the perturbation series expansion, it is found that the first order DWB1 term contains the effects of the atom on the projectile. The effects of the distortion of the atom by the projectile occur naturally in the second term of the perturbation expansion. The second term also contains the effects of absorption of flux into other non-observed channels. Consequently, it can be argued that all of the important physical effects are contained in the first two terms of the perturbation series expansion: the distortion of the projectile by the atom, the distortion of the atom by the projectile and the effects of non-observed processes. While the higher order terms contain additional contributions and corrections from these same physical effects, one must include at least the first two terms if one wishes to include all the potentially important physics.

In this chapter, we will present a brief overview of the theory, discuss some of the problems and considerations involved in performing a practical calculation and finally show some typical results of first and second order calculations for electron–alkali atom scattering.

10.2   Theory

The details of the distorted wave approach may be found elsewhere (see, for example, Madison and Shelton, [369] Madison *et al.* [385, 386]) so only a brief outline will be presented here. For simplicity we shall limit the discussion to electron–alkali atom scattering. For this case the first two terms in the distorted wave series can be expressed as

$$T_{fi} = T_{fi}^{(1)} + T_{fi}^{(2)} , \qquad (10.1)$$

where $T_{fi}^{(1)}$ is the first order amplitude

$$T_{fi}^{(1)} = 2\langle \phi_f^-(0)\Psi_f(1) \mid [V - U_f(0)] \, \mathbf{A} \mid \Psi_i(1)\phi_i^+(0)\rangle . \qquad (10.2)$$

Here Ψ_i (Ψ_f) is the initial (final) bound state atomic wave function for the active electron and V is the full interaction between the projectile electron and the atom. The distorting potential U_f is a final state spherically symmetric approximation for V, and the final state distorted wave ϕ_f^- is a solution of Schrödinger's equation using U_f

$$(K + U_f - E_f)\,\phi_f^- = 0. \tag{10.3}$$

Here K is the kinetic energy operator and E_f is the energy of the final state electron. The initial state distorted wave ϕ_i^+ is obtained similarly using the initial state distorting potential U_i and incident energy E_i. The remaining undefined symbol in Eq. (10.2) is the antisymmetrizing operator

$$\mathbf{A} = \frac{1}{2}\left[1 + (-1)^S P_{01}\right]. \tag{10.4}$$

Here $S = 0$ (1) corresponds to singlet (triplet) scattering and P_{01} is the operator that interchanges particles 0 and 1. In the first order amplitude, the unity operator produces the direct amplitude and the interchange operator produces the exchange amplitude. Consequently, the first order amplitude can be written as

$$T_{fi}^{(1)} = \mathsf{D} + (-1)^S \mathsf{E}, \tag{10.5}$$

where D is the direct scattering amplitude and E is the exchange amplitude. The second order amplitude is given by

$$
\begin{aligned}
T_{fi}^{(2)} = {}& 2\langle\phi_f^-(0)\Psi_f(1) \mid [V - U_f(0)]\,\mathbf{A}\,[E^+ - H_\mathrm{A} - K - U_\mathrm{G}]^{-1}\,\mathbf{A} \\
& \times [V - U_i(0)] \mid \Psi_i(1)\phi_i^+(0)\rangle.
\end{aligned} \tag{10.6}
$$

The operator $[E^+ - H_\mathrm{A} - K - U_\mathrm{G}]^{-1}$ in Eq. (10.6) is the distorted Green's function with H_A being the Hamiltonian for an isolated atom and U_G the distorting potential in the Green's function. The two antisymmetrizing operators in the second order term produce four different second order amplitudes

$$T_{fi}^{(2)} = \mathsf{DD} + (-1)^S \mathsf{DE} + (-1)^S \mathsf{ED} + \mathsf{EE}. \tag{10.7}$$

The second order amplitude corresponds to two collisions between the projectile and atom and for the DD term both collisions are of the direct type; DE (direct-exchange) corresponds to a direct process followed by an exchange collision; ED is the reverse possibility (exchange first and direct second); and EE corresponds to two exchange collisions. Explicit expressions for these amplitudes along with partial wave expansions may be found in Madison *et al.* [385]

10.3   Practical considerations

10.3.1   Bound state wavefunctions

The evaluation of the perturbation series T-matrix requires initial and final state wavefunctions for the projectile and the active electron. The projectile wave function is determined by the distorting potential which will be discussed in the next section. Normally, the bound state wave function for the active electron is chosen to be the asymptotic initial and final state wavefunctions for a static isolated atom. Consequently, analytic wavefunctions are used for hydrogen and numerical Hartree-Fock wavefunctions are used for heavier atoms.

In the previous section, it was noted that the first term of the perturbation series contains the effects of the atom on the projectile and one of the effects contained in the second term is the polarization of the atom by the projectile. Since higher order terms are very difficult to evaluate, it becomes desirable to try to simulate higher order effects in lower order terms. This can be partially accomplished by using bound state wavefunctions which have been distorted by the presence of the projectile. The distorted wave polarized orbital (DWPO) model of McDowell *et al.* [387] represented one of the first practical applications of this idea in a first order model. The effects of polarization are manifested in two ways.

- the distortion of the atom by the projectile

- the back effect of this distortion on the projectile

The first effect can be simulated by using polarized wave functions in the evaluation of the matrix elements. To include the second effect, the polarized wavefunctions must be used in the calculation of the distorting potential for the projectile. Very recently, Madison *et al.* [386] reported a second order calculation for electron–sodium scattering in which elastic scattering and excitation of the $3p$, $4s$ and $4p$ states were studied. In that work, it was found that the back effect of atomic polarization on the projectile was particularly dramatic for excitation of the $4p$ state. This will be demonstrated in section 10.4.

10.3.2   Distorting potentials

The projectile wavefunctions are determined by the choice for the distorting potentials U_i and U_f (see Eq. (10.3)). These distorting potentials are obtained by using the bound state wavefunctions to form a charge density for the atom, and then this charge density is used to calculate a potential in the usual manner. For ease of calculation, the potential obtained from the atomic charge density is averaged over angles to obtain a spherically symmetric distorting potential. Madison *et al.* [388] recently examined the importance of making the spherically symmetric approximation by performing a calculation for excitation of the $2p$ state of hydrogen using the full angle dependent distorting potential. For this case, it was found that the non-spherical part of the distorting potential did not make a significant difference in the energy range for which perturbation series would be expected to be valid.

There have been several different first order distorted wave calculations reported over the years. [369]–[388] The primary difference between various works lies in the distorting potential used to calculate the initial and final state distorted waves. Intuitively, one would expect that since the incoming electron "sees" the initial state of the atom, U_i should be the static potential for the atomic ground state and since the outgoing electron "sees" the final state of the atom, U_f should be the static potential for the final atomic state. However, this is not required by the theory and in fact these potentials may, in principle, be chosen arbitrarily. In the first order many body theory (FOMBT) method, [370, 371],[378]–[381],[383, 384] it is argued that both U_i and U_f should be the static potential for the ground state. There have been several FOMBT calculations reported over the years and it is clear that this choice gives better agreement with experiment than the intuitive one. Madison and Winters [389] performed a computer experiment to determine the choice for the distorting potentials which would give the best agreement with experimental data for excitation of the $2(^1P)$ state of helium. It was found that if one used the potential $[(1/3)U_i+(2/3)U_f]$ to calculate both the initial and final state distorted waves, reasonably good agreement was achieved with the available experimental data. It was also noted that simply using U_f to calculate both the initial and final state distorted waves gave results not very much different from the 1/3, 2/3 combination. In a similar vein, Srivastava *et al.* [377] found that $[U_i + U_f]/2$ also gave good reasonably good agreement with experimental data.

Over the last few years, distorting potentials for several different transitions in many different atoms have been investigated, and it has been found consistently that the intuitive choice (U_i for initial, U_f for final) gives the poorest agreement with experimental data, the FOMBT (U_i for initial, U_i for final) choice gives better agreement and the (U_f for initial, U_f for final) choice gives the best agreement with data of these three possibilities. This observation leads one to speculate that this seemingly unnatural choice must somehow simulate not only the first order amplitude but also some part of the second order amplitude which is now known to be important. So far no theoretical basis for this speculation has been found. Nevertheless, if one wishes to perform only a first order calculation, then experience dictates that the wavefunctions for both the initial and final state of the projectile electron should be calculated using either the final state potential or at least a strong component of this potential.

10.3.3 Exchange distortion

Exchange enters the calculation in two different ways. The first, and most obvious, is through the exchange amplitude. If one is performing a first order calculation, including exchange generally means that the amplitude E of Eq. (10.5) is evaluated or for a second order calculation the amplitudes DE, ED, and EE of Eq. (10.7) are also evaluated. The second, and less obvious, way in which exchange enters a calculation is through the determination of the distorted waves. For static distorting potentials, there is no distinction between a projectile that is an electron and an imaginary equivalent particle that is not an electron. A proper calculation of the distorted wave for an electron should include a non-local exchange potential which depends upon the wavefunctions for all the other electrons in the system. That is, one should perform a Hartree-Fock calculation for all the atomic electrons plus the projectile electron. So far, this type of calculation has not been performed. Nevertheless, the effect of this "exchange potential," which we label "exchange distortion," can be significant and should be included in the calculation of distorted waves. The standard procedure for taking this effect into account is to approximate the non-local exchange potential by a local potential which is then added to the U_i or U_f for calculating the distorted waves. This is the procedure that was used to include "exchange distortion" for the sodium results which are presented in section 10.4.

10.3.4 Relativistic effects

Relativistic effects can be important for scattering of low energy electrons from heavy atoms. The relativistic effects can be manifested through either the target wave function, the projectile wave function or both and experiments can be designed to isolate a particular type of effect (Furst et al. [390]). Madison and Shelton [391] reported the first distorted wave calculation for atomic excitation to include relativistic effects for the projectile. In that work, spin polarizations were calculated for electron–mercury scattering using relativistic distorted waves for the projectile and non-relativistic bound state wavefunctions for the atom and reasonable agreement with experiment was found for both elastic and inelastic scattering. Subsequently, however, more detailed measurements were made for various *J*-states of mercury and the semi-relativistic description worked reasonably well for some states but not others (Bartschat and Madison [372]). Bartschat and Madison [375] reported the first distorted wave calculation for which relativistic effects were included both in the projectile and the atomic wavefunctions. This work dealt with excitation of the allowed and forbidden transitions in the inert gases. Relativistic effects in the atom were approximated by using a *J*-dependent wave function obtained from mixed singlet and triplet states. It was found that relativistic effects in the atom became increasingly important for heavier atoms.

The work of Bartschat and Madison [375] for the inert gases was still semi-relativistic in the sense that two-component wavefunctions were used for both the projectile and atom. Recently, there has developed an increased interest in this problem and as a result suggestions for improving the relativistic description [392]–[395] have been made. The first fully relativistic calculation using four component wavefunctions has recently been reported by the group at York (Zuo *et al.* [392]). Relativistic calculations were performed for mercury and better agreement with experiment was found than was achieved with the earlier semi-relativistic work. [396] The fully relativistic calculations have also been performed for the inert gases, [397] cadmium [398] and calcium, strontium and barium. [399, 400] Typically, the fully relativistic calculations are in qualitative agreement with the previous semi-relativistic work and overall give better agreement with experimental data.

10.3.5 *Second order calculations*

Second order perturbation series calculations for electron–atom scattering have typically been avoided due to the difficulty involved in evaluating the second order amplitude. The biggest problem associated with the second order term lies in the fact that one must sum over the infinite number of intermediate states. Since the intermediate states can be either bound or continuum, this means that one must sum over all bound states and integrate over the continuum states, and it is the continuum states that represent the biggest problem. The standard procedure for avoiding the explicit evaluation of this sum and integral is to invoke the closure approximation. [401]–[405] Madison and Winters [406] examined the various approximations which have been made to simplify the evaluation of the second order amplitude and they found that although there were situations for which the approximations were reasonable, in general approximations should be avoided.

The first second order calculation to explicitly evaluate the sum and integral over intermediate states was a plane wave Born calculation for excitation of the 2*s* state of hydrogen which was performed by Ermolaev and Walters. [407] Madison [408, 409] reported the first second order distorted wave (DWB2) calculation which summed and integrated over the intermediate states to numerical convergence. This calculation was for excitation of the $2p$ state of hydrogen and only the DD term of Eq. (10.7) was evaluated. It was found that second order effects were important and significantly improved agreement with experiment in the intermediate energy range. Since the agreement was still not as good as one might expect for the fundamental scattering problem, it was decided to investigate the importance of second order exchange which was initially expected to be small. Consequently, the DE, ED, and EE terms of Eq. (10.7) were also evaluated and it was found that these terms were important and brought theory and experiment into very good accord for the DCS for energies above about 50 eV. [385] The second order work for hydrogen has now been extended to the alkali atoms [386] and some typical results will be shown in the next section.

10.3.6   Green's function

In addition to the infinite sum and integral over intermediate states, the second order amplitude also contains the Green's function propagator for the projectile electron $[E^+ - \varepsilon - K - U_G]^{-1}$ where ε is the energy of the intermediate atomic state. The distorting potential in the Green's function U_G is, in principle, arbitrary. Consequently, the fundamental question concerns how to choose this potential. The easiest choice is to pick U_G to be zero in which case the distorted Green's function reduces to the free particle Green's function. Madison *et al.* [385] found that for scattering from hydrogen, picking U_G equal to U_{1s} where U_{1s} is the static potential for the ground state of hydrogen gave the best agreement with experimental data. This is the choice that would be suggested by many body perturbation theory. For the case of scattering from sodium, a non-zero U_G produced some resonance-behavior which caused problems in the numerical integration over continuum states. Consequently, the free particle Green's function was used for the initial sodium work. [386] We hope to overcome these numerical difficulties soon so that we can investigate the effect of distortion in the Green's function for scattering from alkali atoms as well.

10.4   Results

10.4.1   First order results

First order DWB1 results are compared with experimental DCS for excitation of several states of cadmium from the ground state in Fig. 10.1.

The experimental data are not absolute and were normalized to the best overall visual fit to the theory. The results seen in this figure represent the general features of the many DWB1 calculations which have been performed for different states of many atoms. The DWB1 results are in reasonable qualitative agreement with the shape of the experimental DCS for all the excited states and some fairly complicated structure is well predicted. For the optically allowed transition, the small angle DWB1 DCS is in reasonable agreement with the shape of the data. This is a general feature for the dominant optically allowed transition. In fact, the elementary first order Born approximation typically yields reasonable small angle DCS results for these transitions. The Born approximation completely fails for large angle DCS and transitions to most other states however. For transitions to states other than the dominant optically allowed ones, the small angle DWB1 results are typically not in good agreement with experiment as is seen in Fig. 10.1. The small angle problem for these states results primarily from the second order effects of atomic polarization and absorption. It should also be noted that for the few cases for which absolute measurements have been made, the agreement between the DWB1 calculations and experiment was not as good as one would have assumed by normalizing to a best visual fit such as has been done for Fig. 10.1.

10.4.2 Second order effects

It is now very clear that a second order calculation is required if quantitative agreement between experiment and theory is desired for the DCS. As was mentioned in section 10.2, one of the strengths of the perturbation series method lies in the fact that the importance of different types of effects can easily be examined. As an example, by either including or excluding various parts of the T-matrix, one can examine the importance of polarization of the atom by the projectile, distortion of the projectile by atomic polarization, the importance of discrete or ionization channels on the process of interest, exchange scattering or exchange distortion, etc. Although it is impractical to present all the various interesting comparisons here, we have selected a couple of the possibilities to illustrate the point. The next three figures contain first and second order results for three different types of transitions in sodium.

Figure 10.2 compares first and second order results with experiment for the dominant optically allowed transition in sodium. The broken and dotted curve gives the DWB1 results and the other curves are two different DWB2 calculations. The dotted curve in Fig. 10.2 was calculated using the second order T-matrix of Eq. (10.7) and static distorting potentials for the projectile so we label this possibility as DWB2S. This calculation contains the effects of polarization of the atom by the projectile, absorption into

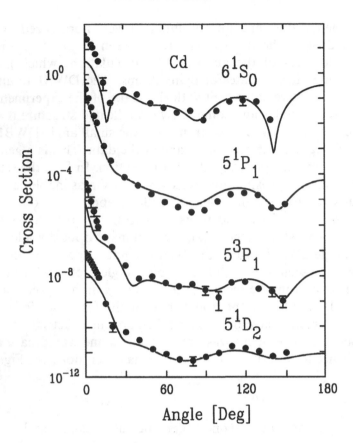

Fig. 10.1. Differential cross section for inelastic scattering of 60 eV electrons from cadmium in units of the Bohr radius squared a_0^2 per steradian: a_0^2/sr. The experimental data are those of Marinkovic *et al.* [410]. The theoretical curves are the DWB1. The $5\,^1P_1$ results have been multiplied by 10^{-4}; the $5\,^3P_1$ by 10^{-5}; and the $5\,^1D_2$ by 10^{-9} .

other channels and exchange scattering. It does not contain the back effects of atomic polarization on the projectile or the effect of exchange distortion on the projectile. The solid curve in Fig. 10.2, which we label DWB2PE, includes these two later effects.

From Fig. 10.2, it is seen that for excitation of this optically allowed state, polarization, absorption and exchange distortion are not very important for small angle scattering. For large angle scattering these effects are important and when included give reasonable agreement with the experimental data over the entire angular range. All the experimental data for sodium shown here and in subsequent figures have been normalized to theory by integrating the data over the angular range of the measurement and then normalizing to the DWB2PE results integrated over the same

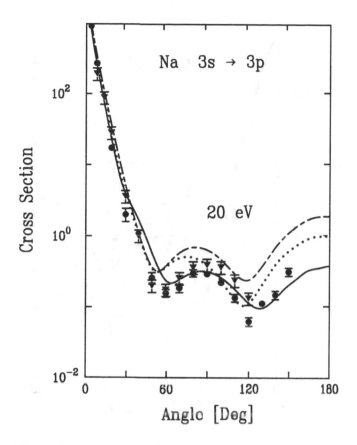

Fig. 10.2. Theoretical and experimental DCS for 20 eV electron excitation of the 3p state of sodium in units of (a_0^2)/sr. The experimental data are: circles, Marinkovic *et al.* [411]; triangles, Srivastava and Vuskovic [412]. The theoretical calculations are: dashed line, DWB1; dotted line, DWB2S; and solid line, DWB2PE.

angular range. This procedure was suggested by Vuskovic [413] and we believe that it is a very good method for normalizing experimental data.

Figure 10.3 contains the same type of comparison for elastic scattering from sodium. Here we see a quite different behavior from thats for the 3p transition. For elastic scattering at small scattering angles, polarization of the atom and absorption are very important (DWB2S). The difference between DWB2S and DWB2PE represents the effects of atomic polarization and exchange distortion on the projectile and it is clear that these effects are not very important for small angles.

At large scattering angles, the situation is reversed—atomic polarization and absorption are not important while the effects on the projectile are very important. Again we see that the full DWB2PE results are in quite

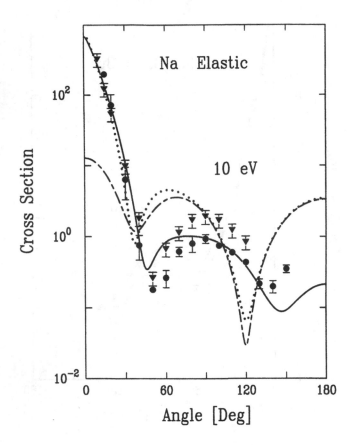

Fig. 10.3. The same as Fig. 10.2 except that here we have elastic scattering of 10 eV incident electrons.

good agreement over the entire scattering range for the fairly low electron energy of 10 eV. Figure 10.4 contains the same comparison for transitions to the 4p state. Again a very interesting, but different type of picture emerges. For this case, atomic polarization and absorption have a small effect for small scattering angles while the effect of polarization on the projectile is enormous and totally changes the small angle behavior such that theory and experiment come into agreement. For large scattering angles, all the effects are important and again when combined give good agreement with experiment except perhaps at the largest angle data point.

10.4.3 Convergence of perturbation series

The rate of convergence of the perturbation series expansion depends on the energy of the incident electron. Although everyone agrees that perturbation series should converge faster with increasing energy, it is

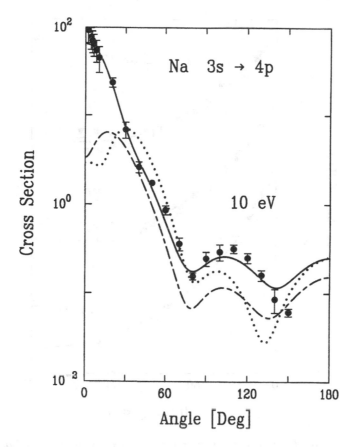

Fig. 10.4. The same as Fig. 10.2 except that here we have excitation of the *4p* state of sodium by 10 eV incident electrons.

not clear what the crucial factor is. It is clear, however, that it is not simply the absolute value of the incident electron energy but rather some combination of the electron energy and a property of the atom. Many feel that the ratio of electron energy to the ionization potential should be an important factor in determining the rate of convergence. In the next few figures, the rate of convergence is illustrated for different types of transitions and different atoms.

Figure 10.5 compares DWB2PE results and experiment for elastic scattering of electrons from hydrogen. Unlike the sodium case, the experimental measurements for hydrogen are absolute. At the lowest energy, the two experiments are in agreement with each other and the DWB2PE for angles out to about 110°. For larger angles, the DWB2PE lies above the data of Williams [414] which exhibits the standard large angle behavior and below the data of Shyn and Cho [415] which rises with angle. By 30

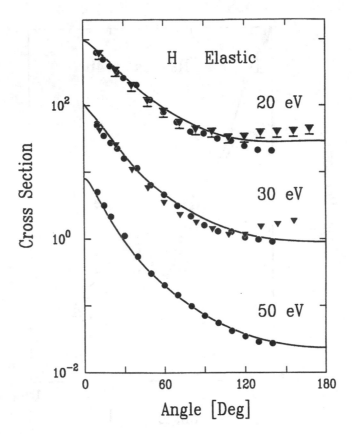

Fig. 10.5. Elastic DCS for scattering from hydrogen in units of a_0^2/sr. The experimental data are: circles, Williams [414] and triangles Shyn and Cho [415]. The theoretical curves are those of the DWB2PE. The 20 eV results have been multiplied by 10^2 and the 30 eV results by 10.

eV, the DWB2PE is in reasonable agreement with the Williams' [414] data and this agreement is even better at 50 eV. Consequently, the DWB2PE is very similar to Williams' data by 20 eV and has converged to it by 30 eV. In terms of this data, one would say that perturbation series had converged to the second term in the vicinity of two times the ionization potential for elastic scattering from hydrogen.

Figure 10.6 shows a similar comparison for excitation of the $2s + 2p$ states of hydrogen. Again the experimental data is absolute. For this case, it is clear that the DWB2PE results are only in qualitative agreement with experiment at 20 eV. At 54 eV, however, the agreement between experiment and theory is excellent. Consequently, these results for excitation are consistent with those for elastic scattering suggesting that the perturbation series converges to the second term for scattering from hydrogen at around 2-3 times the ionization potential.

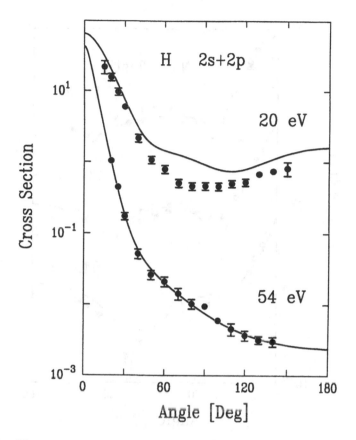

Fig. 10.6. The same as Fig. 10.5 except that here we have excitation of the $2s + 2p$ states. Here the 20 eV experimental data are those of Williams [416] and the 54 eV experimental data are those of Williams and Willis [417]. The 20 eV results have been multiplied by 10.

Figures 10.7, 10.8, and 10.9 contain a similar comparison for scattering from sodium. In Fig. 10.7, the elastic scattering case is presented. Considering the spread in the experimental data (as normalized by the group at Missouri-Rolla), the agreement between the DWB2PE and the data is reasonably good over the entire energy range of 10 - 54 eV. For sodium, 10 eV corresponds to two times the ionization energy so these results are consistent with those for hydrogen. In Fig. 10.8, the same comparison is made for excitation of the $3p$ state of sodium. Although the DWB2PE lies within the experimental data spread at 10 eV, both sets of data indicate some structure near 60° which is not predicted by the theory.

Consequently, it is not clear whether or not one can say that the perturbation series has converged to the second term at 10 eV for this case. However, the agreement between the DWB2PE and experiment is

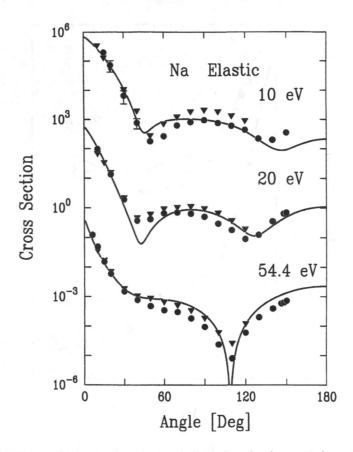

Fig. 10.7.   Theoretical and experimental DCS for elastic scattering of electrons from sodium in units of (a_0^2)/sr. The experimental data are: circles, Marinkovic *et al.* [411] and triangles, Srivastava and Vuskovic [412]. The theoretical calculations are those of the DWB2PE. The 10 eV results have been multiplied by 10^3 and the 54.4 eV results by 10^{-3}.

very good at 20 eV. Consequently, these results indicate that convergence has been achieved at between two and four times the ionization energy for excitation of the 3*p* state.

Finally, the same type of comparison is presented for excitation of the 4*s* state in Fig. 10.9. Once again the conclusions are similar to the previous cases. For 10 eV, there is a large spread in the experimental measurements and the DWB2PE results are in qualitative but probably not quantitative agreement with experiment. By 20 eV, the agreement between experiment and theory is much improved and it can be argued that convergence to the second term of the perturbation series has been achieved at this energy. The possible exception to this statement lies in the fact that the DWB2PE

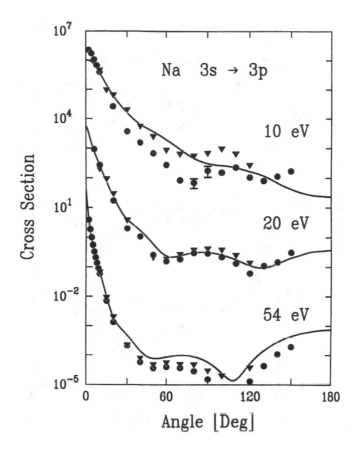

Fig. 10.8. The same as Fig. 10.7 except that here we have excitation of the $3p$ state. The 10 eV results have been multiplied by 10^3 and the 54 eV results by 10^{-3}.

consistently predicts a minimum near 20° which is not seen in the data. On the more positive side, there is a clear shoulder in the experiment at this angle for 54.4 eV scattering so some additional measurements in this angular range for lower energies would be very interesting.

10.4.4 Comparison with close-coupling calculations

Finally we would like to examine how the perturbation series results compare with close-coupling calculations.

In the close-coupling method, one expands the scattering wave function in terms of a complete basis set and the primary difficulty lies in the treatment of the continuum contributions to this basis set. The early close-coupling calculations included a small number of bound states and assumed that the continuum could be ignored (see, for example, the

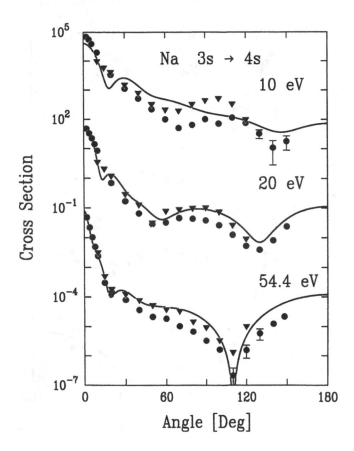

Fig. 10.9. The same as Fig. 10.7 except that here we have excitation of the 4s state. The 10 eV results have been multiplied by 10^3 and the 54.4 eV results by 10^{-3}.

reviews of Bransden and McDowell [367, 368]). Although this procedure is valid near threshold, the continuum states become more important with increasing energy. Consequently, this type of close-coupling calculation is valid for very low energies and becomes less reliable with increasing energy. Most of the work in recent years has concentrated on approximating the effects of the continuum and the techniques which have emerged are the pseudo-state method, the optical potential method and the *R*-matrix method.

In the pseudo-state method, all the bound and continuum states are approximated by a set of wavefunctions which are finite in the region of interest and zero outside this region. The primary idea is to avoid the problems associated with the continuum by using finite ranged wave functions while at the same time providing a good description of the scattering wave function in the region of interest. The different types of basis sets which have been used over the years were reviewed by Callaway. [418] Madison and Callaway [419] and Walters [420, 421] have examined the effective completeness of many of these basis sets. One of the problems associated with pseudostates lies in the fact that one normally could not establish convergence by simply using a larger basis set. Recently, however, Bray and Stelbovics [422] used a Laguerre basis set and they found convergence by increasing the basis size for electron–hydrogen scattering.

In the optical potential method, the complete basis set is divided into two parts which are normally called **P**-space and **Q**-space. **P**-space contains the bound states which are explicitly included in the close-coupling expansion and **Q**-space contains the remaining bound and continuum states. The effects of **Q**-space are approximated in the close-coupling equations through a complex operator called the optical potential. This method has been developed for atomic scattering primarily by McCarthy and his coworkers [422]–[426] over the last several years.

The third close-coupling method that has received considerable attention is the R-matrix method. In the R-matrix method, space is divided into an internal region and an external region. The wave function in the internal region is expanded in terms of a large basis set. In the asymptotic external region, the form of wave function is known and the internal and external wavefunctions are matched at the boundary between the two regions. The R-matrix method for atomic scattering has been developed primarily by Burke and his collaborators at Belfast (see chapter thirteen of this book for a complete description and references to original works). The most recent calculations for hydrogen by Scholz *et al.* [427] have been labeled the intermediate energy R-matrix (IERM) method and the new feature of the IERM method lies in the fact that the expansion for the internal region wave function consists not only of the standard bound-bound and bound-continuum terms but also a large set of continuum-continuum terms.

There have been many close-coupling calculations performed over the years using the techniques discussed above and it is impractical to try to present a detailed comparison with all of them. Consequently, we have chosen to compare the DWB2PE with the most recent R-matrix calculation for hydrogen scattering and a recent optical model calculation for sodium scattering. The DWB2PE results are compared with the IERM results of Scholz *et al.* [427] in Fig. 10.10 for elastic scattering of electrons from hydrogen. To within plotting accuracy, the two calculations are almost indistinguishable for all angles at 50 eV and for angles out to about 120° for 30 eV. Even at 20 eV, there is not much difference between the calculations except for large angles. Even for the largest angles at

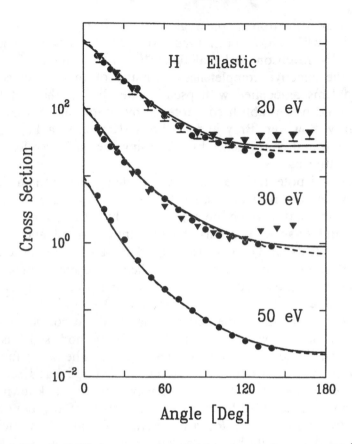

Fig. 10.10. Elastic DCS for scattering from hydrogen in units of $(a_0^2)/\text{sr}$. The experimental data are: circles, Williams [414] and triangles, Shyn and Cho [415]. The theoretical curves are: dashed line, IERM of Scholz *et al.* [427] and solid line, DWB2PE. The 20 eV results have been multiplied by 10^2 and the 30 eV results by 10.

20 and 30 eV, the difference between the DWB2PE and IERM is about the size of the experimental error. One of the significant features of the DWB2 method lies in the fact the continuum states are included exactly, to second order. One of the nice features of a close-coupling approach lies in the fact that the effects of all states included in the close-coupling expansion are included to all orders of perturbation theory. The fact that the IERM and DWB2PE results are almost the same indicates that: (a) perturbation series has converged to the second term and (b) the basis set used in the IERM expansion adequately represents the continuum effects in this energy range.

The DWB2PE and IERM results are compared with experiment for excitation of the $2s + 2p$ states of hydrogen in Fig. 10.11. For this case,

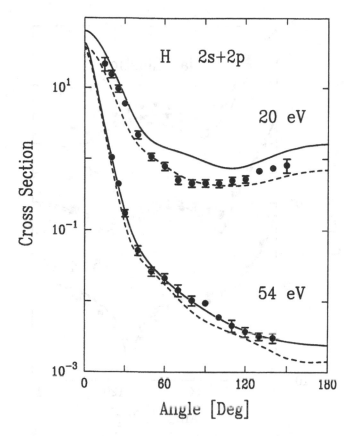

Fig. 10.11. The same as Fig. 10.5 except that here we have excitation of the $2s + 2p$ states. Here the 20 eV experimental data are those of Williams [416] and the 54 eV experimental data are those of Williams and Willis [417]. The 20 eV results have been multiplied by 10.

there is a larger difference between the two theories. At 20 eV, the IERM is in better agreement with experiment than the DWB2PE which lends further support to our earlier conclusion that higher order terms were starting to be important for this energy. For 54 eV, the DWB2PE results are in better agreement with the absolute measurements than the IERM. Assuming that the experiment is correct, this would suggest that the IERM representation of the continuum is not adequate at the higher energy for this process.

In Fig. 10.12, the DWB2PE results are compared with the three state coupled channels optical model (3CCO) results of Bray *et al.* [426] for elastic electron-sodium scattering. As for the results of Fig. 10.10 for elastic scattering from hydrogen, the perturbation theory results and close-coupling results are very similar. At 20 and 54 eV, the difference between

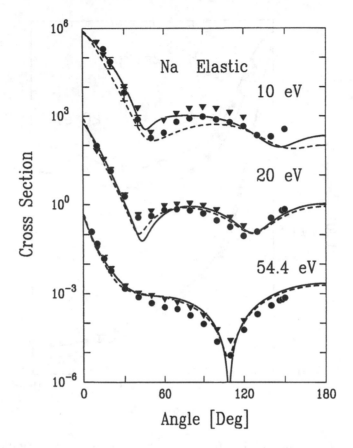

Fig. 10.12.   Theoretical and experimental DCS for elastic scattering of electrons from sodium in units of $(a_0^2)/\text{sr}$. The experimental data are: circles, Marinkovic *et al.* [411] and triangles, Srivastava and Vuskovic [412]. The theoretical curves are: dashed line, 3CCO of Bray *et al.* [425] and solid line, DWB2PE. The 10 eV results have been multiplied by 10^3 and the 54.4 eV results by 10^{-3}.

the two theories is smaller than the experimental uncertainties. Similarly to the hydrogen conclusion, this would imply both that the perturbation series had converged to the second term and that the optical potential had properly accounted for the effects of the continuum in the 20–54 eV energy range. The 10 eV results are interesting. The difference between the theories is again about the same as the difference between the two experiments and it is difficult to say which one is in better agreement with experiment since the data are not absolute. The DWB2PE does appear to be in somewhat better qualitative agreement with the shape of the data however.

Figure 10.13 contains the same comparison for excitation of the $3p$ state. For this case, the difference between the two theories is smaller at

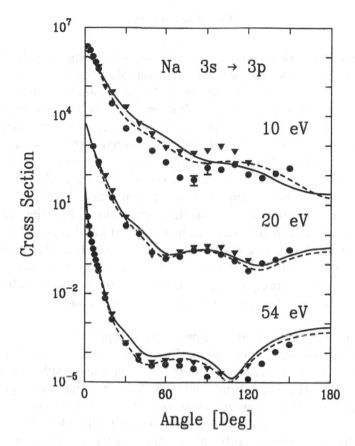

Fig. 10.13. The same as Fig. 10.12 except that here we have excitation of the $3p$ state. The 10 eV results have been multiplied by 10^3 and the 54 eV results by 10^{-3}.

10 eV than was seen on the previous figure for elastic scattering. Again the difference between the two calculations is generally substantially less than the difference between the two measurements. The similarity of the 3CCO and DWB2PE results for 10 eV scattering would indicate that perturbation series had actually converged to the second term in spite of the fact that one would not have concluded this by comparing experiment and theory. Finally, it should be noted that these comparisons between perturbation series and close-coupling results have all been performed for the DCS. Parameters obtained from coincidence measurements or from using spin polarized beams are more sensitive to the theoretical model and for these parameters the perturbation and close-coupling results exhibit larger differences. [422]

10.5   Conclusions

In summary, the DWB1 is very useful for obtaining a qualitative description of the scattering process. Reasonable results can be obtained quickly and easily on any desktop machine particularly for the dominant optically allowed transition and higher energies. In the intermediate energy range (10–200 eV), however, higher order terms in the perturbation series expansion must be evaluated if quantitative agreement with experiment is desired for all scattering angles and all transitions of interest. In the last few years, DWB2 calculations which sum and integrate over all intermediate states to numerical convergence have been performed for electron–hydrogen and electron–sodium scattering. For the differential cross section, these DWB2 calculations exhibited good agreement with experiment for incident energies higher than 2–3 times the ionization energy. For lower energies, it is clear that third and higher order terms become more important. Consequently, it appears that the perturbation series converges to the second term around 2–3 times the ionization energy for the DCS.

This conclusion is further supported by comparing the DWB2 results with recent sophisticated close-coupling calculations. This comparison revealed that both approaches yielded very similar results for energies above 2–3 times the ionization potential. The fact that two significantly different theoretical approaches yield almost the same answer is a strong indication that both approaches have properly included the important physical effects. Consequently, after many years of effort, we may be finally reaching the point where theory and experiment are in agreement at least for some of the elementary scattering problems. Both the DWB2 and close-coupling calculations give rise to significant numerical problems. On the one hand, the close-coupling results have an advantage in that they are valid down to threshold. On the other hand, the perturbation series approach has an advantage in that the importance of different physical effects can easily be isolated and studied. The most sophisticated work so far has been for scattering from the alkali atoms. One can expect the emphasis to shift to atoms with more valence electrons in the near future.

11

Target dependence of the triply differential cross section for low energy (e, 2e) processes

Cheng Pan and Anthony F. Starace

11.1 Introduction

In recent years, experimental studies on low and intermediate energy (*e, 2e*) processes have accumulated large amounts of triply differential cross section data. [428]-[439] These (*e, 2e*) results, in which the energies and angles of both of the outgoing electrons produced in the electron-impact ionization process are specified, [428, 429] display strong electron-correlation effects. Owing to the difficulty involved in describing precisely various electron correlations, in particular, the Coulomb interaction between the two final-state continuum electrons, only approximate theoretical treatments have been carried out. At present, theoretical understanding of these data and the underlying effects are far from complete. [434, 436, 437]

The near-threshold energy dependence of two electrons escaping from a positive ion has been studied theoretically by many authors using a number of methods. [440]-[451] These studies cover the threshold behavior of the total and the differential cross sections for electron-impact ionization of atoms and ions.

In the early 1950's, Wannier [440] applied to this problem the idea that the near-threshold energy-dependence of a reaction could be derived by investigating only the long-range interactions among its final products, without having a detailed knowledge about a small "reaction zone," the size of which is of the order of magnitude of the Bohr radius. [441] He revealed the importance of the configuration $r_1 = -r_2$ for the double escape of slow electrons from a positive ion by using methods of classical mechanics. He concluded that in the zero-energy limit all the orbits leading to the double escape approached asymptotically to this configuration. An assumption he made is that there is no strong selectivity against the two outgoing electrons emerging from the reaction zone in this configuration as compared to other configurations. He obtained a threshold law for the cross section for the $^1S^e$ state of the two final-state continuum electrons. For neutral-atom targets, the electron-impact ionization cross section is predicted to vary as the 1.127th power of the excess energy, which is the difference between the kinetic energy of the incident electron and the threshold energy for ionization.

Later, the double escape problem was treated by Temkin [442] assuming that the dominant configurations leading to double escape are those with one relatively fast outgoing electron and one slow outgoing electron. Asymptotically, the corresponding processes can then be described as the relatively fast electron moving in the dipole field generated by the combination of the positive ion and the slow electron. A different threshold law was obtained which gives an oscillatory behavior very close to the threshold. [442]

The threshold behaviors of the energy partition and the angular distribution of the escaping electrons were studied by Vinkalns and Gailitis [443] using Wannier's technique. The energy partition close to threshold is predicted to be nearly uniform, and so for neutral-atom targets the differential cross section with respect to the energy of one of the two escaping electrons varies as the 0.127th power of the excess energy. The angular distribution for cases in which the two escaping electrons leave in opposite directions is predicted to be independent of the excess energy E_{ex}. The distribution with respect to the mutual angle θ_{12} between the two escaping electrons is centered around $\theta_{12} = \pi$, and the width of this distribution is predicted to vary as $E_{ex}^{1/4}$ for cases in which the charge of the residual ion is one or two.

The quantum mechanical formulation of the Wannier theory was first given by Rau [444](a) and by Peterkop [445](a) in the early 1970's. This theory has been extended to treat all angular momentum states of the two escaping electrons. [446] It has also been extended to treat the distribution of the angular momenta of the outgoing electrons, [447] and it has been generalized to account for the finite mass of the ion and to include processes involving particles with various masses and charges. [448] The Wannier theory and its extensions are often referred to as the Wannier-Peterkop-Rau (WPR) theory, and many experiments have been carried out to verify its predictions. [452] Recent measurements for the spin asymmetry of electron-impact ionization of atoms have indicated the need to further extend the existing threshold theories. [453]

Earlier calculations for cross sections and other theoretical aspects of the electron-impact ionization problem have been reviewed by Rudge [454] and by Peterkop. [335] Rudge pointed out that the methods used in actual calculations (up to 1968) did not take proper account of the Coulomb interaction between the two final-state outgoing electrons, and so they could not be expected to give precise cross sections and angular distributions for the outgoing electrons at low excess energies. Bottcher discussed various numerical methods for including the interaction between the outgoing electrons. [455] Peterkop [456] and Rudge and Seaton [457] gave in the early 1960's a relation specifying momentum-dependent effective charges which can be used to approximate the interaction between the two outgoing electrons, but calculations employing it were not reported until 1989. [458] The exact asymptotic boundary conditions for three charged particles interacting via Coulomb forces were first given in 1973, [343] but calculations which incorporated them also did not come until 1989. [344]

Measurements for the relative triply differential cross sections for (*e*, 2*e*) processes were first reported in 1969 by Ehrhardt *et al.* [459] for the He atom and by Amaldi *et al.* [333] for carbon. In 1972 and 1981, further measurements were made for He down to 6 eV above the ionization threshold. [460] In 1984 Fournier-Lagarde, Mazeau, and Huetz reported measurements for the triply differential cross section for He down to 1 eV above threshold. [461] To study the near-threshold experimental results, Altick used partial wave expansions and a correlation factor to describe the repulsion between the outgoing electrons. [462] The resulting expressions fit the measured results very well. Crothers in 1986 [449] reported an *ab initio* calculation which explained well some of the characteristics of the measured near threshold angular distributions even though only singlet states of the two outgoing electrons were included. He compared the contribution from the configurations emphasized in the WPR theory and the contribution from those emphasized in the theory of Temkin and found that the latter is relatively small. His results for the total and differential cross sections are consistent with the WPR theory.

Since 1987, a large number of (*e*, 2*e*) measurements have been reported for low excess energies [431]-[439]. A number of theoretical studies for the low-energy range have also been reported. [436],[439],[463]-[466] At present, there is no theoretical method which can reproduce the results of all the measurements.

Significant progress in treating the Coulomb interaction between the final-state continuum electrons has been made in the past several years by the use of wave functions which satisfy the asymptotic boundary conditions for three particles interacting via Coulomb forces. In 1989, Brauner, Briggs, and Klar first reported such a calculation for the triply differential cross section of the H atom. [344] Franz and Altick have developed a partial-wave expansion for ionization amplitudes involving wave functions which satisfy the asymptotic boundary conditions. [351] Topics related to this progress are reviewed by P. L. Altick in chapter nine of this book.

The main purpose of this chapter is to review our calculations, [464] whose aim is to explain the target dependence of the triply differential cross sections that has been observed in a number of recent experiments [430]-[432],[435],[438],[439] performed at relatively low energies above the ionization threshold for the two outgoing electrons leaving in opposite directions ($\theta_{12} = \pi$). One reason for focusing on this $\theta_{12} = \pi$ geometry is that it is the most important geometry for near-threshold double-escape processes in the WPR theory. [440, 444, 445] Another reason is that up to now, except for H and He targets, the triply differential cross section for low-energy electron-impact ionization has been measured for Ne, Ar, Kr, and Xe targets only in this geometry. [432, 435] In these experiments, for a given sharing of the excess energy by the two outgoing electrons, the differential cross section was measured, typically for a number of different angles between the incident electron beam and the vector connecting the two final-state outgoing electrons. When plotted against this angle, these measured results show striking target dependence.

Since at asymptotic separations the long-range fields in the final state of these ionization processes are the same, the observed target effects must be related to the short-range interactions between the incident electron and the target and between the final-state outgoing electrons and the residual ion. Therefore, in our calculations we focus mostly on the accurate treatment of such short-range interactions. For the interaction between the two final-state continuum electrons, we use an approximation which employs effective screening potentials [464](a,b) because of the difficulty of treating simultaneously the long-range and the short-range interactions.

The methods we use are essentially distorted-wave methods, but we have also treated some electron-correlation effects perturbatively. Our calculations can be presented in the framework of many-body perturbation theory, and this is another purpose of the present chapter.

Theoretical studies for low-energy triply differential cross sections in various geometrical conditions other than $\theta_{12} = \pi$ have been reported by Altick, [462](a) by Crothers, [449] by Shaw and Altick, [462](b) by Selles, Huetz, and Mazeau, [430, 463] by Altick and Rösel, [462](c) by Brauner et al., [436] by Jones, Madison, and Srivastava, [465] by Botero and Macek, [466] and by Jones and Madison, given in Ref. [439]. Interested readers are referred to these references for details of these studies.

In the next section our theoretical approach is discussed. Specifically, the partial-wave expansions for our approximate wave functions are given. Then, many-body perturbation theory is used to treat the matrix elements between *LS*-coupled wave functions. Finally, the formula for the differential cross section is reduced to a simple form for the $\theta_{12} = \pi$ case.

In a following section low-energy results of experimental and theoretical studies for the $\theta_{12} = \pi$ geometry are briefly reviewed. Calculated triply differential cross sections for H and He targets are compared with experimental results, and the observed difference between H and He is shown to stem from the short-range effects on the s-wave phase shifts of both incident and final-state continuum electrons. The comparisons between the calculated and the measured results for Ne, Ar, Kr, and Xe are also given. Finally, the energy-dependence of the theoretical results for He is compared with available absolute experimental results.

11.2 Theory

Various aspects of the theory of electron-impact ionization of atoms and atomic ions were reviewed extensively by Rudge in 1968 [454] and by Peterkop in 1977. [335] More recently, Brauner, Briggs, and Klar have discussed the use of wave functions that satisfy the asymptotic three-body boundary conditions; [344] Byron and Joachain have reviewed various higher-Born approximation methods; [467] Curran, Whelan, and Walters have discussed a coupled-pseudostate method; [348] Franz and Altick have given a partial-wave expansion for the asymptotic three-body boundary conditions; [351] and Jetzke, Zaremba, and Faisal [458] have discussed the use of the effective charges specified by Peterkop [456] and by Rudge and Seaton. [457]

Chapter 9 by P. L. Altick gives a review of recent progress on theoretical methods. Here we limit our discussion to the partial-wave expansions used for our approximate initial- and final-state wave functions, methods of many-body perturbation theory, and the simplified form of the formula for the triply differential cross section for the coplanar, $\theta_{12} = \pi$ geometry. In this work, we neglect relativistic effects and assume an infinite nuclear mass. Atomic units are used throughout this chapter except when otherwise indicated.

11.2.1 Partial wave expansion

The differential cross section for electron-impact ionization is given by

$$\frac{d\sigma}{d\mathbf{k}_1 d\mathbf{k}_2} = \frac{(2\pi)^4}{k} \, | \, \langle \Psi_f^- \, | \, \Delta H \, | \, \Phi_i^+ \rangle \, |^2 \, \delta(E_f - E_i) \,, \qquad (11.1)$$

or, equivalently, by

$$\frac{d\sigma}{d\mathbf{k}_1 d\mathbf{k}_2} = \frac{(2\pi)^4}{k} \, | \, \langle \Phi_f^- \, | \, \Delta H \, | \, \Psi_i^+ \rangle \, |^2 \, \delta(E_f - E_i) \,, \qquad (11.2)$$

according to formal scattering theory. [468] In Eqs. (11.1) and (11.2), k is the magnitude of the momentum of the incident electron, $\mathbf{k}_1$ and $\mathbf{k}_2$ are the momenta of the two continuum electrons in the final state, and E_i and E_f are the energies of the initial and final states. In Eq. (11.1), the final-state wave function Ψ_f^- is the exact solution of the full Hamiltonian, the initial-state wave function Φ_i^+ is the solution of an approximate Hamiltonian, and the perturbation ΔH is the difference between the exact Hamiltonian and the approximate Hamiltonian used to solve for Φ_i^+. Similarly, in Eq. (11.2), the initial-state wave function Ψ_i^+ is the exact solution of the full Hamiltonian, the final-state wave function Φ_f^- is the solution of an approximate Hamiltonian, and the perturbation ΔH is the difference between the exact Hamiltonian and the approximate Hamiltonian for Φ_f^-. In Eqs. (11.1) and (11.2), the normalization for Ψ_i^+ and Φ_i^+ is assumed to be $\delta(\mathbf{k}' - \mathbf{k})$, the normalization for Ψ_f^- and Φ_f^- is assumed to be $\delta(\mathbf{k}_1' - \mathbf{k}_1)\delta(\mathbf{k}_2' - \mathbf{k}_2)$, and the $+$ and $-$ superscripts on these wave functions denote, respectively, the outgoing-wave and incoming-wave boundary conditions.

In this chapter, Eq. (11.2) is used to calculate the triply differential cross section.* Let E_{ex} denote the excess kinetic energy of the incident electron above the ionization threshold. Then $\delta(E_f - E_i) = \delta(\varepsilon_1 + \varepsilon_2 - E_{ex})$, where $\varepsilon_j = k_j^2/2$ is the kinetic energy of the jth final-state continuum electron. Noting that $d\mathbf{k}_j = k_j d\varepsilon_j d\Omega_j$ and integrating over $d\varepsilon_2$, Eq. (11.2) becomes

$$\frac{d^3\sigma}{d\varepsilon_1 d\Omega_1 d\Omega_2} = \frac{(2\pi)^4}{k} \mid \langle \Phi_f^- \mid \Delta H \mid \Psi_i^+ \rangle \mid^2 k_1 k_2 \,, \qquad (11.3)$$

where $k_2 = (2E_{ex} - 2\varepsilon_1)^{1/2}$. In our calculations, we use a perturbation expansion to represent the initial-state wave function Ψ_i^+ in Eq. (11.3). For the approximate final-state wave function Φ_f^-, it is desirable to choose one that satisfies the asymptotic boundary conditions for three charged particles such that ΔH (*i.e.*, the difference between the exact Hamiltonian and the approximate one satisfied by Φ_f^-) is a short-range interaction. However, we use a more approximate final-state wave function which contains a product of two one-electron continuum wave functions. The interaction between the two final-state continuum electrons is approximated by using an effective screening potential for each of the continuum electrons. While this final-state wave function does not satisfy the exact boundary conditions, its form facilitates the use of partial wave expansions to treat the short-range interactions which govern the target dependence of the cross sections.

The initial-state wave function Ψ_i^+ is characterized by the orbital and

---

* Equation (11.2) is used here for convenience in constructing a many-body perturbation expansion. The first-order calculation in this paper is the same as the one reported in Ref. [464], which employs Eq. (11.1) for the cross section.

spin angular momentum of the N-electron target, denoted by $L_0 M_0 S_0 M_{S_0}$, and by the momentum $\mathbf{k}$ and spin magnetic quantum number m_s of the incident electron. The final state wave function Φ_f^- is characterized by the orbital and spin angular momentum of the $(N-1)$-electron residual ion core, denoted by $L_C M_C S_C M_{S_C}$, and by the momenta, $\mathbf{k}_1$ and $\mathbf{k}_2$, and spin magnetic quantum numbers, m_{s_1} and m_{s_2}, of the two continuum electrons.

A single-electron wave function characterized asymptotically by the momentum $\mathbf{k}$, and spin magnetic quantum number m_s can be expanded as

$$\phi_{\mathbf{k}m_s}^{(\pm)}(\mathbf{r}) = k^{-1/2} \sum_{\ell=0}^{\infty} \sum_{m=-\ell}^{\ell} i^\ell e^{\pm i(\delta_\ell^C + \delta_\ell)} Y_{\ell m}^*(\hat{\mathbf{k}}) u_{\varepsilon\ell m m_s}(\mathbf{r}), \tag{11.4}$$

where the partial-wave states $u_{\varepsilon\ell m m_s}$ are given by

$$u_{\varepsilon\ell m m_s}(\mathbf{r}) = r^{-1} P_{\varepsilon\ell}(r) Y_{\ell m}(\hat{\mathbf{r}}) \chi_{m_s}. \tag{11.5}$$

The $Y_{\ell m}$ functions in Eqs. (11.4) and (11.5) are spherical harmonics. In Eq. (11.5), χ_{m_s} is a two-component spinor, and the radial wave functions $P_{\varepsilon\ell}(r)$ are eigenstates of a single-particle Hamiltonian

$$h_l = -\frac{1}{2}\frac{d^2}{dr^2} - \frac{Z}{r} + \frac{\ell(\ell+1)}{2r^2} + V_\ell(r), \tag{11.6}$$

where Z is the nuclear charge and $V_\ell(r)$ is a radial potential. The function $P_{n\ell}(r)$ has the following asymptotic form

$$P_{\varepsilon\ell}(r) \sim \left(\frac{2}{\pi k}\right)^{\frac{1}{2}} \sin\left[kr - \frac{\ell\pi}{2} + \frac{q}{k}\ln(2kr) + \delta_\ell^C + \delta_\ell\right] \tag{11.7}$$

as $r \to \infty$. In Eq. (11.7), q is the net charge given by the asymptotic value of $Z - rV_\ell(r)$. In Eqs. (11.4) and (11.7), δ_ℓ^C is the Coulomb phase shift, and δ_ℓ is the non-Coulomb phase shift due to short-range interactions.

Using a partial-wave expansion such as that in Eq. (11.4) for the incident electron, we can couple each partial-wave function $u_{\varepsilon\ell m m_s}$ to the wave function of the target atom, forming antisymmetrized wave functions $\Psi_i(L_T M_T S_T M_{S_T})$. The initial-state wave function Ψ_i^+ for the target atom plus an incident electron with specified momentum can be expanded in terms of such LS-coupled states for the $(N+1)$-electron complex. Similarly, using partial wave expansions for the two final-state continuum electrons, we can couple each pair of partial-wave functions, $u_{\varepsilon_1\ell_1 m_1 m_{s_1}}$ and $u_{\varepsilon_2\ell_2 m_2 m_{s_2}}$, to form a two-electron function characterized by quantum numbers $LMSM_S$. This wave function can then be coupled to the wave function of the residual ion, forming antisymmetrized wave functions $\Phi_f(L_T M_T S_T M_{S_T})$. The final-state wave function Φ_f^- for the residual ion plus two continuum electrons, each with well-defined momentum, can be expanded in terms of such LS-coupled states. In what follows, we use unsuperscripted symbols Ψ and Φ to denote these LS-coupled wave functions.

Then, the matrix element $\langle \Phi_f^- \mid \Delta H \mid \Psi_i^+ \rangle$ in Eq. (11.3) can be expanded in terms of the matrix elements of the LS-coupled states, $\langle \Phi_f \mid \Delta H \mid \Psi_i \rangle$.

11.2.2   *MBPT for* $\langle \Phi_f \mid \Delta H \mid \Psi_i \rangle$

Methods using many-body perturbation expansions for the optical potential [469]-[471] have been applied to electron-atom scattering problems, [469],[472]-[474] including the calculation of electron-impact ionization cross sections, [475] by evaluating the imaginary part of the matrix element for the optical potential. Since our main concern here is the triply differential cross section for electron-impact ionization of atoms, we focus instead on the expansion of the matrix element $\langle \Phi_f \mid \Delta H \mid \Psi_i \rangle$.

To develop a perturbation expansion [476] for the matrix element $\langle \Phi_f \mid \Delta H \mid \Psi_i \rangle$, we use the Hartree-Fock (HF) method [174] to calculate the approximate final-state wave function Φ_f as well as the basis functions (orbitals) for expanding the initial-state wave function Ψ_i. Specifically, we use the self-consistent HF approximation for the wave function of the target atom with N electrons. Thus we obtain as basis functions for the occupied one-electron states in the target atom the solutions of a set of single-particle Hamiltonians, each in the form of Eq. (11.6). For orbitals ($\varepsilon\ell$) describing the incident continuum electron (as well as for excited orbitals), the general form [175]-[178] of the single-particle potential in Eq. (11.6) is given by

$$V_\ell = V_{g,\ell} + (1 - \mathbf{P}_\ell)(V_{\varepsilon\ell} - V_{g,\ell})(1 - \mathbf{P}_\ell) , \qquad (11.8)$$

where $\mathbf{P}_\ell = \sum_i \mid n_i\ell \rangle \langle n_i\ell \mid$, with the summation running over the occupied states of the target, and where $V_{g,\ell}$ is the potential for the occupied states. The right hand side of Eq. (11.8) reduces to $V_{\varepsilon\ell}$ when there is no occupied one-electron state with the same ℓ in the target atom. Calculating occupied and excited states of a given angular momentum quantum number ℓ in the same Hermitian potential guarantees orthogonality.

For each partial wave ($\varepsilon\ell$) of the incident electron, $V_{\varepsilon\ell}$ is derived according to[†]

$$\delta\langle \Phi_i \mid H \mid \Phi_i \rangle = 0 \qquad (11.9)$$

by keeping those terms with the factor $\delta P_{\varepsilon\ell}$. Using the HF wave functions for the target atom and for each partial wave of the incident electron, we can construct a zeroth-order initial-state wave function Φ_i corresponding

---

[†] In Eq. (A2) of Ref. [464](b), Ψ_f and Ψ_i should both be $\Psi_{i,f}$. The subscripts were mistyped.

to a zeroth-order Hamiltonian which is a sum of $N+1$ single-particle Hamiltonians.

We can define then the approximate final-state wave function Φ_f. For the one-electron states in the residual ion, we use the basis functions obtained from the self-consistent solution for the initial-state target atom. For each one-electron partial wave $(\varepsilon_j \ell_j, \; j = 1, 2)$ of the two final-state continuum electrons, a HF potential $V_{\varepsilon_j \ell_j}$ is derived by neglecting the radial integrals involving four continuum orbitals in the matrix element $\langle \Phi_f \mid H \mid \Phi_f \rangle$ and by keeping those terms with the factor $\delta P_{\varepsilon_j \ell_j}$ in the equation

$$\delta \langle \Phi_f \mid H \mid \Phi_f \rangle = 0 \, . \tag{11.10}$$

The potential derived this way does not include the interaction between the two continuum electrons in the final state, but we approximate this interaction by adding to this potential an effective screening potential. [464](a,b) We use the relation derived by Peterkop [456] and by Rudge and Seaton [457] for the effective charges of the screening potential at large distances,

$$- \frac{Z_T - \Delta_1}{k_1} - \frac{Z_T - \Delta_2}{k_2} = - \frac{Z_T}{k_1} - \frac{Z_T}{k_2} + \frac{1}{\mid \mathbf{k}_2 - \mathbf{k}_1 \mid} , \tag{11.11}$$

where Z_T is the net charge of the residual ion, and Δ_1 and Δ_2 are the effective charges. For the configuration considered here in which $\hat{\mathbf{k}}_1 = \hat{\mathbf{k}}_2$, Eq. (11.11) can be satisfied by using the following effective charges: [458]

$$\Delta_i = \frac{k_i^2}{(k_1 + k_2)^2} , \quad (i = 1, 2) \, . \tag{11.12}$$

Here also, Eq. (11.8) is used to ensure the orthogonality between each of the partial waves with the occupied one-electron states in the target atom having the same orbital angular momentum ℓ. Since, for a given ℓ, the one-electron partial wave for the two outgoing electrons in the final state and that for the incident electron are calculated using different potentials, they are not orthogonal to each other, although they are both orthogonal to the occupied state with the same ℓ in the target atom.

The difference between the exact Hamiltonian and the approximate Hamiltonian for the final state is

$$\Delta H = \sum_{i<j=1}^{N+1} \frac{1}{\mid \mathbf{r}_i - \mathbf{r}_j \mid} - \sum_{j=1}^{N+1} V_{\ell_j}(r_j) \, . \tag{11.13}$$

The difference between the exact Hamiltonian and the zeroth-order Hamiltonian for the initial state is also given by Eq. (11.13) for the corresponding single-particle HF potentials V_{ℓ_j}. This latter difference,

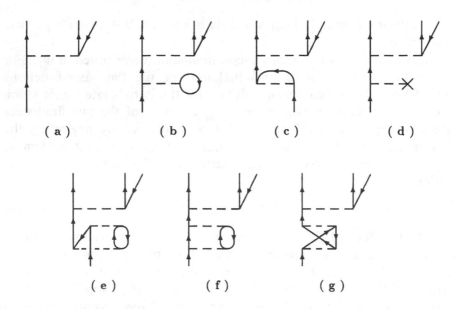

Fig. 11.1. Some diagrams contributing to the matrix element $\langle \Phi_f \mid \Delta H \mid \Psi_i \rangle$. Higher-order diagrams (b)–(g) are treated by defining an appropriate potential for the radial wave function of the incident electron. Diagrams (f) and (g) can only be approximately included by using Hermitian potentials.

ΔH, is treated as a perturbation for expanding the initial state wave function Ψ_i.

A perturbation expansion for the matrix element $\langle \Phi_f \mid \Delta H \mid \Psi_i \rangle$ can be obtained when the initial-state wave function Ψ_i is expanded, and terms of this perturbation expansion can be represented by many-body diagrams. Examples of the diagrams are shown in Figs. 11.1 – 11.3. In these diagrams, each horizontal dashed line with both ends connected to lines with arrows represents the Coulomb interaction term in Eq. (11.13), and each dashed line with a cross at one end denotes the interaction with the negative of the potential term in Eq. (11.13). Lines with arrows drawn downward represent vacancies (holes) in the ground-state target atom, and lines with arrows drawn upward denote continuum or bound excitations (particles). Time runs from bottom to top in these diagrams, and the order of a diagram refers to the number of horizontal dashed lines (interactions) in it. The open particle line on the bottom of each diagram represents the zeroth-order initial-state wave function, and the open hole line and the two open particle lines on top of each diagram denote the approximate final-state wave function. Notice that we do not include diagrams in which the two open particle lines are connected to two other particle lines at the ends of one Coulomb interaction line; these continuum-continuum interactions are considered to be accounted for by the effective screening potentials which are added to the potentials for the orbitals represented by the two open particle lines on top of each diagram.

Shown in Fig. 11.1 are some diagrams that we treat, either exactly or approximately, by using different potentials for the incident electron. The diagram of Fig. 11.1(a) gives the first-order matrix element. The diagram of Fig. 11.1(b) includes the correction to the first-order result due to the direct interaction between the incident electron and the bound electrons in the target atom, and the diagram of Fig. 11.1(c) includes the correction due to the corresponding exchange interaction. The diagram of Fig. 11.1(d) includes the correction due to the interaction with the negative of the single-particle potential used for calculating the wave function for the incident electron. When we use the HF potential derived using Eq. (11.9) for the wave function of the incident electron, the three diagrams in Figs. 11.1(b)–11.1(d) sum to zero. If we neglect the exchange terms in the HF potential for the incident electron, then only the diagrams of Figs. 11.1(b) and 11.1(d) sum to zero. The diagram of Fig. 11.1(e) can be treated by adding second-order terms to the HF potential used for the incident electron. The diagrams of Figs. 11.1(f) and 11.1(g) describe effects due to the interactions between the elastic scattering channel and inelastic scattering channels including impact ionization channels, and each of them comprises both real and imaginary parts. A complete treatment of such inter-channel interactions is in general difficult. However, the effects of these interactions on the elastic scattering channel can be estimated by calculating a second-order matrix element, which contains both a real part and an imaginary part. [475] For the He target and for the incident energies of interest here, the imaginary part is small relative to the real part. So, in this work we neglect the imaginary parts of the diagrams of Figs. 11.1(f) and 11.1(g) and treat the real parts of these diagrams by adding second-order terms to the HF potential used for the incident electron. The second-order potential terms that we use to treat Figs. 11.1(e)–11.1(g) are dependent on the incident energy.

Shown in Fig. 11.2 are diagrams which describe the effects of the interactions of each final-state continuum electron with the bound electrons in the residual ion and with the negatives of the single-particle potentials used for calculating the wave functions of the continuum electrons. When we use the potentials derived according to Eq. (11.10) for the final-state continuum electrons, all the diagrams in Fig. 11.2 sum to zero for targets having closed-subshell residual ions. In general, inter-channel interactions between final-state wave functions with different intermediate angular momentum couplings are nonzero, so that

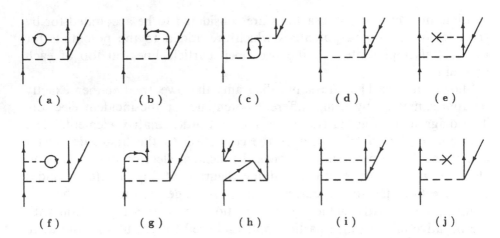

(a)   (b)   (c)   (d)   (e)

(f)   (g)   (h)   (i)   (j)

Fig. 11.2. Some second-order diagrams contributing to $\langle \Phi_f \mid \Delta H \mid \Psi_i \rangle$. For target atoms having closed-subshell residual ions, these diagrams sum to zero when potentials (of Hartree-Fock type) derived according to Eq. (11.10) are used for the radial wave functions of the outgoing electrons in the final state.

not all the diagrams in Fig. 11.2 can be cancelled by simply choosing appropriate single-particle potentials. To include inter-channel interactions for the case of helium targets, we use two-channel close-coupling solutions [477] for the one-electron continuum wave functions associated with the singlet and the triplet states of the two final-state continuum electrons.

Shown in Fig. 11.3 are second-order diagrams having intermediate states in which two target electrons are excited. These diagrams correspond to corrections to the single-configuration HF description of the target atom. We use Figs. 11.3(e) and 11.3(f) to represent second-order perturbation terms which include overlap integrals between the continuum wave functions of the incident electron and one of the two outgoing electrons. Notice that all the first-order diagrams that include overlap integrals (not shown here) sum to zero because we use the general form of potential defined in Eq. (11.8).

11.2.3   *Triply differential cross section for the $\theta_{12} = \pi$ case*

For the special case treated in this chapter of $\hat{\mathbf{k}}_1 = -\hat{\mathbf{k}}_2$, the differential cross section in Eq. (11.3) can be reduced to a simple form in which geometrical and dynamical dependences are clearly separated.

Once the wave functions Ψ_i^+ and Φ_f^- in Eq. (11.3) are expanded using the antisymmetrized, LS-coupled wave functions Ψ_i and Φ_f, Eq. (11.3)

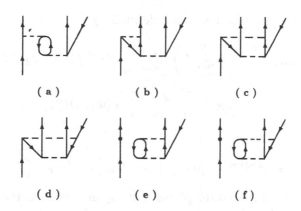

Fig. 11.3. Some other second-order diagrams contributing to $\langle \Phi_f \mid \Delta H \mid \Psi_i \rangle$. These diagrams have intermediate states in which two target electrons are excited; they correct the single-configuration Hartree-Fock description of the target atom. Diagrams (e) and (f) are used here to represent second-order terms which include overlap integrals (denoted by solid dots).

can be simplified by using the relation

$$\sum_{m_1, m_2} Y^*_{\ell_1 m_1}(\hat{\mathbf{k}}_1) Y^*_{\ell_2 m_2}(-\hat{\mathbf{k}}_1) \langle LM \mid \ell_1 m_1 \ell_2 m_2 \rangle =$$

$$(-1)^{\ell_1} \left(\frac{[\ell_1][\ell_2]}{4\pi} \right)^{1/2} Y^*_{LM}(\hat{\mathbf{k}}_1) \begin{pmatrix} \ell_1 & \ell_2 & L \\ 0 & 0 & 0 \end{pmatrix}, \qquad (11.14)$$

where the symbol $[x]$ is defined by $[x] \equiv 2x + 1$. The resulting expression is then summed over the final-state magnetic quantum numbers m_{s_1}, m_{s_2}, M_{S_C}, and M_C and averaged over the initial-state magnetic quantum numbers M_0, M_{S_0}, and m_s, i.e.,

$$\sigma^{(3)} = (2[L_0][S_0])^{-1} \sum_{M_0, M_{S_0}, m_s} \sum_{m_{s_1}, m_{s_2}, M_{S_C}, M_C} \frac{d^3\sigma}{d\varepsilon_1 d\Omega_1 d\Omega_2}. \qquad (11.15)$$

The result is [464](b)

$$\sigma^{(3)} = \frac{(-1)^{L_0 + L_C} \pi}{4k^2 [L_o]} \sum_{\substack{L_T, L'_T, L, L' \\ \ell, \ell', S_T, S}} A(\ell, L, S, L_T, S_T) A^*(\ell', L', S, L'_T, S_T)$$

$$\times \sum_\lambda [\lambda] P_\lambda(\hat{\mathbf{k}}_1 \cdot \hat{\mathbf{k}}) [L_T][L'_T] ([L][L'][\ell][\ell'])^{1/2}$$

$$\times \begin{pmatrix} L & L' & \lambda \\ 0 & 0 & 0 \end{pmatrix} \begin{pmatrix} \ell & \ell' & \lambda \\ 0 & 0 & 0 \end{pmatrix} \begin{Bmatrix} L_T & L'_T & \lambda \\ L' & L & L_C \end{Bmatrix} \begin{Bmatrix} L_T & L'_T & \lambda \\ \ell' & \ell & L_0 \end{Bmatrix}. $$

$$(11.16)$$

In Eq. (11.16), the amplitudes A are defined by

$$A(\ell, L, S, L_T, S_T) = \left(\frac{[S_T]}{2[S_0]}\right)^{\frac{1}{2}} \sum_{\ell_1,\ell_2}(-1)^{\ell_1+L}f(\ell,\ell_1,\ell_2)([\ell_1][\ell_2])^{1/2}$$

$$\times \begin{pmatrix} \ell_1 & \ell_2 & L \\ 0 & 0 & 0 \end{pmatrix} \langle\Phi_f|\Delta H|\Psi_i\rangle \,, \tag{11.17}$$

where

$$f(\ell,\ell_1,\ell_2) = i^{\ell+\ell_1+\ell_2}\exp[i(\delta_\ell + \delta_{\ell_1}^C + \delta_{\ell_1} + \delta_{\ell_2}^C + \delta_{\ell_2})] \,. \tag{11.18}$$

In Eq. (11.18), the initial-state phase shift δ_ℓ depends on the state of the target atom and the angular momentum coupling between the incident electron and the target atom, and in general, the final-state phase shifts δ_{ℓ_1} and δ_{ℓ_2} depend on the state of the residual ion, the coupling between the two continuum electrons, and the coupling between the residual ion and the two-electron partial-wave state.

We now discuss the relatively simple cases of H and He targets. For the case of H, $L_0 = L_C = S_C = 0$, $L = \ell = L_T$, $L' = \ell' = L'_T$, and $S_T = S$ in Eqs. (11.16) and (11.17). For the case of He, $L_0 = S_0 = L_C = 0$, $L = \ell = L_T$, $L' = \ell' = L'_T$, and $S_T = S_C = 1/2$. For these cases Eq. (11.16) becomes

$$\sigma^{(3)} = \frac{\pi}{4k^2}\sum_{L,L'}\sum_{S}A(LS)A^*(L'S)\sum_{\lambda}[\lambda]P_\lambda(\hat{\mathbf{k}}_1 \cdot \hat{\mathbf{k}})[L][L']\begin{pmatrix} L & L' & \lambda \\ 0 & 0 & 0 \end{pmatrix}^2 \,. \tag{11.19}$$

For the lowest-order (first-order) calculations, the explicit form for the amplitude A in Eq. (11.19) is

$$A(LS) = C\delta_{\ell,L}2^{-1}[S]^{1/2}\sum_{\ell_1,\ell_2}(-1)^{\ell_1}f(\ell,\ell_1,\ell_2)[\ell_1][\ell_2]\begin{pmatrix} \ell_1 & \ell_2 & \ell \\ 0 & 0 & 0 \end{pmatrix}^2$$

$$\times \{[\ell_1]^{-1}R^{\ell_1}(\varepsilon_1\ell_1,\varepsilon_2\ell_2;1s,\varepsilon\ell) + (-1)^S[\ell_2]^{-1}R^{\ell_2}(\varepsilon_2\ell_2,\varepsilon_1\ell_1;1s,\varepsilon\ell)\} \,, \tag{11.20}$$

where $C = 1$ for H and $C = (-1)^{S+1}\sqrt{2}$ for He. In Eq. (11.20) R^{ℓ_1} and R^{ℓ_2} are Slater integrals defined by

$$R^\lambda(1,2;3,4) = \int\int P_1(r)P_2(r')\frac{r_<^\lambda}{r_>^{\lambda+1}}P_3(r)P_4(r')drdr' \,, \tag{11.21}$$

where $r_<$ and $r_>$ are respectively the smaller and greater of r and r'.

For convenience we may rewrite the triply differential cross section $\sigma^{(3)}$ in Eq. (11.16) in terms of a doubly differential cross section $\sigma^{(2)}$ and asymmetry parameters β_λ. Let $\hat{\mathbf{k}} = \hat{\mathbf{z}}$ and write $\hat{\mathbf{k}}_1 \cdot \hat{\mathbf{k}} = \cos\theta_1$. Then $\sigma^{(2)}$

is defined by

$$\sigma^{(2)} = \int_0^{2\pi} d\phi_1 \int_{-1}^{+1} \sigma^{(3)} d(\cos\theta_1), \qquad (11.22)$$

and the parameters β_λ are determined by comparing with Eq. (11.16) the equation

$$\sigma^{(3)} = \frac{\sigma^{(2)}}{4\pi} \left[1 + \sum_{\lambda \geq 1} \beta_\lambda P_\lambda(\cos\theta_1) \right]. \qquad (11.23)$$

We use the expression in the brackets on the right hand side of Eq. (11.23) to define the relative triply differential cross section, $4\pi\sigma^{(3)}/\sigma^{(2)}$.

11.3 Results and discussion

In this section, we discuss our theoretical results for the triply differential cross section (TDCS) for low-energy electron-impact ionization of atoms in the $\theta_{12} = \pi$ (i.e., $\hat{\mathbf{k}}_1 = -\hat{\mathbf{k}}_2$) geometry. That is, the TDCS for cases in which the two final-state continuum electrons leave in opposite directions. Figure 11.4 shows the theoretical [436],[464](a,b),[465] and experimental [432, 436] relative TDCS for H for 2 eV excess energy shared equally by the two outgoing electrons. The U-shaped angular distribution for H was first reported by Schlemmer *et al.* in 1989 for equal-sharing of 4eV excess energy. [431] The overall agreement between the three sets of theoretical results and the experimental results of Rösel *et al.* [432] is very good.

Notice that because these relative experimental results have been fitted to the calculated results of Pan and Starace, [464](a,b) the comparisons with the theoretical calculations of Brauner *et al.* [436] and Jones, Madison, and Srivastava [465] are shown at a disadvantage. Although in Fig. 11.4 the three sets of theoretical results more or less agree qualitatively for the relative TDCS, there are significant quantitative discrepancies for the absolute values. The results of Brauner *et al.* were calculated according to Eq. (11.1) using an asymptotic wave function, which satisfies the exact asymptotic three-body boundary conditions, for Ψ_f^- and using a plane wave for Φ_i^+. [436] Their results are about two orders of magnitude smaller than those of the other two calculations, and they have attributed the small values of their absolute results near threshold to the wavefunction normalization they used. The other two sets of results are of the same order of magnitude. Jones, Madison, and Srivastava also used effective screening potentials to approximate the interaction between the two outgoing electrons, but they used a different ansatz to choose the effective charges. [465] They did not include the effects due to exchange between

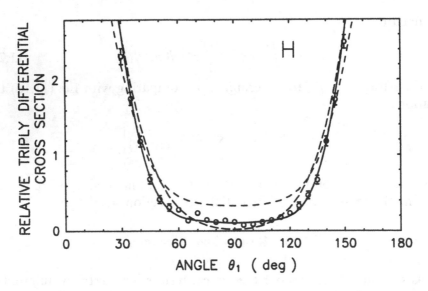

Fig. 11.4. Relative triply differential (*e*, 2*e*) cross sections for a H-atom target for final states having $\theta_{12} = \pi$ and the continuum electrons sharing 2 eV excess energy equally. Open circles: experimental results of Rösel *et al.* Ref. [432]. Curves are theoretical results. Solid curve: Pan and Starace, Ref. [464](b). Long-broken curve: Brauner *et al.*, Ref. [436]. Short-broken curve: Jones, Madison, and Srivastava, Ref. [465]. This figure is reproduced from Ref. [464](b).

the incident electron and the target electron. The first-order calculations of Pan and Starace [464](a,b) include the diagrams of Figs. 11.1(a)–11.1(d).

Figure 11.5 shows the theoretical [439],[449],[464](a,b),[465] and experimental [438, 439] TDCS for He for 2 eV excess energy shared equally by the two outgoing electrons. The W-shaped angular distribution in the $\theta_{12} = \pi$ geometry for He for low excess energies was revealed in the measurements of Schubert *et al.* in 1981 [460] and of Fournier-Lagarde, Mazeau, and Huetz in 1984 [461] and was discussed in detail by Selles, Huetz, and Mazeau in 1987. [430] The first measurement for the absolute TDCS at near-threshold excess energies was reported by Rösel *et al.* in 1992 for He. [438, 439] The theoretical results of Pan and Starace [464](b) agree very well with the absolute results of Rösel *et al.* [438, 439] Those of Jones and Madison given in Ref. [439] also agree reasonably well, although they are somewhat lower than experiment near $\theta_1 = 90°$. The main differences [439] of these two calculations are that Jones and Madison use a local approximation to the exchange interactions compared to an exact treatment of exchange in Ref. [464]; also, the two calculations employ different values for the effective screening charges describing the interaction between the final-state continuum electrons. The earlier calculation of Jones, Madison, and Srivastava [465] did not include any effects

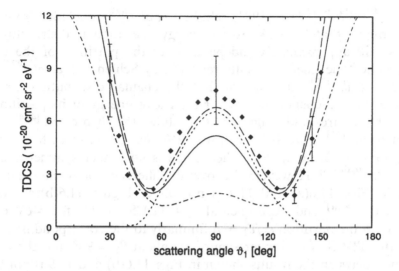

Fig. 11.5. Triply differential $(e, 2e)$ cross sections for a He-atom target for final states having $\theta_{12} = \pi$ and the continuum electrons sharing 2 eV excess energy equally. Solid diamonds: experimental results of Rösel *et al.*, Ref. [438]. Curves are theoretical results. Solid curve: Jones and Madison, given in Ref. [439]. Broken curve: Pan and Starace, Ref. [464](b). Dotted curve: Crothers, Ref. [449]. Broken and dotted curve: Jones, Madison, and Srivastava, Ref. [465]. This figure is reproduced from Ref. [439].

due to the exchange interactions between the continuum electrons and the bound target electrons; their results do not show a prominent peak at $\theta_1 = 90°$. The theoretical results of Crothers [449] did not include contributions from the triplet states of the two outgoing electrons. However, contributions from these triplet states are zero at $\theta_1 = 90°$, and his results agree well with the experimental results for the peak centered at $90°$.

The striking difference between the low-energy angular distributions for H and He (shown respectively in Figs. 11.4 and 11.5) is due to short-range interactions since the asymptotic conditions are the same for both cases. As is shown in the calculations done by Pan and Starace, [464](a,b) by Jones, Madison, and Srivastava, [465] and by Jones and Madison, [439] the difference can be explained in part by including the effects of the charge distributions of the target atom and the residual ion on the continuum electrons. Inclusion in addition of the effects due to exchange of the continuum electron and the bound target electrons leads then to good agreement with experiment for both cases. [439, 464](a,b)

According to the WPR theory, in the limit of zero excess energy, the angular distribution for the $\theta_{12} = \pi$ case is independent of the excess energy and is almost independent of the partition of the excess energy. [443],[444](b),[450] In 1981, Schubert *et al.* [460] concluded from their measurements that at 6 eV excess energy, the angular distribution for He was still approximately independent of the partition of the excess energy. The experimental results reported by Schlemmer *et al.* [431] and by Rösel *et al.* [432] have shown that the angular distribution for H is approximately independent of either the excess energy or its partition up to 4eV above threshold. Figure 11.6(a) shows the theoretical [464](a) and experimental [431] TDCS for H for 4 eV excess energy with the partition $\varepsilon_1/\varepsilon_2 = 7$. Since $\varepsilon_1 \neq \varepsilon_2$, the TDCS is no longer symmetric about $\theta_1 = 90°$. [464](a,b) However, the overall difference between the results shown in Figs. 11.6(a) and 11.4 is not large. Figure 11.6(b) shows the theoretical [464](a) and experimental [431] TDCS for He for 4 eV excess energy with the partition $\varepsilon_1/\varepsilon_2 = 7$. Similarly to the corresponding results for H, the TDCS is no longer symmetric about $\theta_1 = 90°$, but the overall difference between the results shown in Figs. 11.6(b) and 11.5 is not large.

Figure 11.7 compares our present theoretical results for the TDCS including electron-correlation effects with those of our first-order calculation (Ref. [464](b)) and the results of experimental measurements. [430, 438, 439] Starting from the first-order calculation, which includes effectively the diagrams of Figs. 11.1(a)–11.1(d), we first add the inter-channel interaction between the singlet and triplet states of the two outgoing electrons induced by the He^+1s core, *i.e.*, the interaction described by the matrix element $\langle 1s[\varepsilon_1\ell_1\varepsilon_2\ell_2(^1L)](^2L) \mid \sum r_{ij}^{-1} \mid 1s[\varepsilon_1\ell_1\varepsilon_2\ell_2(^3L)](^2L)\rangle$. With the inter-channel interaction added, all the diagrams in Fig. 11.2 are now included. Then, we include the diagrams of the type shown in Fig. 11.1(e) and, using the approximation discussed in Sec. 11.2.2, the diagrams of the types shown in Figs. 11.1(f) and 11.1(g). The diagram of Fig. 11.1(e) is found to have a relatively small effect as compared to the other two diagrams in Figs. 11.1(f) and 11.1(g). Finally, we include the second-order diagrams in Fig. 11.3, which correspond to corrections to the single-configuration HF description for the wave function of the He atom. Note that except for the diagrams involving overlap integrals (which are found to be relatively small for the present case), these corrections have been treated previously by Pan and Starace. [464](b) One can see in Fig. 11.7 that the overall change in the TDCS is not very large after including these electron-correlation effects. The theoretical results are improved appreciably near $\theta_1 = 50°$ and $130°$, where the TDCS displays minima, although the peak at $\theta_1 = 90°$ becomes somewhat lower. The effect of the inter-channel interaction is relatively strong.

In 1990, Selles, Mazeau, and Huetz reported measurements of TDCS for Ne, Ar, and Kr targets for equal-sharing of 0.5, 1, 2, and 4 eV excess energies. [435] These results were compared with the results for He,

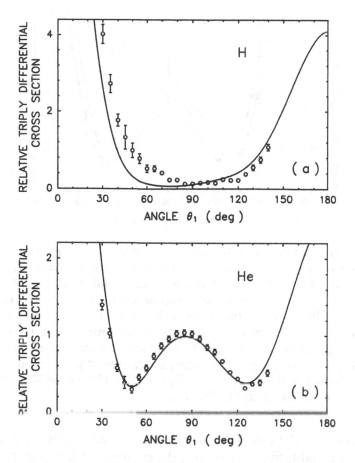

Fig. 11.6. Relative triply differential (*e, 2e*) cross sections for (a) H and (b) He targets for final states having $\theta_{12} = \pi$ and the continuum electrons having $1/2\ k_1^2 = 3.5$ eV and $1/2\ k_2^2 = 0.5$ eV. Open circles: experimental results of Schlemmer *et al.*, Ref. [431]. Solid curves: theoretical results of Pan and Starace, Ref. [464](a). This figure is reproduced from Ref. [464](a).

and the angular distributions for all four of these targets were shown to become stable as the excess energy was lowered, in agreement with the prediction of the WPR theory. [443],[444](b),[450] Nevertheless, the upper limits of the excess energy below which the angular distribution curves are stable are different for these targets. The upper limit is found to change from about 2 eV to below 1 eV when following the He to Kr chain. This is presumably due to the stronger electron-correlation effects in the heavier target atoms. Rösel *et al.* also measured the TDCS for Kr and Xe targets for equal-sharing of 2 eV excess energy. [432] The results for these heavy rare-gas targets also show strong target dependence. Pan and Starace [464](b) have performed first-order calculations for these targets;

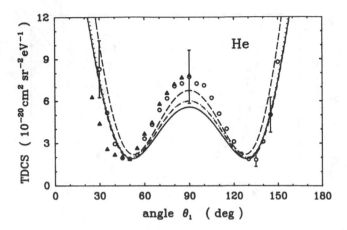

Fig. 11.7. Triply differential $(e, 2e)$ cross sections for a He-atom target for final states having $\theta_{12} = \pi$ and the continuum electrons sharing 2 eV excess energy equally. Open circles: absolute experimental results of Rösel *et al.*, Ref. [438]. Solid triangles: relative experimental results of Selles, Huetz, and Mazeau, Ref. [430], fitted to the result of Rösel *et al.* at $\theta_1 = 50°$. Curves are the present theoretical results. Broken curve: first-order results for the case in which the diagrams of Figs. 11.1(b)–11.1(d) sum to zero. Dotted curve: results also including the final-state inter-channel interaction. Broken-and-dotted curve: results also including the diagrams of Figs. 11.1(e)–11.1(g). Solid curve: results also including all the diagrams in Fig. 11.3.

for the most part, the calculated and the measured results for Ne, Ar, Kr, and Xe agree qualitatively. However, the comparison indicates the need in future calculations to include electron-correlation effects. Figures 11.8(a, b) show the theoretical [464](b) and experimental [435] TDCS for Ne and Ar targets for equal-sharing of 0.5 eV excess energy, which is the lowest excess energy used in the measurements. One can see a rough qualitative agreement between theory and experiment. However, the calculated results for Kr do not agree with the measured ones at 0.5 eV excess energy. [464](b) Figures 11.9(a) and 11.9(b) show the theoretical [464](b) and experimental [432, 435] TDCS for Kr and Xe for equal-sharing of 2 eV excess energy. Good agreement between the calculated and the measured results for Xe can be seen for this excess energy.

The H atom and the He atom are the two simplest target atoms, and the theoretical explanation for the difference between the TDCS for H and He in the $\theta_{12} = \pi$ geometry turns out to be relatively simple. [464] The difference between the TDCS for either H or He and the TDCS for any other target atom is in general more complicated; the Li atom can be used as an example to demonstrate this point. The Li atom has an outer subshell which is isoelectronic to the H atom, and so Eqs. (11.19)

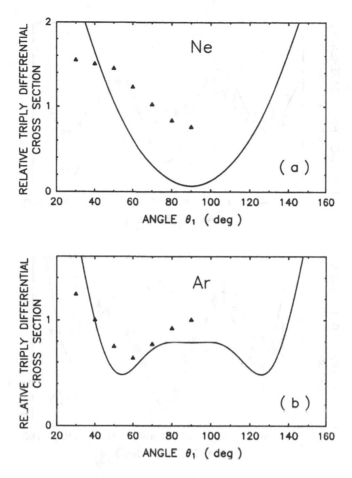

Fig. 11.8. Relative triply differential (e, $2e$) cross sections for (a) Ne and (b) Ar targets for final states having $\theta_{12} = \pi$ and the continuum electrons sharing 0.5 eV excess energy equally. Solid triangles: experimental results of Selles, Mazeau, and Huetz, Ref. [435]. Solid curves: theoretical results of Pan and Starace, Ref. [464](b). This figure is reproduced from Ref. [464](b).

and (11.20) can be used to calculate its TDCS provided that the 1s labels in Eq. (11.20) are changed to 2s labels. Figure 11.10 shows the theoretical TDCS for Li for equal-sharing of 2 eV excess energy. [464](c) One immediately sees that the angular distribution for Li is more complex than the ones for H and He. Table 11.1 presents the parameters $\sigma^{(2)}$ and β_λ that describe the angular distributions for H, He, and Li targets for equal-sharing of 2eV excess energy. [464](c) One notices that β_2 is the largest asymmetry parameter for H, β_4 is the largest for He, and both β_6 and β_8 are very large for Li.

Table 11.2 compares the relative magnitudes and phases of the partial-

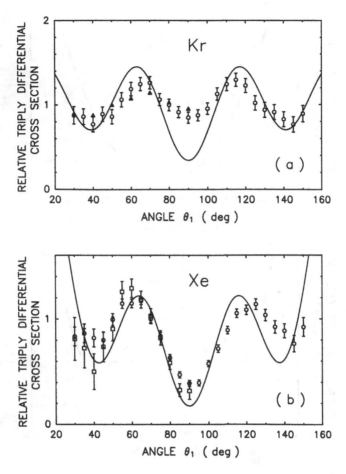

Fig. 11.9.   Relative triply differential (*e, 2e*) cross sections for (a) Kr and (b) Xe targets for final states having $\theta_{12} = \pi$ and the continuum electrons sharing 2 eV excess energy equally. Solid triangles: experimental results of Selles, Mazeau, and Huetz, Ref. [435]. Open circles: experimental results of Rösel *et al.*, Ref. [432], for 2 eV above the $Xe^+(^2P_{3/2})$ threshold. Open squares: experimental results of Rösel *et al.*, Ref. [432], for 2 eV above the $Xe^+(^2P_{1/2})$ threshold. Solid curves: theoretical results of Pan and Starace, Ref. [464](b). This figure is reproduced from Ref. [464](b).

wave amplitudes $A(LS)$ (cf. Eq. (11.20)) for H, He, and Li targets for equal-sharing of 2eV excess energy.[464](c) One can see that the relative magnitudes of these amplitudes are very similar for H and He. The arguments of these amplitudes are also very similar for H and He except for the case of the $^1S^e$ amplitudes, which differ by more than 1.5 rad. This difference affects the TDCS in Eq. (11.19) primarily via the interference terms [*i.e.*, $A(^1S^e)A^*(^1D^e) + A(^1D^e)A^*(^1S^e)$] between the $L = 0$ and $L = 2$

Table 11.1. Parameters $\sigma^{(2)}$ and β_λ determining the triply differential cross section $\sigma^{(3)}$ [cf. Eq. (11.23)] for $E_{ex} = 2$ eV for H, He, and Li targets. Dashed lines indicate values smaller than 0.003.

Parameter	H [a]	He [b]	Li [c]
$\sigma^{(2)}$ (a.u.)	3.73	0.781	34.5
β_2	3.090	1.082	1.638
β_4	2.367	1.868	0.856
β_6	0.855	0.338	1.881
β_8	0.115	0.023	2.036
β_{10}	0.008		0.924
β_{12}			0.216
β_{14}			0.029

[a] Ref. [464](b), Table III.
[b] First-order results corresponding to the long-dashed curves in Figs. 11.5 and 11.7.
[c] Ref. [464](c), Table I.

partial waves, which contribute to the asymmetry parameter β_2. The interference terms for H and for He largely explain the observed difference between the angular distributions for the two targets. [464](a) The difference between the phases of the $^1S^e$ amplitudes for the two targets stems mainly from the differences between the one-electron s-wave phase shifts for the corresponding continuum electrons in the two cases due to short-range interactions.

While for H and He targets the phase shifts of the $\ell \geq 1$ partial waves of the continuum electrons are all close to zero, this is no longer true for heavier targets. Also, for heavier targets, the ejected electron in the low-energy ionization process is not the 1s electron. As a result, more partial-wave amplitudes for the two outgoing electrons are important in describing the TDCS than for H and He targets. One can see in Table 11.2 that six partial-wave amplitudes for Li are significant, but only four amplitudes for H and He are significant.

It is shown in Fig. 11.5 for He that when $\theta_{12} = \pi$, $\theta_1 = 90°$, and $E_{ex} = 2$ eV, the calculations by Crothers, [449] by Pan and Starace, [464](b) and by Jones and Madison [439] all agree fairly well with the experimental TDCS. However, the threshold behaviors of the TDCS for the $\theta_{12} = \pi$ case given by the theoretical methods used in these calculations are very different. On the one hand, Crothers' results for the total electron-impact ionization cross section in the range $0 \leq E_{ex} \leq 6$ eV have an energy dependence $E_{ex}^{1.127}$, and his result for the TDCS for the $\theta_{12} = \pi$ case varies according

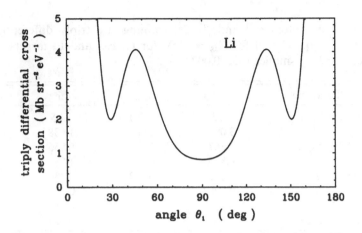

Fig. 11.10. Triply differential $(e, 2e)$ cross sections for a Li-atom target for final states having $\theta_{12} = \pi$ and the continuum electrons sharing 2 eV excess energy equally. Solid curve: theoretical results of Pan and Starace, Ref. [464](c). This figure is reproduced from Ref. [464](c).

Table 11.2. Relative amplitude and phase for electron-impact ionization scattering amplitudes $A(LS)$ for H, He, and Li targets for final-state electron kinetic energies $1/2\, k_1^2 = 1/2\, k_2^2 = 1$ eV. Only the first six partial waves for each target are shown.

Partial wave	Relative amplitude $|A(LS)|/|A(^1S^e)|$			arg $A(LS)$ (rad)		
$^{2S+1}L^\pi$	H^a	He^b	Li^a	H^a	He^b	Li^a
$^1S^e$	1.000^c	1.000^c	1.000^c	2.95	4.52	2.85
$^3P^o$	0.471	0.466	0.236	1.10	1.37	4.13
$^1D^e$	0.504	0.577	0.121	2.32	2.58	0.88
$^3F^o$	0.324	0.211	0.251	1.27	1.20	5.48
$^1G^e$	0.036	0.029	0.192	1.28	1.26	0.59
$^3H^o$	0.023	0.009	0.164	0.80	0.46	5.84

^{a} Ref. [464](c), Table II.
^{b} First-order results corresponding to the long-dashed curves in Figs. 11.5 and 11.7.
^{c} The values of $|A(^1S^e)|$ for H, He, and Li targets are respectively 0.3430, 0.2076, and 0.9003.

to $E_{ex}^{1.127-3/2}$. [449] These energy dependences agree with the WPR theory in the limit $E_{ex} \to 0$. [443],[445](b) The diverging behavior of this TDCS as $E_{ex} \to 0$ is related to the rapid narrowing of the width of the distribution with respect to θ_{12} centered around $\theta_{12} = \pi$, which varies as $E_{ex}^{1/4}$. On the other hand, the TDCS for the $\theta_{12} = \pi$ case given by calculations using the

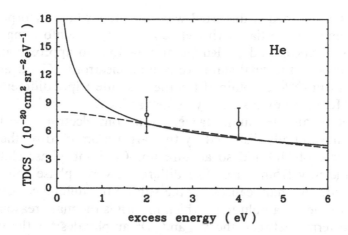

Fig. 11.11. Triply differential $(e, 2e)$ cross sections for a He-atom target for final states having $\theta_{12} = \pi$ and $\theta_1 = \theta_2 = 90°$ and the continuum electrons sharing equally the excess energy in the range 0–6 eV. Open circles: experimental results of Rösel *et al.*, Ref. [438]. Solid curve: theoretical results of Crothers, Ref. [449]. Broken curve: the first-order theoretical results of Pan and Starace.

effective-charge approximation [456]-[458],[464],[465] is independent of the excess energy in the $E_{ex} \to 0$ limit. This threshold behavior can be obtained by analyzing the wave functions used in such calculations. Figure 11.11 compares the results of our first-order calculations and the results of Crothers [449] with the measured TDCS for He for the case of $\theta_{12} = \pi$ and $\theta_1 = 90°$ in the energy range $0 \leq E_{ex} \leq 6$ eV. Crothers explicitly gives the TDCS results only for 1 and 2 eV excess energies. Here we extract results for other energies using both Fig. 1 and Eqs. (74) and (86) of Ref. [449]; the TDCS as a function of the excess energy E_{ex} for the case of $\theta_{12} = \pi$ and $\theta = 90°$ is given by $8.9(E_{ex}/eV)^{-0.373} \times 10^{-20} \mathrm{cm}^2 \mathrm{sr}^{-2} \mathrm{eV}^{-1}$. The available absolute experimental results are also plotted. [438] One sees that although the two theoretical curves agree well with the available experimental results at and above 2 eV excess energy, they depart from each other as the excess energy decreases below 2 eV. At present, the range of validity of the WPR theory is not known. However, it is clear that as the excess energy approaches zero the absolute TDCS results calculated by using approximations employing effective screening potentials [456]-[458],[464],[465] cannot be relied upon.

11.4 Concluding remarks

We have discussed theoretical and experimental results for low-energy electron-impact ionization of atoms in the $\theta_{12} = \pi$ geometry. Our own calculations have used a distorted-wave method which incorporates effective screening potentials (with effective charges conforming to those specified by Peterkop and by Rudge and Seaton) to approximate the interaction between the final-state continuum electrons. Good agreement with experiment [438] is obtained for the absolute triply differential cross section of He for an excess energy as low as 2 eV.

For H atom and He atom targets, only one-electron $\ell = 0$ partial waves are substantially distorted by the target atom (and by the residual ion in the case of He) and so acquire non-Coulomb phase shifts which differ significantly from zero. The different s-wave phase shifts for the two targets are the main causes for the substantial difference between the arguments of the $^1S^e$ amplitudes, which in turn is the main reason that the interference terms between the $^1S^e$ and $^1D^e$ amplitudes for the two cases differ, causing thereby the β_2 parameters to differ. Thus the observed difference between the angular distributions for H and He [430, 431, 432] is largely explained.

For H and He, inclusion of electron-correlation effects changes the angular distributions for the $\theta_{12} = \pi$ case only slightly from those obtained from first-order calculations which include distortion effects arising from both the direct and the exchange interactions. For He, comparison with experiment is improved near $\theta_1 = 50°$ and $130°$, where the angular distribution has minima, but is made worse near $\theta_1 = 90°$ where there is a local maximum. Amplitudes for $L > 3$ two-electron partial waves are not important for describing the angular distributions for H and He, as was found empirically. [431]

In comparison to H and He, in the case of heavier target atoms, [432, 435] more one-electron partial waves are substantially distorted, and consequently, more two-electron partial-wave amplitudes are important for describing the angular distributions. Also, for heavier targets electron-correlation effects become very important.

12

Overview of Thomas processes
for fast mass transfer

J. H. McGuire, Jack C. Straton,
and T. Ishihara

12.1 Introduction

Transfer of mass is a quasi-forbidden process. It does not occur very simply. So new insights into how masses interact in our environment may be discovered by considering this challenging little puzzle of how a quasi-forbidden process operates. The simple pickup of a stationary mass, M_2, by a moving mass, M_1, is forbidden by conservation of energy and momentum. Specifically, if $M_1 < M_2$ then M_1 rebounds, if $M_1 = M_2$ then M_1 stops and M_2 continues on, and if $M_1 > M_2$ then M_2 leaves faster than M_1. In none of these cases do M_1 and M_2 leave together. Thus, mass transfer may not occur in a single elastic collision. L.H. Thomas [478] understood this in 1927 and further realized that transfer of mass occurs only when a third mass is present and all three masses interact. The simplest allowable process is a two-step process now called a Thomas process. It was not until the work of Shakeshaft and Spruch 52 years later [479] that the significance of Thomas processes was properly understood. There is evidence, for example, that Bohr [480] did not understand the significance of the failure of the first Born approximation to reduce to the correct classical limit.

The resolution of Bohr's dilemma lies in the unusual feature that the second Born term for mass transfer is larger than the first Born term at high collision velocity, v. Recognition of this result by Drisko [481] in 1955 raised the inconvenient specter of a Born series in which each Born term becomes successively larger. Fortunately, this particular theoretical nightmare does not exist in the theory of mass transfer. In general, the second Born term is the largest Born term at high velocity and corresponds to the simplest allowed classical process, namely the Thomas process. The first Born term for transfer corresponds to the classically forbidden process described above. While this classically forbidden first Born term is not zero

287

(owing to the Uncertainty Principle), the first Born cross section varies at high v as v^{-12} in contrast to the second Born cross section, which varies as v^{-11} and so dominates. Higher Born terms correspond to multi-step processes that are unlikely in fast collisions where there is not enough time for complicated processes. The higher Born terms ($n > 2$) are also smaller than the second Born term.

So the two-step Thomas process, described below, is the simplest and the dominant process in collisions in which mass is transferred to (or from) a fast projectile.

12.2　The classical Thomas process

Transfer of mass occurs when a mass M_2 leaves a target of mass $M_3 + M_2$ and joins a projectile of mass M_1. The classical Thomas process may be easily understood by an undergraduate student who has taken one semester of an introductory physics course using algebra (even the quantum version may be largely understood without calculus).

The basic Thomas process is shown in Fig. 12.1. Here the entire collision is coplanar since particles 1 and 2 go off together (which is what is meant by mass transfer). We assume that all the masses and the incident velocity, $\mathbf{v}$, are known. Then there are six unknowns, $\mathbf{v}', \mathbf{v}_f$ and $\mathbf{v}_3$, each vector having two components. Conservation of momentum gives two equations of constraint for each collision. Conservation of overall energy gives a fifth constraint, and conservation of energy in the intermediate state gives a sixth constraint. With six equations of constraint, all six unknowns may be completely determined.

The allowed values of $\mathbf{v}', \mathbf{v}_f$ and $\mathbf{v}_3$ depend on the masses, M_1, M_2 and M_3. For example, in the case of the transfer of an electron from atomic hydrogen to a proton, i.e. $p^+ + \mathrm{H} \rightarrow \mathrm{H} + p^+$, it is easily verified that (in the notation of Fig. 12.1) $\alpha = \dfrac{M_2}{M_1} \sin 60°, \beta = 60°$, and $\gamma = 120°$ where $m' = M_2 = m_e$ (the mass of an electron), and $M_1 = M_3 = 1836 m_e = M_p$ (the mass of a proton).

Another case is formation of a bound state of an electron and an anti-electron (denoted Ps and called positronium) by transfer of an electron in hydrogen to an incident anti-electron, e^+ (called a positron). This reaction is denoted by $e^+ + \mathrm{H} \rightarrow \mathrm{Ps} + p^+$. Here, $M_1 = M_2 = m_e$ (the electron mass and the anti-electron mass are the same), and $M_3 = M_p$. Here m' may be either M_2 or M_1, both of which equal m_e. Then it may be easily verified that $\alpha = 45°, \beta = 45°$, and $\gamma = 90°$.

In general the intermediate mass, m', may equal M_1, M_2, or M_3. We shall regard these as different Thomas processes, and label them B, A and

$$I + (2,3) \longrightarrow (1,2) + 3$$

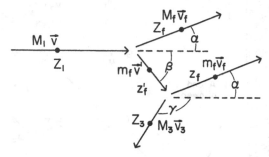

Fig. 12.1. Diagram for mass transfer via a Thomas two-step process. A projectile mass M_1, interacts with a target particle (M_2 or M_3). Either the projectile or the target particle leaves the system with velocity $\mathbf{v}_f$. The other particle, the "intermediate" particle of mass m', then interacts with the second target particle (M_3 or M_2). M_3 then leaves with velocity $\mathbf{v}_3$ and the remaining particle leaves with velocity $\mathbf{v}_f$.

C respectively. The standard Thomas process (the one actually considered by Thomas in 1927) is case A and corresponds to the first example given above. The Thomas processes A, B, and C are illustrated in a figure due to Lieber [482] (see Fig. 12.2). In the Lieber diagram, mass regions in which solutions exist for processes A, B, and C are shown. (An equivalent diagram was given earlier by Detmann and Liebfried. [483]) There are some regions in which two-step processes are forbidden. In these regions the theory of the mass transfer is not fully understood at present.

The main point of this overview of Thomas processes is to understand the simplest processes (smallest number of collisions possible) for mass transfer. As an aside, it is interesting to note that at high collision velocities there is a limit to the largest number of collisions that may occur. [481] This upper limit suggests that quantum Born terms beyond this limit do not contribute as much as the lower "allowed" Born terms.

12.3 Quantum mechanics

In quantum mechanics, energy conservation in the intermediate states may be violated within the limits of the Uncertainty Principle, namely, $\Delta E \geq \hbar/(\Delta t)$, where Δt is the uncertainty in time of the mass transfer. In quantum mechanics it is not possible to determine if mass transfer actually occurs at the beginning, in the middle or at the end of the collision. Thus we choose $\Delta t = \bar{r}/\bar{v}$, where $\bar{r}$ is the size of the collision region and $\bar{v}$ is

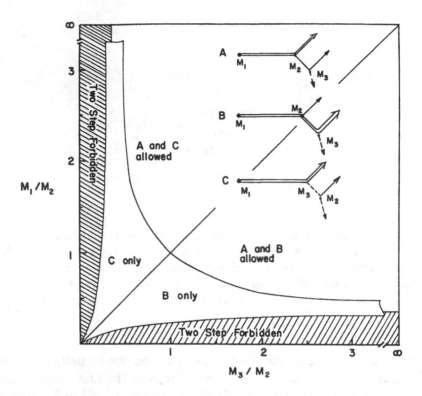

Fig. 12.2. Lieber's diagram for mass transfer. The three possible Thomas processes are shown diagramatically and labeled A, B and C. The allowed regions of the mass ratios are labeled for each Thomas process. It is not clear what is the minimum number of steps for transfer in the two-step forbidden regions.

the mean collision velocity. Taking $\bar{r} \approx r_0/Z_{target}$ and $\bar{v} \approx v$, the projectile velocity, we have $\Delta E \approx \dfrac{\hbar}{\Delta t} = \dfrac{\hbar \bar{v}}{\bar{r}} = \dfrac{\hbar v}{r_0/Z_{target}} = \dfrac{\hbar v Z_{target}}{r_0}$. Here r_0 is the Bohr radius (5.3×10^{-11} meters) and Z_{target} is the nuclear charge of the target in units of the electron charge. Within this uncertainty ΔE the constraint of energy conservation in the intermediate state may be relaxed.

Returning to Fig. 12.1, let us consider the influence of conservation of overall energy and momentum. Let us for the moment ignore the intermediate state characterized by $\mathbf{v}'$ and m'. Conservation of overall energy and momentum then gives three equations of constraint on the four unknowns comprised by $\mathbf{v}_f$ and $\mathbf{v}_3$, namely,

$$M_1 \mathbf{v} = (M_f + m_f)\mathbf{v}_f + M_3 \mathbf{v}_3, \qquad (12.1)$$

$$\tfrac{1}{2}M_1 v^2 = \tfrac{1}{2}(m_f + m_f)v_f^2 + \tfrac{1}{2}M_3 v_3^2, \qquad (12.2)$$

where $M_f(m_f)$ is the mass of the upper (lower) particle in the final state of the bound system shown in Fig. 12.1, in which m' is the mass of the intermediate particle (M_1, M_2 or M_3). From Eqs. (12.1) and (12.2) it is easily shown (see the Appendix in Sec. 12.7) that the velocity of the recoil particle is constrained by the condition,

$$2\hat{v}_3 \cdot \hat{v} = 2\cos(\gamma) = \frac{M_1 + M_2 + M_3}{M_1} \frac{v_3}{v} - \frac{M_2}{M_3} \frac{v}{v_3}, \tag{12.3}$$

Thus, the magnitude and the direction of v_3 are not independent. Specifically, specifying either v_3 or $\hat{v}_3$ is sufficient, together with the equations of constraint, to determine the energies and directions of all particles in the final state. (Similarly one may express the equations of constraint in terms of v_f or $\hat{v}_f$.)

The effect of the constraints of conservation of overall energy and momentum may be seen in Fig. 12.3, where a sharp ridge is clearly evident

in the reaction,

$$p^+ + \text{He} \rightarrow \text{H} + \text{He}^{++} + e^-, \tag{12.4}$$

where $M_1 = M_p, M_2 = M_3 = m_e$. Here v_3 is the speed of the recoiling ionized target electron, and the target nucleus is not directly involved in the reaction. The width of the sharp ridge is due to the momentum spread of the electrons in helium and may be regarded as being caused by the Uncertainty Principle since this momentum (or velocity) spread corresponds to $\Delta p = \dfrac{\hbar}{\Delta r}$ where Δr is taken as the radius of the helium atom.

In a classical two-step process, the projectile hits a particle in the target and the intermediate mass, m', then propagates and subsequently undergoes a second collision. Quantum mechanically, this corresponds to a second Born term represented by $V_1 G_0 V_2$, where V represents an interaction and G_0 is the propagator of the intermediate state. Now, it may be shown from the mathematics of complex variables that

$$G_0 = (E - H_0 + i\eta)^{-1} = -i\pi\delta(E - H_0) + \mathrm{P}\frac{1}{E - H_0}, \tag{12.5}$$

where P is the principal part of G_0 which excludes the singularity at $E = H_0$. The singularity at $E = H_0$ corresponds to conservation of energy in the intermediate state. It is this singularity that gives rise to the weaker secondary ridge in Fig. 12.3 at $v_3 = v$. The width of the secondary ridge is given approximately by $\Delta E = \hbar/\Delta t$, discussed above. The intersection of the ridges is the Thomas peak. At very high collision velocities, the Thomas peak dominates the total cross section (or counting rate) for mass transfer.

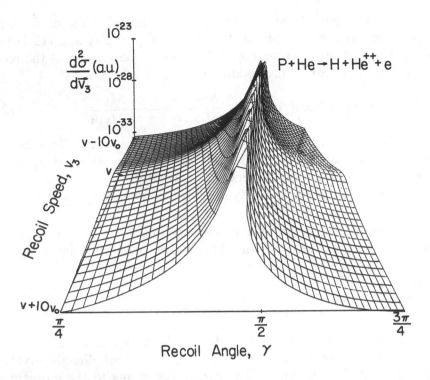

Fig. 12.3. Counting rate (or cross section) on the vertical axis versus recoil speed, v_3, and recoil angle, γ, for a Thomas process in which a proton picks up an electron from helium; the captured electron then bounces off the second target electron in the helium target. These are the results of a four-body second Born quantum calculation by Ishihara and McGuire.

The constraint imposed by conservation of intermediate energy may be expressed in an elegant way by replacing the speed of the recoil particle v_3 by the scaled variable $K = M_3 v_3 / m' v$. Then it may be shown, after some algebra (see the Appendix in Sec. 12.7), that the conservation of intermediate energy may be expressed as,

$$(M_3 v_3 / m' v)^2 \equiv K^2 = 1. \tag{12.6}$$

The constraints of conservation of overall energy and momentum, *i.e.* Eq. (12.3), may be easily written in terms of K as

$$2\cos(\gamma) = r \frac{m'}{M_2} K - \frac{M_2}{m'} \frac{1}{K}, \tag{12.7}$$

where $r = (M_1 + M_2 + M_3) M_2 / M_1 M_3$.

The locus of the sharp ridge in Fig. 12.3, corresponding to conservation of overall energy and momentum, is given by Eq. (12.7). The locus of the weaker ridge in Fig. 12.3, corresponding to conservation of energy in the

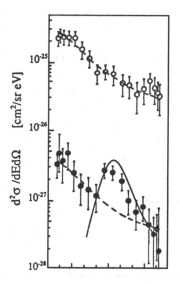

Fig. 12.4. Observation of a slice of the Thomas ridge structure in $p^+ + \text{He} \rightarrow$ $\text{H} + \text{He}^{++} + e^-$ at 1 Mev by Palinkas *et al.* The top data are a control. The Thomas effect, given by the solid curve, is confirmed by the data in the lower half of the figure. The variable on the x axis is γ in Fig. 12.3 at $v_3 = v$.

intermediate state, is given by Eq. (12.6). The intersection of these two loci gives the unique classical result suggested by Thomas. The width of these ridges may be estimated from the Uncertainty Principle, as described above.

Experimental evidence for the double ridge structure has been reported by Palinkas *et al.* [486] corresponding to the calculations given in Fig. 12.3, but at a collision energy of 1 MeV, as shown in Fig. 12.4.

The data in Fig. 12.4 corresponds to a slice across the sharp ridge shown in Fig. 12.3 at $v = v_3$. The solid line is a second Born calculation [485] at 1 MeV. The bump of data above a smooth background is the indication of the ridge structure. The origin of the smooth background is not fully understood, although it may be due to a competing process in which mass transfer to the projectile and ionization of the second electron by the projectile happen independently. At the time of writing, there is no experiment which gives data for the entire double ridge structure. We recommend that such experiments be done.

12.4 Interpretations

Let us now consider some interpretations of the second Born Thomas quantum amplitudes, and how this analysis may be applied to the experi-

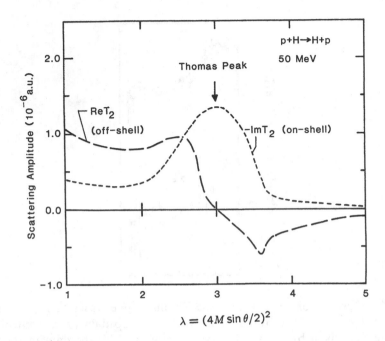

Fig. 12.5. Energy-conserving (on-shell) and energy-nonconserving contributions to the second Born scattering amplitude, f_2, from Simony [487]. These amplitudes satisfy a dispersion relation. Anomalous dispersion [490] is evident in the vicinity of the Thomas peak at $\lambda = 3$. The angle θ here is the same as α in Fig. 12.1.

mental data. The second Born amplitude corresponds to $V_1 G_0 V_2$ terms in which an interaction, V_1, is followed by propagation of the intermediate state via the Green's function, G_0, and then the second interaction, V_2. In this interpretation we shall focus on the significance of the Green's function, G_0.

As we noted in Eq. (12.5) above, the Green's function G_0 contains an energy-conserving term, $i\pi\delta(E - H_0)$ and an energy-nonconserving term, $P\dfrac{1}{E - H_0}$. The energy-conserving term is imaginary, while the energy-nonconserving term is real. The energy non-conserving term does not occur classically; it is permitted by the Uncertainty Principle and represents the contribution of virtual (energy-nonconserving) states within $\pm\Delta E = \hbar/\Delta t$ about the classical value $E = H_0$. It is also possible to show [488] that this quantum term represents the effect of time-ordering in the second Born amplitude. In plane wave second Born calculations the energy-nonconserving (off-energy-shell) term gives the real part of the scattering amplitude, f_2, and the energy-conserving (on-shell) term gives the imaginary part of f_2. These two contributions are shown in Fig. 12.5.

It is evident from Fig. 12.5 that the energy-nonconserving part of the am-

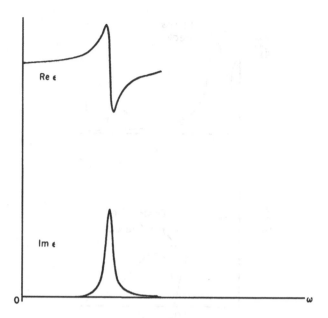

Fig. 12.6. Real and imaginary parts of the dielectric constant, ϵ, in the neighborhood of a resonance.

plitude is not small. In fact it may be shown for electron transfer to a proton that the energy-nonconserving contribution makes precisely the same contribution to the total cross section as the energy-conserving contribution. Thus half of the Thomas peak comes from energy-nonconserving contributions which are not included in a classical description. Also, the energy-nonconserving contribution plays a significant role in determining the shape of the standard Thomas peak, which has been observed. [489] In this problem it is necessary to use quantum mechanics to obtain a fully correct result. (Nonetheless, we feel that it would be an understatement of Thomas' contribution to this problem to say that he got it only half right.)

McGuire and Weaver recognized [490] that Fig. 12.5 bore a striking resemblance to the real and imaginary parts of the dielectric constant, ϵ, near a resonance, as discussed in the electrodynamics text of Jackson, [165]; see Fig. 12.6.

In the case of the dielectric constant ϵ it is well known that the real and imaginary parts are related by a dispersion relation, namely the Kramers-Kronig relation. [165] The similarity between Fig. 12.6 and Fig. 12.5 is not accidental, as we shall now show.

Because of the form of the Green's function of Eq. (12.5) the second Born contribution to the scattering amplitude has a single pole in the

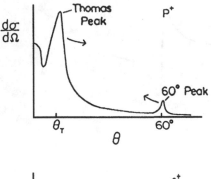

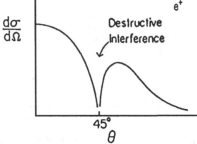

Fig. 12.7. Change of position and nature of the Thomas peaks with decreasing projectile mass. Upper figure, p^+; lower figure, e^+. Here θ is the same as α in Fig. 12.1.

lower half of the complex plane, *i.e.*

$$f_2 = \frac{c}{\lambda - \lambda_{\text{Thomas}} + i\eta},$$

$$\lambda = (4M_1 \sin(\alpha/2))^2, \tag{12.8}$$

where α is given in Fig. 12.1 and α_{Thomas} is the Thomas angle for diagram A of Fig. 12.2.

Because f_2 is analytic in the upper half plane it satisfies the following dispersion relation, [490]

$$\text{Re}f_2(\lambda) = -\frac{1}{\pi}\text{P}\int_{-\infty}^{+\infty} d\lambda' \frac{\text{Im}f_2(\lambda')}{\lambda - \lambda'},$$

$$\text{Im}f_2(\lambda) = \frac{1}{\pi}\text{P}\int_{-\infty}^{+\infty} d\lambda' \frac{\text{Re}f_2(\lambda')}{\lambda - \lambda'}. \tag{12.9}$$

Thus the energy-nonconserving part of f_2 is related to an integral over the energy-conserving part, and vice versa. Mathematically the resonant structures of f_2 in Fig. 12.5 and ϵ in Fig. 12.6 are similar and lead to dispersion relations with the characteristic structure seen in Figs. 12.5 and 12.6.

Normally we think of a resonance as a function of some energy. The width of such a resonance gives the lifetime τ of the resonance. Classically τ is the time for which the projectile orbits the target before it leaves, corresponding to a delay or shift in time of the projectile during the interaction. If the width of the resonance is ΔE, then the lifetime is $\tau = \hbar/\Delta E$. In quantum mechanics, energy E and time τ are conjugate variables. The Thomas peak is a resonance in a scattering angle, corresponding to a well defined momentum transfer of the projectile. The conjugate variable to the momentum transfer is the impact parameter of the projectile. It has been shown [491] that Thomas resonances thus correspond to a shift in the impact parameter of the scattering event. However, unlike energy resonances, our Thomas resonance in the scattering angle seems to have no classical analogue.

12.5   Destructive interference of Thomas amplitudes

It was noted in the introduction that the location of the Thomas peaks depends on the mass of the collision partners. For $p^+ +$atom $\rightarrow$ atom$^+ +$H there are two separate Thomas peaks [479, 492] corresponding to cases A and B in the Lieber diagram (Fig. 12.2). Experimental evidence exists for both peaks. The standard Thomas peak occurs at small forward angles, [489] and the second peak [493] occurs about 60°. If the mass of the projectile is reduced, the positions of these Thomas peaks move toward one another [494] as illustrated in Fig. 12.7. When $M_1 = M_2$, then both Thomas peaks occur at 45°. This occurs in positronium formation where $M_1 = M_2 = m_e$, i.e.,

$$e^+ + \text{He} \rightarrow \text{Ps} + \text{He}^{++} + e^-. \tag{12.10}$$

In cases A and B in Fig. 12.2 the two $V_1 G_0 V_2$ second Born terms are of opposite sign because V_2 is of opposite sign in diagrams A and B. This leads to destructive interference for $1s$–$1s$ electron capture (which is dominant at high velocities) as was first discussed by Shakeshaft and Wadehra. [494] Consequently, the observed Thomas peak structure is expected [495, 496] to be quite different for e^+ impact than for impact of p^+ or other projectiles heavier than an electron. The double ridge structure for transfer-ionization of helium by e^+ is expected to differ significantly from the structure shown in Fig. 12.3. Understanding such destructive interference between resonant amplitudes is likely to give deeper insight into the physical nature of the intermediate states in this special few-body collision system. We encourage experimental investigation.

12.6 Summary

Mass transfer is a quasi-forbidden process, that provides an opportunity to understand some new physics (albeit in a process of small overall importance). The points we have considered include the following.

- The second Born amplitude is dominant at high collision velocities.

- The second Born amplitude corresponds to a classical two-step mechanism first given by Thomas in 1927. Mass transfer in a simple single collision is forbidden by conservation of energy and momentum.

- Mass transfer may not occur in a single elastic collision. The minimum number of steps in which mass transfer may occur is two.

- Two-step processes are allowed only for certain mass ratios of the participants.

- There is a maximum number of steps that are classically allowed.

- There are energy-nonconserving contributions in the intermediate state to the quantum amplitude that are as significant in the $1s$–$1s$ capture amplitude as the classically allowed energy-conserving amplitudes.

- The quantum and the classical cross sections for electron transfer in atoms are different (they have a different coefficient).

- The energy non-conserving and energy-conserving parts of the second Born amplitude are inter-related by a dispersion relation.

- The Thomas amplitude is a resonance amplitude in the momentum transfer (or scattering angle). This resonance corresponds to a shift in the impact parameter of the projectile during the collision. No classical analogue for this shift has been given.

- Thomas peaks have been observed in the scattering cross section versus angle of the projectile and quantum calculations are in good agreement with the data, for the most part.

- There are three different ways in which a Thomas two-step process may occur classically. These give rise to observable singularities at different scattering angles and energies.

- If the mass of the projectile and the mass of the transferred particle are the same, two of these Thomas processes may occur at the same angle. The quantum amplitudes for these processes can interfere destructively. This effect is qualitatively observable in the formation of positronium, but experimental data has not yet been taken.

- A crossing double ridge structure has been predicted for recoil electrons in transfer-ionization. The loci of these ridges is given by conservation of overall energy and momentum, and by conservation of intermediate energy. The widths of these ridges may be estimated from the Uncertainty Principle. Approximate second Born quantum calculations have been partially confirmed by experiment. Additional data would be useful.

12.7   Appendix

Here we derive the expressions for the loci for the double ridge structure in the recoil velocities illustrated in Fig. 12.3. Specifically we shall first derive the constraint imposed by conservation of overall momentum and energy, Eq. (12.3), giving the locus of the sharp ridge. Then we shall impose the classical Thomas condition of conservation of intermediate energy and derive the more difficult result of Eq. (12.6), corresponding to the broader Thomas ridge which intersects the sharp ridge at the Thomas peak. Our mass conditions suffice for the transfer of a single arbitrary mass in a two-step collision, but do not allow the mass to change in the second step of the collision. We follow the notation of Fig. 12.1.

12.7.1   Overall momentum and energy conservation: the sharp ridge

Let us derive the relatively simple condition that gives the locus of the sharp ridge in Fig. 12.3, namely Eq. (12.3) or equivalently, Eq. (12.7).

From Fig. 12.1 with $m' = m_f$ or M_3, one has for *momentum conservation*,

$$M_1 \mathbf{v} \;=\; m_f \mathbf{v}' + M_f \mathbf{v}_f \,,$$

$$m' \mathbf{v}' = m_f \mathbf{v}_f + M_3 \mathbf{v}_3 \,,$$

which gives

$$M_1 \mathbf{v} = (M_f + m_f) \mathbf{v}_f + M_3 \mathbf{v}_3 \,. \qquad (P)$$

For *overall energy conservation*,

$$\tfrac{1}{2} M_1 v^2 = \tfrac{1}{2} (M_f + m_f) v_f^2 + \tfrac{1}{2} M_3 v_3^2. \qquad (E)$$

From (P),

$$(M_f + m_f) \mathbf{v}_f = M_1 \mathbf{v} - M_3 \mathbf{v}_3, \qquad (+)$$

and from (E),

$$(M_f + m_f)v_f^2 = M_1 v^2 - M_3 v_3^2. \tag{++}$$

Then from $(+)$ and $(++)$,

$$
\begin{aligned}
v_f^2 &= \frac{M_1}{M_f + m_f} v^2 - \frac{M_3}{M_f + m_f} v_3^2 \\
&= \frac{1}{(M_f + m_f)^2} [M_1 \mathbf{v} - M_3 \mathbf{v_3}]^2 \\
&= \frac{1}{(M_f + m_f)^2} \left[M_1^2 v^2 + M_3^2 v_3^2 - 2 M_1 M_3 \mathbf{v} \cdot \mathbf{v_3} \right]
\end{aligned}
$$

$$\frac{2 M_1 M_3}{M_f + m_f} \mathbf{v} \cdot \mathbf{v_3} = v_3^2 \left[\frac{M_3^2}{M_f + m_f} + M_3 \right] + v^2 \left[\frac{M_1^2}{M_f + m_f} - M_1 \right]$$

$$2 \mathbf{v_3} \cdot \mathbf{v} = \frac{M_3 + M_f + m_f}{M_1} v_3^2 - \frac{M_f + m_f - M_1}{M_3} v^2.$$

This gives Eq. (12.3), namely,

$$2 \cos(\gamma) = \frac{M_1 + M_2 + M_3}{M_1} \frac{v_3}{v} - \frac{M_2}{M_3} \frac{v}{v_3},$$

or, with

$$K \equiv \frac{M_3 v_3}{m_f v},$$

we have Eq. (12.7), namely

$$2 \cos(\gamma) = \frac{(M_1 + M_2 + M_3) m_f}{M_1 M_3} K - \frac{M_2}{m_f} \frac{1}{K}.$$

This gives the locus of the sharp peak in Fig. 12.3.

12.7.2   Conservation of intermediate energy: the broad ridge

Next, let us consider the more difficult condition that gives the locus of the broad ridge in Fig. 12.3, namely Eq. (12.6). Imposing *energy conservation in the intermediate state* at the first collision,

$$\frac{1}{2} m' v'^2 = \frac{1}{2} M_1 v^2 - \frac{1}{2} M_f v_f^2, \tag{E'}$$

and momentum conservation overall, by squaring (P),

$$2 M_3 \mathbf{v_f} \cdot \mathbf{v_3} = \frac{1}{M_f + m_f} \left[M_1^2 v^2 - M_3^2 v_3^2 \right] - [M_f + m_f] v_f^2,$$

and substituting these into the equation for intermediate state momentum conservation applied at the second collision, which we also square,

$$m'^2 v'^2 = m_f^2 v_f^2 + 2m_f M_3 \mathbf{v}_f \cdot \mathbf{v}_3 + M_3^2 v_3^2,$$

one has

$$\begin{aligned} m' M_1 v^2 - m' M_f v_f^2 &= m'^2 v'^2 \\ &= m_f^2 v_f^2 + \frac{m_f}{M_f + m_f} \left[M_1^2 v^2 - M_3^2 v_3^2 \right] \\ &\quad - m_f \left[M_f + m_f \right] v_f^2 + M_3^2 v_3^2. \end{aligned}$$

Finally, applying overall energy conservation, (E), to eliminate v_f, and multiplying by the denominator, $M_f + m_f = M_1 + M_2$, one obtains

$$\left[M_1^2(m' - m_f) + M_1 M_2 m' \right] v^2 + M_f(m_f - m') \left[M_1 v^2 - M_3 v_3^2 \right]$$
$$= M_3^2 M_f v_3^2 .$$

Collecting terms, one has

$$\frac{v_3^2}{v^2} = \frac{M_1 m_f (M_2 + m' - m_f)}{M_3 M_f (M_3 - m' + m_f)} .$$

Then from the definition of K, Eq. (12.6), one has for any intermediate mass m'

$$\begin{aligned} K^2 \equiv \left(\frac{M_3 v_3}{m' v} \right)^2 &= \frac{M_3 M_1 m_f (M_2 + m' - m_f)}{M_f m'^2 (M_3 - m' + m_f)} \\ &= \frac{M_3 m_f^2 (M_2 + m' - m_f)}{M_2 m'^2 (M_3 - m' + m_f)} . \end{aligned}$$

There are two possibilities. The first is $m' = m_f$, where m_f is either M_1 or M_2, giving

$$K^2 = \frac{M_3 M_2}{M_2 M_3} = 1 .$$

The second possibility is $m' = M_3$, in which case $m_f = M_2$, so that

$$K^2 = \frac{M_3 M_2^2 M_3}{M_2 M_3^2 M_2} = 1 .$$

In this way we have shown that applying energy conservation to the intermediate state gives $K^2 = 1$ for all three cases, A, B, and C in Fig. 12.2.

Part 3

ATOMIC SCATTERING:
C. All-order applications

13

R-matrix theory: some recent applications

Philip G. Burke

13.1 Introduction

The *R*-matrix theory was first introduced by Wigner [497] and Wigner and Eisenbud [498] to describe nuclear resonance reactions. This theory was subsequently developed and applied by many workers in nuclear physics, and comprehensive reviews have been written by Lane and Thomas, [499] Breit, [500] and Barrett *et al.* [501]

In recent years, it has been increasingly realised that the *R*-matrix method can also be used as an *ab initio* approach to describe a broad range of atomic and molecular structure and collision processes. These include: electron–atom and –ion excitation and ionization processes, [502, 503] electron–molecule scattering, [504]-[506] positron–atom and –molecule scattering, [507, 508] atomic and molecular photoionization processes, [160, 509] free–free transitions, [510] atomic and molecular bound state energies and oscillator strengths, [511, 512] atomic polarizabilities, [513] atom–molecule reactive scattering, [514] dissociative attachment and recombination processes [515] and atomic multiphoton processes. [516, 517] Reviews of the theory and applications to atomic and molecular processes have been written by Burke and Robb [518] and Burke and Berrington. [519]

The present chapter is organised as follows. In the next section the basic *R*-matrix theory of atomic and molecular processes is summarised. In the following section some recent applications of this basic theory are presented. Three examples are discussed: (i) low energy electron scattering by ions of Fe; (ii) electron scattering by atoms and ions at intermediate energies and (iii) multiphoton processes.

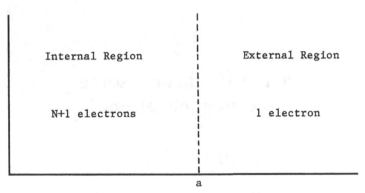

Fig. 13.1. Partitioning of configuration space in R-matrix theory.

13.2   Basic theory

We summarise in this section R-matrix theory describing low energy electron impact excitation of light atoms and ions where relativistic effects can be neglected.* This theory provides the basis for extensions to other atomic and molecular processes, some of which are discussed in the following section.

The basic theory commences by partitioning configuration space into two regions by a sphere of radius $r = a$, where r is the relative coordinate of the scattered electron and the target nucleus, as illustrated in Fig. 13.1. The radius a is chosen so that the charge distribution of the target states of interest are just contained within this radius. In the internal region, electron exchange and correlation effects between the scattered electron and the N–electron target are important and a configuration interaction expansion of the $(N + 1)$–electron complex is appropriate. In the external region, electron exchange between the scattered electron and the target is negligible and the scattered electron moves in the long-range multipole potential of the target. The wave functions in these two regions are related by the R-matrix, or inverse logarithmic derivative, on the boundary.

In order to calculate the R-matrix we must solve the collision problem in the internal region. We start from the time-independent non-relativistic Schrödinger equation

$$H_{N+1}\Psi = E\Psi, \tag{13.1}$$

where E is the total energy of the electron–atom (ion) system and the

---

* The inclusion of relativistic effects using the Dirac R-matrix theory is discussed in Chapter 14.

Hamiltonian H_{N+1} is given by

$$H_{N+1} = \sum_{i=1}^{N+1} \left(-\frac{\nabla_i^2}{2} - \frac{Z}{r_i} \right) + \sum_{i>j=1}^{N+1} \frac{1}{r_{ij}}, \tag{13.2}$$

where Z is the nuclear charge and $r_{ij} =| \mathbf{r}_i - \mathbf{r}_j |$, $\mathbf{r}_i$ and $\mathbf{r}_j$ being the vector coordinates of electrons i and j respectively. The origin of coordinates is taken to be the target nucleus, which is assumed to have infinite mass, and we use atomic units where $\hbar = m_e = |e| = 1$, m_e and e being the mass and charge of the electron.

We now introduce a set of target eigenstates and pseudostates Φ_i together with their corresponding energies E_i^N by the equation

$$\langle \Phi_i | H_N | \Phi_j \rangle = E_i^N \delta_{ij}, \tag{13.3}$$

where H_N is the target Hamiltonian defined by Eq. (13.2) with $(N + 1)$ replaced by N. These target states are usually represented in terms of a set of basis configurations ϕ_j by the expansion

$$\Phi_i = \sum_j \phi_j a_{ji}, \tag{13.4}$$

where the coefficients a_{ji} are determined by diagonalizing H_N as in Eq. (13.3).

In order to solve Eq. (13.1) in the internal region we introduce the Bloch operator [520] L_{N+1} defined by

$$L_{N+1} = \sum_{i=1}^{N+1} \frac{1}{2} \delta \left(r_i - a \right) \left(\frac{d}{dr_i} - \frac{b-1}{r_i} \right). \tag{13.5}$$

This operator ensures that $H_{N+1} + L_{N+1}$ is Hermitian over the internal region by cancelling the surface terms which arise from the kinetic energy operators in H_{N+1}. The constant b can be chosen arbitrarily and is usually taken to be zero. Eq (13.1) can then be formally solved, by adding $L_{N+1}\Psi$ to each side, to yield the equation

$$\Psi = (H_{N+1} + L_{N+1} - E)^{-1} L_{N+1}\Psi. \tag{13.6}$$

The operator $(H_{N+1} + L_{N+1} - E)^{-1}$ in this equation can be expanded in terms of the basis functions

$$\Psi_k^\Gamma (\mathbf{X}_{N+1}) = A \sum_{ij} \overline{\Phi}_i^\Gamma (\mathbf{x}_1, \cdots, \mathbf{x}_N ; \hat{\mathbf{r}}_{N+1}, \sigma_{N+1}) r_{N+1}^{-1} u_j (r_{N+1}) a_{ijk}^\Gamma$$

$$+ \sum_i \chi_i^\Gamma (\mathbf{X}_{N+1}) b_{ik}^\Gamma. \tag{13.7}$$

Here $\mathbf{X}_{N+1} \equiv \mathbf{x}_1, \cdots, \mathbf{x}_{N+1}$ where $\mathbf{x}_i \equiv (\mathbf{r}_i, \sigma_i)$ is the space and spin coordinates of the ith electron. The $\overline{\Phi}_i^\Gamma$ are channel functions formed by coupling the target states Φ_i included in the calculation to the spin-angle functions of the scattered electron to give functions belonging to the conserved quantum numbers

$$\Gamma \equiv \gamma L S M_L M_S \Pi, \tag{13.8}$$

where L and S are the total orbital and spin angular momenta, M_L and M_S are their z components, Π is the total parity and γ represents any other quantum numbers necessary to define the channel. The u_j are radial basis functions describing the scattered electron. These functions are non-zero on the boundary of the internal region and provide the link to external region solutions through the R-matrix, as described below. The χ_i^Γ are quadratically integrable functions which, like the target states Φ_i, vanish by the boundary of the internal region. They are usually formed from the same orbital basis as the Φ_i. The operator $\mathbf{A}$ antisymmetrises the first expansion in accordance with the Pauli exclusion principle. Finally the expansion coefficients a_{ijk}^Γ and b_{ik}^Γ are determined by diagonalizing $H_{N+1} + L_{N+1}$ in the basis Ψ_k^Γ giving

$$\left\langle \Psi_k^\Gamma \left| H_{N+1} + L_{N+1} \right| \Psi_{k'}^\Gamma \right\rangle = E_k^\Gamma \, \delta_{kk'}, \tag{13.9}$$

where the radial integrals in this equation are confined to the internal region.

We can now rewrite Eq. (13.6) in terms of the basis Ψ_k^Γ as

$$| \Psi^\Gamma \rangle = \sum_k \frac{| \Psi_k^\Gamma \rangle \langle \Psi_k^\Gamma |}{E_k^\Gamma - E} L_{N+1} | \Psi^\Gamma \rangle. \tag{13.10}$$

Projecting this equation onto the channel functions $\overline{\Phi}_i^\Gamma$ and evaluating on the boundary $r = a$ of the internal region yields

$$F_i^\Gamma (a) = \sum_j R_{ij}^\Gamma (E) \left(a \frac{dF_j^\Gamma}{dr} - b F_j^\Gamma \right)_{r=a}, \tag{13.11}$$

where we have defined the R-matrix by

$$R_{ij}^\Gamma (E) = \frac{1}{2a} \sum_k \frac{w_{ik}^\Gamma \, w_{jk}^\Gamma}{E_k^\Gamma - E}, \tag{13.12}$$

the reduced radial wave functions by

$$a^{-1} F_i^\Gamma (a) = \left\langle \overline{\Phi}_i^\Gamma \left| \Psi^\Gamma \right\rangle \right._{r_{N+1}=a}, \tag{13.13}$$

and the surface amplitudes by

$$a^{-1} w_{ik}^\Gamma = \left\langle \overline{\Phi}_i^\Gamma \left| \Psi_k^\Gamma \right\rangle \right._{r_{N+1}=a}. \tag{13.14}$$

Eqs. (13.11) and (13.12) are the basic equations describing the solution of the scattering problem in the internal region. We see that the R-matrix is obtained at all energies by a single diagonalization of $H_{N+1} + L_{N+1}$ for each set of conserved quantum numbers Γ.

It is appropriate at this point to consider the choice of the radial basis functions u_j in Eq. (13.7). In principle, members of any linearly independent set of functions satisfying appropriate boundary conditions at $r = 0$ and $r = a$ could be adopted. However a careful choice will enable the convergence of the first expansion in Eq. (13.7) to be more rapid. A basis which has been used in many electron–atom and electron–ion scattering calculations is obtained by solving the following zero-order radial equation

$$\left(\frac{d^2}{dr^2} - \frac{\ell(\ell+1)}{r^2} + V(r) + k_i^2 \right) u_i(r) = \sum_j \lambda_{ij} P_j(r), \qquad (13.15)$$

subject to the homogeneous boundary conditions

$$u_i(0) = 0, \quad \left(\frac{a}{u_i} \right) \left. \frac{du_i}{dr} \right|_{r=a} = b. \qquad (13.16)$$

The Lagrange multipliers λ_{ij} are chosen so that the orthogonality conditions

$$\int_0^a u_i(r) \, P_j(r) \, dr = 0 \,, \text{all } i \text{ and } j, \qquad (13.17)$$

are satisfied, where the $P_j(r)$ are the reduced radial parts of the occupied orbitals of the target. Finally, the potential $V(r)$ is usually chosen to be the spherical part of the static potential of the target ground state. A straightforward extension of this approach has also been developed for electron–molecule collision calculations. [521]

Since the basis functions, defined by Eqs. (13.15)-(13.17), satisfy homogeneous boundary conditions at $r = a$, a correction, first proposed by Buttle, [522] must be included to allow for the high lying omitted poles in the R-matrix expansion of Eq. (13.12). On the other hand, the fact that R-matrix theory can be derived from a variational principle [523]-[525] has been used by Greene [526] to develop an approach which enables continuum basis functions satisfying arbitrary boundary conditions at $r = a$ to be used. This obviates the need to include a correction to the R-matrix and gives rapid convergence for low energy collisions.

The final step in the calculation is to solve the collision problem in the external region. Since the radius a is chosen so that electron exchange and correlation with the target is negligible in this region we can expand the total wavefunction in an unsymmetrised form consistent with the expansion of Eq. (13.7), by writing

$$\Psi^{\Gamma}(\mathbf{X}_{N+1}) = \sum_i \overline{\Phi}_i^{\Gamma}(\mathbf{x}_1, \cdots, \mathbf{x}_N; \hat{\mathbf{r}}_{N+1}, \sigma_{N+1}) r_{N+1}^{-1} F_i^{\Gamma}(r_{N+1}), \qquad (13.18)$$

where the $\overline{\Phi}_i^{\Gamma}$ are the same channel functions retained in the expansion of Eq. (13.7) and the F_i^{Γ} correspond to the reduced radial wave functions introduced in Eq. (13.13). Substituting the expansion of Eq. (13.18) into the Schrödinger equation (13.1) and projecting onto the channel functions $\overline{\Phi}_i^{\Gamma}$ yields the following coupled differential equations satisfied by the functions F_i^{Γ}

$$\left(\frac{d^2}{dr^2} - \frac{\ell_i(\ell_i+1)}{r^2} + \frac{2(Z-N)}{r} + k_i^2 \right) F_i^{\Gamma}(r) = 2\sum_{j=1}^n V_{ij}^{\Gamma}(r) F_j^{\Gamma}(r),$$

$$i = 1, \cdots, n, \ r \geq a, \qquad (13.19)$$

where $k_i^2 = 2(E - E_i^N)$ and we have assumed that n channel functions have been retained in the expansions of Eqs. (13.7) and (13.18). The potential matrix V_{ij}^{Γ} has a particularly simple form being expressible in inverse powers of r, the leading term in r^{-2} giving rise in second-order to the long-range polarizability. Eq. (13.19) can be integrated outwards, subject to the R-matrix boundary condition of Eq. (13.11) at $r = a$ using one of a number of standard techniques [514, 527, 528] and then fitted to an asymptotic expansion. Usually the following K-matrix asymptotic boundary conditions are adopted

$$\lim_{r\to\infty} F_{ij}^{\Gamma}(r) \sim k_i^{-\frac{1}{2}} \left(\sin\theta_i \delta_{ij} + \cos\theta_i K_{ij}^{\Gamma} \right), \text{ open channels,}$$

$$(13.20)$$

$$\lim_{r\to\infty} F_{ij}^{\Gamma}(r) \sim 0, \text{ closed channels,}$$

where the second index j on F_{ij}^{Γ} distinguishes the n_a linearly independent solutions of Eq. (13.19), where n_a is the number of open channels, and where $k_i^2 \geq 0$ and $\theta_i = k_i r - \frac{1}{2}\ell_i\pi$ for a neutral target. In this way we can determine the $n_a \times n_a$ real symmetric K-matrix. The $n_a \times n_a$ unitary symmetric S-matrix is then given by the matrix equation

$$\underline{S}^{\Gamma} = \frac{1 + i\underline{K}^{\Gamma}}{1 - i\underline{K}^{\Gamma}}. \qquad (13.21)$$

The scattering observables can be obtained from the S-matrix elements. In particular, the cross section for a transition from an initial target state denoted by the quantum numbers $\gamma_i L_i S_i$ to a final target state denoted by

$\gamma_j L_j S_j$ is given by

$$\sigma(i \to j) = \frac{\pi}{k_i^2} \sum_{\substack{\ell_i \ell_j \\ LS\Pi}} \frac{(2L+1)(2S+1)}{2(2L_i+1)(2S_i+1)} \left| S_{ij}^{\Gamma} - \delta_{ij} \right|^2, \qquad (13.22)$$

in units of a_0^2, where γ_i and γ_j represent the quantum numbers, in addition to the orbital and spin angular momenta, necessary to completely define the initial and final states of the target. In addition we can introduce a collision strength Ω_{ij} which is defined in terms of the cross section by

$$\Omega_{ij} = k_i^2 (2L_i+1)(2S_i+1) \sigma(i \to j). \qquad (13.23)$$

It follows from the symmetry of the S-matrix that the collision strength is symmetric, $\Omega_{ij} = \Omega_{ji}$. It can also be seen from Eqs. (13.22) and (13.23) that it is dimensionless.

It follows from the above discussion that while the R-matrix is determined at all energies by a single diagonalization, the coupled equations of Eq. (13.19) must be solved for $r \geq a$ to yield the solution F_{ij}^{Γ} and hence the K-matrix, S-matrix and cross sections for each energy of interest. Although the coupled equations in the external region have a very simple form the necessity to solve them at each energy often means in practice that the time taken for the calculation in the external region exceeds that taken in the internal region.

13.3 Recent applications

In this section we consider three recent applications of the above theory.

13.3.1 Electron scattering by ions of Fe

Lines of low ionization stage ions of Fe, such as Fe^+ and Fe^{2+}, are present in many astrophysical spectra and this has motivated several major calculations of electron impact excitation cross sections and rate coefficients. However these ions have a large number of closely spaced low lying energy levels which makes accurate calculations very extensive. For example, in the case of Fe^+, we show in Table 13.1 that the lowest six configurations

$$3d^6 4s, \; 3d^7, \; 3d^5 4s^2, \; 3d^6 4p, \; 3d^5 4s 4p, \; 3d^5 4p^2, \qquad (13.24)$$

which can be expected to be important in low energy collisions, have a total of 446 energy levels in LS coupling. Furthermore, these levels are strongly split by the fine-structure interaction. Some of the lowest lying levels, illustrating the complexity of the spectra, are shown in Fig. 13.2. These complex spectra make realistic collision calculations very difficult

Table 13.1. Number of target states belonging to the first six configurations of Fe^+ in *LS*-coupling

Configuration	Target spin				Totals
	1/2	3/2	5/2	7/2	
$3d^64s$	15	8	1	0	24
$3d^7$	6	2	0	0	8
$3d^54s^2$	11	4	1	0	16
$3d^64p$	41	24	3	0	68
$3d^54s4p$	74	56	14	1	145
$3d^54p^2$	103	66	15	1	185
Totals	250	160	34	2	446

Table 13.2. Channels coupled to the configurations listed in Table 13.1 for each *LS* and Π

Parity Π	Total Spin S				
	0	1	2	3	4
$(-1)^L$	891	1413	615	96	3
$(-1)^{L+1}$	873	1401	615	88	1

to perform, as illustrated in Table 13.2, which gives the channels coupled to the above configurations for each *L*, *S* and Π combination. If we assume that a minimum of 15 continuum basis functions in each channel are required to obtain convergence, then Hamiltonian matrices of order more than 20,000 must be diagonalized in the absence of fine-structure effects.

A further computational difficulty presents itself in the case of ions of Fe. This is the very large number and complexity of the angular integrals which have to be calculated in setting-up the Hamiltonian matrix elements. In the computer program package currently used, the angular integrals for each matrix element are calculated individually using the method of Fano [529] which has been programmed by Burke [530] and Hibbert. [531] However it is clear that for target states defined by the configurations given by Eq. (13.24) this procedure involves considerable duplication. A new algorithm and computer program package [532]-[534] has therefore

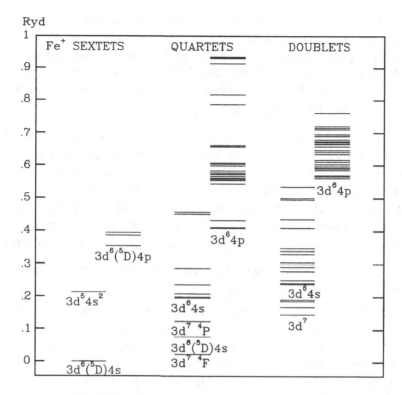

Fig. 13.2.   Lowest lying energy levels of Fe⁺.

bccn dcvclopcd which is cnabling these angular integrals to be evaluated much more efficiently.

In view of the difficulties in performing realistic coupled channel calculations, the earliest e^-–Fe⁺ studies used the distorted-wave approximation (Ref. [535, 536]). The first R-matrix calculation, in which the lowest four terms of the Fe⁺ ion were included, was performed in LS coupling by Baluja *et al.* [537] This was extended by Berrington *et al.*, [538] who included relativistic effects[†] using the Breit-Pauli Hamiltonian, [539, 540] to obtain the fine-structure collision strengths between the lowest 16 fine-structure levels. The collision strengths showed considerable resonant enhancement which was not found in the earlier distorted-wave studies.

Recently, the availability of supercomputers such as the Cray Y-MP has enabled Pradhan and Berrington [541] to include many more target states of Fe⁺. This work calculates collision data for more transitions and examines the effect of truncating the close coupling expansion in the earlier

---

[†] The inclusion of relativistic effects using the Dirac R-matrix theory is discussed in Chapter 14.

16-level (4 *LS* states) calculation on the results for transitions between the lowest lying states. Two calculations were carried out: one with 38 states in *LS* coupling including all $3d^7$, $3d^64s$ and $3d^64p$ quartet and sextet spin states, yielding collision strengths for 703 transitions between *LS* states; the second using the Breit-Pauli Hamiltonian with 41 fine-structure levels belonging to the 10 quartet and sextet states of the form $3d^6\left(^5D\right)4s$, $3d^6\left(^5D\right)4p$, and $3d^7$, giving a total of 820 fine-structure transitions. These new calculations are the first to include the coupling between the low lying even parity states and the higher lying odd parity $4p$ states. The results show that the extra coupling and resonance structures included have a dramatic effect on the energy dependence of the collision strengths even for transitions between low lying levels. This is illustrated in Fig. 13.3 which shows the forbidden transitions $a\left(^6D\right) \to a\left(^4F\right)$, $a\left(^6D\right) \to a\left(^4D\right)$ and $a\left(^6D\right) \to a\left(^4P\right)$ compared with the distorted wave results of Nussbaumer *et al.* [535, 536] at the three energies (0.15, 0.20 and 0.25 Ryd) where they tabulated their results. The extensive resonance structures, particularly close to threshold, make a significant enhancement to the low temperature rate coefficients even though the non-resonant background appears to be lower than the distorted wave results.

It is expected that these results for Fe^+ and other ions of Fe will be progressively extended as these new computer programs are developed and as the power of supercomputers increases. An international collaboration, called the "Iron Project" collaboration, has recently been formed to carry out these calculations in response to the many demands from astronomers for accurate atomic collision data.

13.3.2   *Electron scattering at intermediate energies*

The expansion of Eq. (13.7) can give accurate cross sections at low electron impact energies where only a finite number of channels are open and the corresponding channel functions can all be included explicitly in this expansion. However at intermediate energies, defined to be close to and above the ionization threshold, where an infinite number of channels including ionizing channels are open, this expansion needs to be extended. Although the inclusion of pseudostates in this expansion can approximately allow for the loss of flux into the ionizing channels [542, 543] there is considerable difficulty in choosing a set of pseudostates which give convergent results at intermediate energies and also in choosing a pseudostate basis for complex targets.

Recently a new intermediate energy *R*-matrix (IERM) approach has been introduced by Burke *et al.* [544] which enables a complete set of

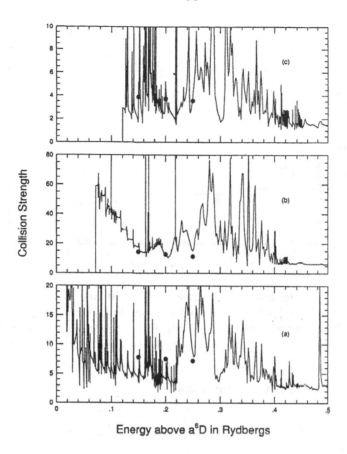

Fig. 13.3. e^-–Fe^+ collision strengths for excitation of the a^6D ground state to the metastable states: (a) a^4F, (b) a^4D and (c) a^4P. The full circles are distorted wave results [535, 536] and the full lines are 38-state R-matrix results [541] (Figure 1 from Pradhan and Berrington [541]).

pseudostates to be defined for arbitrary complex targets. As an example we first consider the scattering of electrons by atomic hydrogen. In this case, the R-matrix basis states defined in the internal region are expanded in terms of two-electron basis functions where the target electron and the scattered electron are both represented by the same one-electron basis consisting of bound hydrogen orbitals together with members of a complete set of continuum orbitals $u_{n\ell}$. The R-matrix expansion replacing the expansion of Eq. (13.7) can then be written, for each L, S and Π symmetry, as

$$\Psi_k^{LS\Pi}(\mathbf{r}_1, \mathbf{r}_2) = \left[1 + (-1)^S P_{12}\right] \sum_{n_1=1}^{n_{1,\max}} \sum_{n_2=1}^{n_{2,\max}} \sum_{\ell_1=0}^{\ell_{1,\max}} \sum_{\ell_2=0}^{\ell_{2,\max}} \qquad (13.25)$$
$$\times r_1^{-1} u_{n_1\ell_1}(r_1) r_2^{-1} u_{n_2\ell_2}(r_2) Y_{\ell_1\ell_2 LM_L}(\hat{\mathbf{r}}_1, \hat{\mathbf{r}}_2) a_{n_1 n_2 \ell_1 \ell_2 k}^{LS\Pi},$$

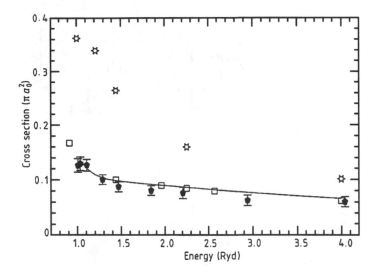

Fig. 13.4. e^-–H integrated $1s$–$2s$ cross sections at intermediate energies. Full curve, IERM results [545]; open squares, pseudostate results [547]; six point stars, $1s$–$2s$–$2p$ close coupling results [548]; full pentagons, experiment [549, 550]. Error bars represent two standard deviations. (Figure 8 from Scott *et al.* [545]).

where the $u_{n\ell}(r)$ satisfy the equation

$$\left(\frac{d^2}{dr^2} - \frac{\ell(\ell+1)}{r^2} + V(r) + k_{n\ell}^2\right) u_{n\ell}(r) = \sum_{n'=1}^{n_{\text{bound}}} \lambda_{n\ell n'\ell}\, u_{n'\ell}(r), \qquad (13.26)$$

subject to the boundary conditions

$$u_{n\ell}(0) = 0, \qquad \frac{du_{n\ell}}{dr}\Big|_{r=a} = 0. \qquad (13.27)$$

The radius a of the internal region is chosen so that all the orbitals with $n \leq n_{\text{bound}}$, which correspond to the physical target's bound orbitals of interest in the calculation, have effectively vanished by the boundary of the region, while the remaining orbitals $n > n_{\text{bound}}$ represent the continuum and are non-zero on this boundary. The Lagrange multipliers $\lambda_{n\ell n'\ell}$ ensure that all the $u_{n\ell}(r)$ are orthogonal over the internal region. The $Y_{\ell_1\ell_2 LM_L}$ in Eq. (13.25) are angular functions which are eigenstates of L^2, ℓ_1^2 and ℓ_2^2 while P_{12} is the space-exchange operator which ensures that eigenfunctions with $S = 0$ are symmetric and those with $S = 1$ are antisymmetric with respect to exchange of their space coordinates.

The first application of this approach [222, 427, 545] was to obtain $1s$–$1s$, $1s$–$2s$ and $1s$–$2p$ cross sections for e^-–H scattering at intermediate energies. The radius a of the internal region was chosen to be 25 a.u., n_{bound} was set equal to 2 and both $n_{1,\text{max}}$ and $n_{2,\text{max}}$ were chosen to be 30.

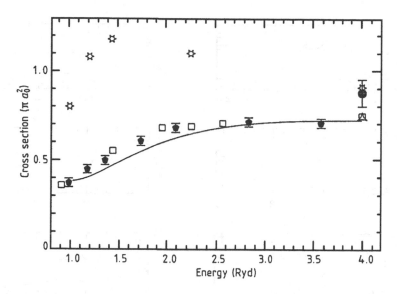

Fig. 13.5. e^-–H integrated $1s$–$2p$ cross section at intermediate energies. Symbols in common with Fig. 13.4 have the same meaning except the full pentagons which gives the data of Long *et al.* [550] normalised to theory at 200 eV; full circle experiment [551]; three point star at 4 Ryd, pseudostate result [552] (Figure 9 from Scott *et al.* [545]).

The main limitation was on the number of (ℓ_1, ℓ_2) combinations which could be included in order that the dimension of the Hamiltonian matrix was ≤ 3500, the limit imposed by the computer available to these workers. A further approximation was to only include the $1s$, $2s$ and $2p$ channels in expansion of Eq. (13.18) in the external region while neglecting all the "continuum channels" in this region. This lead to pseudo-resonance structure in the T-matrix elements which had to be averaged over to yield the physical T-matrix elements. [546]

We show in Figs. 13.4 and 13.5 the integrated $1s$–$2s$ and $1s$–$2p$ excitation cross sections respectively, calculated using this IERM approach, compared with pseudostate calculations of Callaway *et al.* [547] and with $1s$–$2s$–$2p$ close coupling calculations. [548] It is seen that the IERM results are in good agreement with experiment, also shown in these figures, and with the pseudostate results but that the $1s$–$2s$–$2p$ calculations considerably overestimate the cross sections at low energies. The latter calculations do not include coupling with the continuum and hence do not allow for loss of flux into the continuum channels which is important close to the ionization threshold. These results indicate that the IERM theory and the pseudostate theory can give reliable results at intermediate energies at least for these transitions.

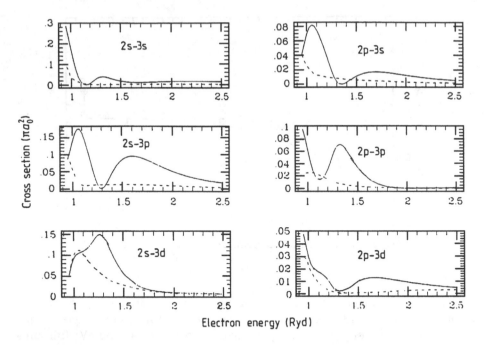

Fig. 13.6. e^-–H $n=2$ to $n=3$ excitation cross sections in the $^1S^e$ state. Full curve, six state R-matrix results; dashed curve, reduced IERM results (Figure 2 from Scott and Burke [553]).

Recently, the IERM approach has been extended to treat transitions in e^-–H scattering involving the $n = 3$ levels. [553] In this case a boundary radius $a = 38$ a.u. is necessary to fully enclose the $n = 3$ orbitals. However, in order to reduce the size of the calculation $n_{1,\text{max}}$ was set equal to 35, which enabled the calculation to span the energy range of interest, and the convergence of the result for smaller values of $n_{2,\text{max}}$ was tested. It was found that the results were little changed when $n_{2,\text{max}}$ was increased beyond 10 and hence a series of calculations were carried out with $n_{2,\text{max}}$ set equal to 12. Also $\ell_{1,\text{max}}$ and $\ell_{2,\text{max}}$ were both set equal to 3 when $L = 0$. We show in Fig. 13.6 the cross section for the six $n = 2$ to $n = 3$ transitions in the $^1S^e$ state for this reduced IERM calculation compared with six state results obtained by including just the $1s$, $2s$, $2p$, $3s$, $3p$ and $3d$ states in a close coupling expansion. We see that all the cross sections are considerably reduced in magnitude from the six state results. These calculations are now being extended to all partial waves of importance at these energies in order to obtain reliable total cross sections for use in applications. [554]

The extension of this approach to calculate transitions for states with higher values of n rapidly becomes prohibitive computationally since the

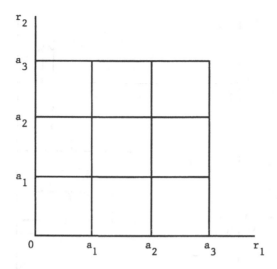

Fig. 13.7. Partitioning of (r_1, r_2) plane in the 2-dimensional R-matrix propagator approach.

computer time goes as n^{12}. A more appropriate approach for high n is to adopt a two-dimensional R-matrix propagator [555] in which the region $0 \leq r_1 \leq u$, $0 \leq r_2 \leq u$ is subdivided into a number of small rectangular subregions, as illustrated in Fig. 13.7. In each subregion an R-matrix expansion analogous to Eq. (13.25) is adopted. The Hamiltonian is then diagonalized in each subregion in this basis and the R-matrix on the boundaries of these subregions propagated out from $r_1 = 0$ and $r_2 = 0$ to $r_1 = a$ and $r_2 = a$. The advantage of this approach is that the number of basis functions which need to be retained in each subregion is proportional to the area of the subregion. Hence the size of the Hamiltonian matrices which need to be diagonalized is much reduced from that required if a single region is adopted. On the other hand, for each energy of interest, the R-matrix must be propagated across the subregions to the boundary $r_1 = a$ and $r_2 = a$.

Recently, this 2-dimensional propagator approach has been applied by LeDourneuf *et al.* [555] to the e^-–H s-wave model where only the $\ell_1 = 0$ and $\ell_2 = 0$ terms are retained in the expansion of Eq. (13.25). Although this model problem is much reduced in size from the physical e^-–H scattering problem it has many of its features including a discrete infinity of thresholds converging to the ionization continuum. In this work, the R-matrix was propagated out to 200 a.u. using subregions of dimension 10 a.u. × 10 a.u. giving converged results for high n values. We show in Fig. 13.8, the calculated $ns \rightarrow (n+1)s$ collision strengths for $n = 1, \cdots, 4$ for energies up to the ionization threshold. The results, which

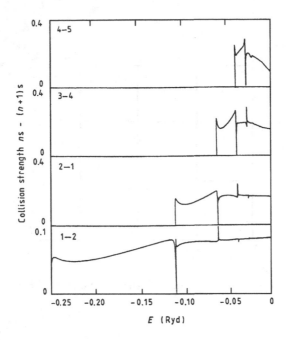

Fig. 13.8. e^-–H $ns \to (n+1)s$ collision strengths below the ionization threshold using the 2 dimensional propagator for the s-wave model (Figure 2 from LeDourneuf *et al.* [555]).

show evidence of narrow resonances near the excitation thresholds, have converged in this energy range. Excitation and ionization cross sections have also been calculated above the ionization threshold for $n \leq 5$ showing that this approach is capable of yielding accurate excitation and ionization cross sections for high Rydberg states at intermediate energies. This work is now being extended to include summations over ℓ_1 and ℓ_2 enabling physically meaningful e^-–H cross sections to be calculated.

The extension of the IERM theory to complex atoms and ions has been described by Burke *et al.* [544] and the development of a general computer program is underway. [534]

13.3.3 Multiphoton processes

Recently a unified R-matrix-Floquet theory of multiphoton ionization and laser-assisted electron scattering for a general atom or ion has been introduced. [516, 517] This theory is non-perturbative and can be applied to an arbitrary atom or ion. It takes advantage of the R-matrix division of configuration space into internal and external regions, as illustrated in Fig. 13.9, to treat the interaction between the laser field and the atomic system in the most appropriate gauge in each region. It also

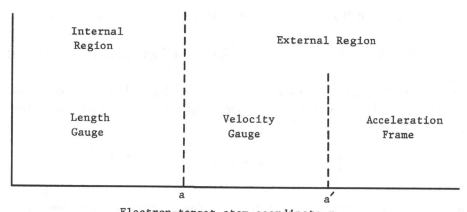

Fig. 13.9. Partitioning of configuration space in *R*-matrix Floquet theory of multiphoton processes.

allows computer program packages developed for the field-free case to be modified in a straightforward way to treat this problem.

In the non-relativistic limit an atomic or ionic system composed of a nucleus of atomic number Z and $N + 1$ electrons in an intense laser field is described by the time-dependent Schrödinger equation

$$i\frac{\partial}{\partial t}\Psi(\mathbf{X}_{N+1}, t) = \left(H_{N+1} + \frac{1}{c}\sum_{i=1}^{N+1}\mathbf{A}(t)\cdot\mathbf{p}_i\right.$$
$$\left. + \frac{N+1}{c^2}\mathbf{A}^2(t)\right)\Psi(\mathbf{X}_{N+1}, t), \quad (13.28)$$

where H_{N+1} has been defined by Eq. (13.2) and where the Coulomb gauge for the electromagnetic field has been adopted so that the vector potential satisfies $\nabla\cdot\mathbf{A} = 0$. The laser field is assumed to be monochromatic, monomode, linearly polarized and spacially homogeneous (*i.e.* its wavelength is large compared with the size of the atom). Hence

$$\mathbf{A}(t) = \hat{\epsilon}\,A_0\sin\omega t, \quad (13.29)$$

where $\hat{\epsilon}$ is the polarization vector and where A_0 is related to E_0, the electric field strength, and ω, the angular frequency, by $A_0 = -cE_0/\omega$.

In the internal region, Eq. (13.28) is transformed to the dipole length gauge, defined by the unitary transformation

$$\Psi(\mathbf{X}_{N+1}, t) = \exp\left(-\frac{i}{c}\sum_{i=1}^{N+1}\mathbf{A}(t)\cdot\mathbf{r}_i\right)\Psi_L(\mathbf{X}_{N+1}, t). \quad (13.30)$$

The wave function $\Psi_L(\mathbf{X}_{N+1}, t)$ then satisfies the Schrödinger equation

$$i\frac{\partial}{\partial t}\Psi_L(\mathbf{X}_{N+1}, t) = \left(H_{N+1} + \sum_{i=1}^{N+1}\mathbf{E}(t)\cdot\mathbf{r}_i\right)\Psi_L(\mathbf{X}_{N+1}, t), \qquad (13.31)$$

where $\mathbf{E}(t) = \hat{\epsilon}E_0\cos\omega t$. In the external region, Eq. (13.28) is transformed to the dipole velocity gauge for the ejected electron, defined by the unitary transformation

$$\Psi(\mathbf{X}_{N+1}, t) = \exp\left(-\frac{i}{c}\sum_{i=1}^{N}\mathbf{A}(t)\cdot\mathbf{r}_i - \frac{i}{2c^2}\int^t\mathbf{A}^2(t')\,dt'\right)\Psi_V(\mathbf{X}_{N+1}, t).$$
$$(13.32)$$

The wave function $\Psi_V(\mathbf{X}_{N+1}, t)$ then satisfies the Schrödinger equation

$$i\frac{\partial}{\partial t}\Psi_V(\mathbf{X}_{N+1}, t) = \left(H_{N+1} + \frac{1}{c}\mathbf{A}(t)\cdot\mathbf{p}_{N+1} + \sum_{i=1}^{N}\mathbf{E}(t)\cdot\mathbf{r}_i\right)\Psi_V(\mathbf{X}_{N+1}, t). \quad (13.33)$$

If laser assisted electron–atom scattering cross sections are required the wave function is transformed to the acceleration frame at a very large value of $r = a'$, the electron–target coordinate.

In both regions the Floquet-Fourier expansion

$$\Psi_R(\mathbf{X}_{N+1}, t) = e^{-iE_Rt}\sum_{n=-\infty}^{\infty}e^{in\omega t}\Psi_n^R(\mathbf{X}_{N+1}), R = L \text{ or } V \qquad (13.34)$$

is adopted. Substituting this equation into Eqs. (13.31) or (13.33) then yields an infinite set of coupled equations for the $\Psi_n^R(\mathbf{X}_{N+1})$. In order to solve these equations, $\Psi_n^L(\mathbf{X}_{N+1})$ is expanded in the internal region in terms of an R-matrix basis defined by Eq. (13.7). The coefficients in this expansion are determined by diagonalizing the resultant Floquet Hamiltonian from which the R-matrix in the length gauge on the boundary of the internal region is obtained. This R-matrix is then transformed from the length gauge to the velocity gauge using Eqs. (13.31) and (13.32). The transformed R-matrix then provides the boundary condition for integrating the coupled equations in the velocity gauge in the external region using a modification of the Light-Walker propagator method. [514] The final step in the case of multiphoton ionization is to match the solution in the velocity gauge at a large radius to outgoing wave boundary conditions, rather than to the K-matrix boundary conditions given by Eq. (13.20) which are appropriate to scattering. The multiphoton ionization rate is then proportional to the negative imaginary part of the resultant complex energy eigenvalue. A general computer program package has been written which implements this theory en-

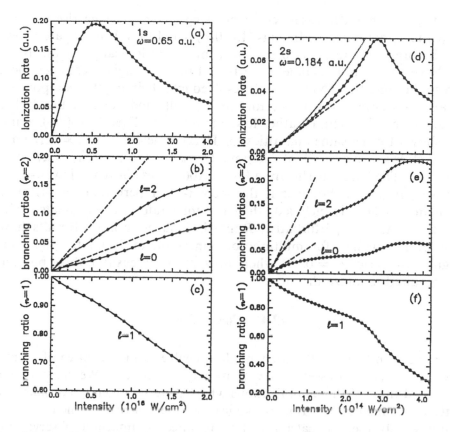

Fig. 13.10. Total multiphoton ionization rates and branching ratios into the dominant ionization channels (the ATI channels $n = 1$ and $n = 2$) for (a)-(c) H(1s) in a field of angular frequency $\omega = 0.65$ a.u. and (d)-(f) H(2s) in a field of angular frequency $\omega = 0.184$ a.u. both versus intensity. Perturbation theory results are given by the dashed lines (Figure 1 from Dörr *et al.* [556]).

abling multiphoton ionization rates to be calculated for a general atom or ion.

Using the above theory and computer program package, multiphoton ionization rates have recently been obtained for H, H$^-$ and He. [556, 557] In the case of H, calculations were carried out for ionization both from its 1s ground state and from its metastable 2s state. Results were first obtained in the high frequency regime where the photon energy ω is larger than the binding energy of the initial state. In Fig. 13.10 (a-c) results are shown versus laser intensity for H(1s) in a field of angular frequency $\omega = 0.65$ a.u. In Fig. 13.10(a), which shows the total ionization rate, perturbative behaviour (linear with intensity) is observed at low intensity. However as the intensity is increased a peak is observed at about 10^{16}W cm^{-2}

and then the rate decreases with increasing intensity giving rise to "high frequency stabilization." In Figs. 13.10(b) and 13.10(c) results are shown for the branching ratios into the most important ionization channels, *i.e.* the lowest ionization channel ($n = 1$) and the next highest channel above threshold ionization (ATI) ($n = 2$) resolved into their angular components. In all cases deviation from perturbation results, indicated by the dashed lines, can be observed at rather low intensities. In Figs. 13.10(d-f) results are shown for the same quantities for H(2s) in field where $\omega = 0.184$ a.u. (the KrF laser wavelength of 248 nm). The same stabilization behaviour is observed. The total rate shown in Fig. 13.10(d) rises above the lowest-order perturbative result, however if the second-order perturbative correction (thin dotted line) is included, then the total rate is overestimated with respect to the nonperturbative *R*-matrix-Floquet results. Results have also been obtained for multiphoton ionization of H(1s) for the same angular frequency $\omega = 0.184$ where now, at low intensities, three photons are required to ionize the atom.

13.4   Conclusions

We have seen that the *R*-matrix method enables a broad range of atomic and molecular processes to be accurately calculated. We have briefly discussed three of these applications where there is currently considerable activity. Computer program packages have been written which are now being widely used to calculate data required to interpret observations particularly of astronomical objects and laboratory plasmas. Work is now underway to develop new versions of these programs which can take advantage of the currently emerging generation of massively parallel computers which are expected to be orders of magnitude faster than the current vector processors. In addition, the need to make these programs more user friendly so that they can be easily used by scientists in other fields has recently been recognised. [558] Thus, in conclusion, we can expect that the *R*-matrix method will continue to play an important role in future research in atomic and molecular physics.

14

Electron scattering from atomic targets: application of Dirac R-matrix theory

Wasantha Wijesundera, Ian Grant, and Patrick Norrington

14.1 Introduction

The Dirac R-matrix method is a relativistic generalization of the usual (nonrelativistic) R-matrix method as described by Burke *et al.*, [502] Burke and Robb, [518] and by Burke in the previous chapter. A nonperturbative relativistic treatment of electron-ion (atom) scattering is expected to be economical and reliable when the target ion (atom) has atomic number Z greater than about 30.

A preliminary computer program for the Dirac R-matrix method for electron-ion scattering has been described by Chang. [559, 560] Numerous enhancements and corrections to this program yielded the first version of the JJMTRX package. The JJMTRX suite of programs has been applied to the calculation of collision parameters for electron scattering by Ne II [561] and Fe XXIII and Fe VII [562] targets. JJMTRX has recently been revised extensively with an emphasis on the efficient use of computational resources for large collision problems.

14.2 Theory

We will discuss the Dirac R-matrix theory briefly. A more detailed description may be found elsewhere. [559, 562] Hartree atomic units are used throughout this chapter.

A relativistic many-body Hamiltonian, H^R, may be constructed from one-body Dirac operators $H^D(i)$, and two-body operators — here taken to be those due to the Coulomb interaction — $H^C(ij)$,

$$H^R = \sum_i H^D(i) + \sum_{i \neq j} H^C(ij). \qquad (14.1)$$

The operator H^R commutes with the ionic angular momentum operator,

J, the *z* component of this operator, J_z, and the ionic parity operator
$\hat{\Pi}$. Configuration state functions (CSFs) are eigenfunctions of the latter
three operators with eigenvalues $J(J+1)$, M, and Π, respectively. They
are linear combinations of Slater determinants of relativistic orbitals,[*]

$$\phi(\mathbf{r}) = \frac{1}{r}\begin{pmatrix} P(r)\,\chi_{\kappa m}(\mathbf{r}/r) \\ i\,Q(r)\,\chi_{-\kappa m}(\mathbf{r}/r) \end{pmatrix}. \tag{14.2}$$

Here $P(r)$ and $Q(r)$ are, respectively, the large- and small-component
radial wavefunctions; κ is the relativistic angular quantum number: $\kappa = \pm(j+1/2)$ for $\ell = j \pm 1/2$; the two components have the spin orbit
functions $\chi_{\kappa m}(\mathbf{r}/r)$ with orbital quantum numbers $\ell = j \pm 1/2$; and m is
the *z* component of the total orbital angular momentum $j(j+1)$.

Approximate many-body wavefunctions may be constructed as linear
combinations of CSFs. When the orbital radial functions are independent of this configurational expansion, this procedure is known as the
configuration-interaction (CI) approach.

The *R*-matrix approach to scattering involves partitioning configuration
space into two regions by a spherical surface centered on the target. Its
radius, *a*, known as the *R*-matrix boundary, is chosen so that exchange
interactions between the scattering electron and the *N*-electron target are
negligible in the outer region $(r > a)$.

14.2.1 Inner region

In the inner region $(r < a)$, the state of the $(N+1)$-electron system is
approximated by a CI expansion. The set of continuum orbitals available
for this expansion is augmented by basis functions obtained from the
solution of radial Dirac equations of the form

$$\begin{aligned} \frac{dq_i}{dr} - \frac{\kappa}{r}q_i + \frac{1}{c}(\varepsilon_i - V(r))p_i &= \frac{1}{c}\sum_l \lambda_{il}\bar{P}_l \\[2mm] -\frac{dp_i}{dr} - \frac{\kappa}{r}p_i + \frac{1}{c}(2c^2 + \varepsilon_i - V(r))q_i &= \frac{1}{c}\sum_l \lambda_{il}\bar{Q}_l, \end{aligned} \tag{14.3}$$

subject to the usual relativistic boundary conditions at the origin $(r = 0)$
(for instance, Grant and Quiney [563]). In Eq. (14.3) ε_i is the diagonal
Lagrange multiplier. The off-diagonal Lagrange multipliers λ_{il} ensure that
each basis function *i* is orthogonal to all target orbitals *l* of the same
symmetry. The model potential $V(r)$ is usually chosen to be the frozen
core potential of the target. Denoting the radial reduced Hamiltonian for

---

[*] Cf. Eq. (2.12).

angular symmetry κ by

$$h_\kappa = \begin{pmatrix} V(r) & c\left(-\dfrac{d}{dr}+\dfrac{\kappa}{r}\right) \\ c\left(\dfrac{d}{dr}+\dfrac{\kappa}{r}\right) & -2c^2+V(r) \end{pmatrix} \tag{14.4}$$

Eq. (14.3) may be rewritten as

$$h_\kappa \begin{pmatrix} p_i(r) \\ q_i(r) \end{pmatrix} = \varepsilon_i \begin{pmatrix} p_i(r) \\ q_i(r) \end{pmatrix} - \sum_l \lambda_{il} \begin{pmatrix} \bar{P}_l(r) \\ \bar{Q}_l(r) \end{pmatrix}. \tag{14.5}$$

Orthogonality among the basis functions is ensured by the relativistic R-matrix boundary condition (Norrington and Grant [561]) which may be easily obtained from Eq. (14.5) by considering

$$\int_0^a \left[\begin{pmatrix} p_j(r) \\ q_j(r) \end{pmatrix}^\dagger h_\kappa \begin{pmatrix} p_i(r) \\ q_i(r) \end{pmatrix} - \begin{pmatrix} p_i(r) \\ q_i(r) \end{pmatrix}^\dagger h_\kappa \begin{pmatrix} p_j(r) \\ q_j(r) \end{pmatrix} \right] dr$$

$$= (\varepsilon_j - \varepsilon_i) \int_0^a \left[\begin{pmatrix} p_j(r) \\ q_j(r) \end{pmatrix}^\dagger \begin{pmatrix} p_i(r) \\ q_i(r) \end{pmatrix} \right] dr$$

$$= 0$$

(for $i \neq j$) which holds for all pairs of basis functions orthogonal to the target orbitals. The left hand side of the above equation is an exact integral, and we find

$$c\left[p_i(r)q_j(r) - q_i(r)p_j(r) \right]_{r=0}^{r=a} = 0.$$

This implies for basis functions of a given angular symmetry, κ, which obey regular boundary conditions at the origin ($r = 0$) [563] that

$$c\frac{q_i(r)}{p_i(r)}\bigg|_{r=a}$$

is independent of i. We therefore choose R-matrix boundary conditions[561]

$$2ac\frac{q_i(r)}{p_i(r)}\bigg|_{r=a} = b + \kappa. \tag{14.6}$$

Here b is a constant. This boundary condition agrees with the usual nonrelativistic expression [502] in the nonrelativistic limit.

An approximate wavefunction of the $(N+1)$-electron system may be written as

$$\Psi = \sum_{il} c_{il} A(\Phi_i u_{il}) + \sum_m d_m \Theta_m. \tag{14.7}$$

Here the Φ_i are target functions constructed from the N-electron target wavefunctions. The u_{il} are basis orbitals of the form of Eq. (14.2). The subscript i in u_{il} is used to show that the scattering electron is coupled to the target function, Φ_i, to form a scattering channel i with given J and parity, Π. Antisymmetry of the $(N+1)$-electron wavefunctions is ensured by the operator $\mathbf{A}$. The quadratically integrable functions Θ_m are constructed using target orbitals only and represent important contributions to bound states of the $(N+1)$-electron system.

We may compute all possible matrix elements of the relativistic Hamiltonian of Eq. (14.1) for $N+1$ particles among and between the functions $\mathbf{A}(\Phi_i u_{il})$ and Θ_m. The diagonalization of the resulting Hamiltonian matrix $\mathbf{H}^{N+1}$ yields the eigenvalues, e_k, orthonormal wavefunctions, Ψ_k, of the $(N+1)$-electron system in the inner region $r < a$, and the c_{ilk} and d_{mk} of Eq. (14.5) for Ψ_k. The solution Ψ_E at an arbitrary energy E can be expanded in $r < a$ as

$$\Psi_E = \sum_k \Psi_k A_{kE}. \tag{14.8}$$

Thus

$$A_{kE} = \langle \Psi_k | \Psi_E \rangle. \tag{14.9}$$

Since

$$\left(H^{N+1} - E \right) \Psi_E = \left(H^{N+1} - e_k \right) \Psi_k$$
$$= 0$$

so that

$$(E - e_k)\langle \Psi_k | \Psi_E \rangle = \left\langle \Psi_k | H^{N+1}\Psi_E \right\rangle - \left\langle H^{N+1}\Psi_k | \Psi_E \right\rangle$$
$$= \sum_{j=1}^{n} c\left[-p_{jk}(r)q_{jE}(r) + q_{jk}(r)p_{jE}(r) \right]_{r=0}^{r=a}. \tag{14.10}$$

Here $p_{jk}(r)$ and $q_{jk}(r)$ are given by the first sum of Eq. (14.7) for Ψ_k

$$\sum_l c_{jlk} u_{jl}(\mathbf{r}) = \frac{1}{r}\left(\begin{array}{c} p_{jk}(r)\,\chi_{\kappa m}(\mathbf{r}/r) \\ i\,q_{jk}(r)\,\chi_{-\kappa m}(\mathbf{r}/r) \end{array} \right). \tag{14.11}$$

In Eq. (14.10) $p_{jE}(r)$ and $q_{jE}(r)$ are radial functions of the scattering electron at energy E in channel j and the n is the number of channels. Since the radial amplitudes $p_{jk}(r)$ and $q_{jk}(r)$ satisfy the R-matrix boundary condition given in Eq. (14.6), we obtain from Eqs. (14.9) and (14.10)

$$A_{kE} = \frac{1}{(e_k - E)}\sum_j \frac{1}{2a}p_{jk}(a)\left[2acq_{jE}(a) - (b+\kappa)p_{jE}(a) \right]. \tag{14.12}$$

It follows from Eq. (14.8) that

$$p_{iE}(a) = \sum_k p_{ik}(a)A_{kE}.$$

Inserting the expansion in Eq. (14.12) in the above equation, we obtain

$$p_{iE}(a) = \sum_k \sum_j \frac{p_{ik}(a)p_{jk}(a)}{2a(e_k - E)}\left[2acq_{jE}(a) - (b + \kappa)p_{jE}(a)\right]. \quad (14.13)$$

We define the R-matrix at the energy E,

$$R_{ij}(E) = \frac{1}{2a}\sum_k \frac{p_{ik}(a)p_{jk}(a)}{e_k - E}. \quad (14.14)$$

The p_{jk} are surface amplitudes, related to the amplitudes of the large-component basis functions at $r = a$ through the c_{ij} (see Eq. (14.11)). At $r = a$, the large-component radial function of the scattering electron in channel i at energy E given by Eq. (14.13) may be rewritten as

$$p_{iE}(a) = \sum_{j=1}^{n} R_{ij}(E)(2acq_{jE}(a) - (b + \kappa)p_{jE}(a)), \quad (14.15)$$

where $p_{jE}(a)$ ($q_{jE}(a)$) is the large (small) radial component of the wave-function of the scattering electron in channel j.

14.2.2 Outer region

In the outer region $r > a$, we may neglect all exchange interactions between the scattering electron and the N target electrons. The radial Dirac equations for the scattered electron are then

$$\frac{dQ_i}{dr} - \frac{\kappa}{r}Q_i + \frac{1}{c}(E_i - V(r))P_i = \frac{1}{c}\sum_{j=1}^{n} v_{ij}P_j$$

$$-\frac{dP_i}{dr} - \frac{\kappa}{r}P_i + \frac{1}{c}(2c^2 + E_i - V(r))Q_i = \frac{1}{c}\sum_{j=1}^{n} v_{ij}Q_j. \quad (14.16)$$

Here $E_i = E - E_i^N$ (where E_i^N is the target energy in channel i and E is the total energy of the $(N + 1)$-electron system) is the energy of the scattering electron in channel i. $V(r)$ is the effective nuclear potential. The wavefunctions of the scattered electron in different channels are coupled only by long-range potentials, v_{ij}. [559]

For open channels with $E_i > 0$ there are two types of solutions of Eq. (14.16) with asymptotic behaviour [564]

$$S1 : \lim_{r \to \infty}\begin{pmatrix} P_{S1} \\ Q_{S1} \end{pmatrix} \sim \begin{pmatrix} \mu\sin\theta(r) \\ \mu^{-1}\cos\theta(r) \end{pmatrix}$$

and

$$S2 : \lim_{r \to \infty}\begin{pmatrix} P_{S2} \\ Q_{S2} \end{pmatrix} \sim \begin{pmatrix} \mu\cos\theta(r) \\ \mu^{-1}\sin\theta(r) \end{pmatrix},$$

where

$$\mu \;\;=\;\; \left(\frac{1 + \frac{E_i}{2c^2}}{2E_i}\right)^{\frac{1}{4}}$$

$$\theta(r) \;\;=\;\; kr + v \ln 2kr + \delta_k$$

$$k^2 \;\;=\;\; 2E_i\left(1 + \frac{E_i}{2c^2}\right)$$

$$\delta_k \;\;=\;\; \rho - \frac{1}{2}\pi\sigma - \arg\Gamma(\sigma + iv)$$

$$\sigma \;\;=\;\; \sqrt{k^2 - \alpha^2 Z^2}$$

$$e^{2i\rho} \;\;=\;\; -\frac{k + iv}{\sigma + iv'}$$

$$v \;\;=\;\; \frac{ZE_i}{c^2 k}$$

$$v' \;\;=\;\; \frac{Z}{k}.$$

For closed channels with $E_i < 0$ there are bound type solutions of Eq. (14.16) with asymptotic behaviour [564]

$$\text{B} \;:\;\; \lim_{r\to\infty}\binom{P_B}{Q_B} \;\sim\; e^{-(|k|r + v\ln 2|k|r)}.$$

The scattering amplitude in channel i resulting from interaction with channel j is therefore

$$u_{ij} \;\;=\;\; (S1)_i\delta_{ij} + (S2)_iK_{ij} \qquad i,j = 1,\ldots,n_C$$

in open channels and

$$u_{ij} \;\;=\;\; B_i\widetilde{K}_{ij} \qquad \begin{aligned} i &= 1,\ldots,n_B \\ j &= 1,\ldots,n_C \end{aligned}$$

in closed channels. Here n_C is the number of open channels and n_B is the number of closed channels for a given symmetry (*i.e.* given J and parity, Π) of the $(N+1)$-electron system. The matrices K and $\widetilde{K}$ are the K-matrix and its extension for closed channels. The $\widetilde{K}$ matrices are not needed for the scattering calculations.

Thus the radial function of the scattering electron at energy E in $r > a$ may be written as

$$P(E) \;=\; \binom{P_{S1} + P_{S2}K}{P_B\widetilde{K}}, \qquad Q(E) \;=\; \binom{Q_{S1} + Q_{S2}K}{Q_B\widetilde{K}} \qquad (14.17)$$

in a matrix representation. In Eq. (14.17) P_{S1}, Q_{S1}, P_{S2} and Q_{S2} are $n_C \times n_C$ diagonal marices. The P_B and Q_B are $n_B \times n_B$ diagonal matrices. The K and $\widetilde{K}$ are $n_C \times n_C$ and $n_B \times n_C$ matrices.

14.2.3 *Matching of inner and outer region solutions*

The inner solution at $r = a$ given by Eq. (14.15) may be rewritten in a matrix representation as

$$[I + R(E)(b + \kappa)] P(E) = 2ac\underline{R}(E)Q(E), \qquad (14.18)$$

where the matrix I is a $n \times n$ unit matrix. Now we partition $\underline{R}(E)$ and $\underline{I}$ matrices into blocks similar to Eq. (14.17). Let

$$\underline{W}(E) = \frac{1}{2ac} [\underline{I} + \underline{R}(E)(b + \kappa)]$$

and define the $n_C \times n_C$ matrices

$$\underline{A} = (W)_{n_C \times n_C}(P_{S2})_{n_C \times n_C} - (R)_{n_C \times n_C}(Q_{S2})_{n_C \times n_C},$$

$$\underline{F} = (W)_{n_C \times n_C}(P_{S1})_{n_C \times n_C} - (R)_{n_C \times n_C}(Q_{S1})_{n_C \times n_C},$$

the $n_C \times n_B$ matrix

$$\underline{B'} = (W)_{n_C \times n_B}(P_B)_{n_B \times n_B} - (R)_{n_C \times n_B}(Q_B)_{n_B \times n_B},$$

the $n_B \times n_C$ matrices

$$\underline{C} = (W)_{n_B \times n_C}(P_{S2})_{n_C \times n_C} - (R)_{n_B \times n_C}(Q_{S2})_{n_C \times n_C},$$
$$\underline{E} = (W)_{n_B \times n_C}(P_{S1})_{n_C \times n_C} - (R)_{n_B \times n_C}(Q_{S1})_{n_C \times n_C},$$

and the $n_B \times n_B$ matrix

$$\underline{D} = \left((W)_{n_B \times n_B}(P_B)_{n_B \times n_B} - (R)_{n_B \times n_B}(Q_B)_{n_B \times n_B} \right)^{-1}.$$

Then the K-matrix is

$$\underline{K} = -\left(\frac{F - B'DE}{A - B'DC} \right). \qquad (14.19)$$

The scattering matrix $\underline{S}$ is given in terms of the K-matrix by

$$\underline{S} = \frac{1 + i\underline{K}}{1 - i\underline{K}} \qquad (14.20)$$

and the collision strength for a transition from initial state r to final state s is given by

$$\Omega_{rs}^{J\Pi} = \frac{2J + 1}{2} \sum_{j_r j_s} |S_{rs} - \delta_{rs}|^2. \qquad (14.21)$$

We denote the total angular momentum quantum numbers of the scattered electron in the initial and final states respectively by j_r and j_s, and the total angular momentum and parity of the $(N + 1)$-electron system by J and Π, $(+,$ even; $-,$ odd). The total collision strength, at a given scattering energy, Ω_{rs}, is obtained by summing $\Omega_{rs}^{J\Pi}$ over all important $J\Pi$ combinations.

The cross section for a transition from initial state r to final state s is given in units of πa_0^2 (where a_0 is Bohr radius) by

$$\sigma_{rs}^{J\Pi} = \frac{2J+1}{2k_r^2(2J_r+1)} \sum_{j_r j_s} | S_{rs} - \delta_{rs} |^2, \qquad (14.22)$$

where the kinetic energy of the scattered electron in the initial state by k_r^2. The quantity J_r is the total angular momentum of the initial target state. The total cross section, at a given scattering energy, σ_{rs}, is obtained by summing $\sigma_{rs}^{J\Pi}$ over all important $J\Pi$ combinations.

14.3 Method of calculation and results

14.3.1 Kr XXIX

Target states. The GRASP^2 multiconfiguration Dirac-Fock (MCDF) program [565] was used to obtain approximate wavefunctions and level energies of the Kr XXIX target. All levels arising from the configurations $2s_{1/2}^x 2p_{1/2}^y 2p_{3/2}^z$ (where $x + y + z = 6$ and the closed shell $1s_{1/2}^2$ is omitted from the description of configurations) of Kr XXIX were included in an average level (AL) [566] calculation. Only the N-body Dirac-Coulomb Hamiltonian was used in the calculation: the transverse electromagnetic interaction (for instance, Grant, McKenzie *et al.* [88] or Grant and McKenzie [567]) was omitted. This is consistent with our collision calculation.

We present the resulting level energies and both jj-coupled and LSJ-coupled compositions in Table 14.1. With the exception of levels 7 and 9 all levels given in Table 14.1 are well-described by jj coupling. Levels 7 and 9 exhibit strong mixing in this scheme. No appreciable simplification is obtained upon transformation to an LSJ coupled basis (see Table 14.1, columns 6 and 7).

Both the relativistic (jj-coupled) and nonrelativistic (LS-coupled) levels shown in Fig. 14.1 arise from the $2s^2 2p^4$, $2s^2 2p^5$ and $2p^6$ configurations. The relativistic energies were taken from Table 14.1 whilst the nonrelativistic energies were taken from Reed and Henry. [568] We have also calculated the nonrelativistic level energies by solving nonrelativistic limiting forms of the Dirac equations using GRASP^2. Our calculated nonrelativistic energies are in very good agreement with those of Reed and Henry. [568] The relativistic and nonrelativistic level schemes shown in Fig. 14.1 are seen to be very different.

Table 14.1. Levels arising from the configurations $2s_{1/2}^x 2p_{1/2}^y 2p_{3/2}^z$ (where $x + y + z = 6$) of Kr XXIX (oxygen-like krypton). Tabulated are the calculated energies (in Ryd) relative to the calculated ground state, $2s_{1/2}^2 2p_{1/2}^2 2p_{3/2}^2$ (98.26%)+ $2s_{1/2}^2 2p_{1/2} 2p_{3/2}^3$ (1.74%), total angular momentum quantum numbers, J, parities, Π (+, even; − odd), and jj-coupled and LSJ-coupled compositions of the levels. The closed 1s shell is omitted from the configuration descriptions. Contributions with weights less than 1.0% are not shown

Level index	Energy	J^Π	jj-coupled composition	Weights	LSJ-coupled composition	Weights
1	0.0000	2^+	$2s_{1/2}^2 2p_{1/2}^2 2p_{3/2}^2$	(98.26%)	$2s^2 2p^4\ (^3P_2^e)$	(78.50%)
			$2s_{1/2}^2 2p_{1/2} 2p_{3/2}^3$	(1.74%)	$2s^2 2p^4\ (^1D_2^e)$	(21.60%)
2	1.3854	0^+	$2s_{1/2}^2 2p_{1/2}^2 2p_{3/2}^2$	(96.25%)	$2s^2 2p^4\ (^3P_0^e)$	(50.69%)
			$2s_{1/2}^2 2p_{3/2}^4$	(3.19%)	$2s^2 2p^4\ (^1S_0^e)$	(48.72%)
3	3.8999	1^+	$2s_{1/2}^2 2p_{1/2} 2p_{3/2}^3$	(100.00%)	$2s^2 2p^4\ (^3P_1^e)$	(100.00%)
4	4.8878	2^+	$2s_{1/2}^2 2p_{1/2} 2p_{3/2}^3$	(98.26%)	$2s^2 2p^4\ (^1D_2^e)$	(78.50%)
			$2s_{1/2}^2 2p_{1/2}^2 2p_{3/2}^2$	(1.74%)	$2s^2 2p^4\ (^3P_2^e)$	(21.60%)
5	9.3712	0^+	$2s_{1/2}^2 2p_{3/2}^4$	(96.10%)	$2s^2 2p^4\ (^1S_0^e)$	(49.84%)
			$2s_{1/2}^2 2p_{1/2}^2 2p_{3/2}^2$	(2.92%)	$2s^2 2p^4\ (^3P_0^e)$	(49.28%)
6	15.4118	2^-	$2s_{1/2} 2p_{1/2}^2 2p_{3/2}^3$	(100.00%)	$2s\, 2p^5\ (^3P_2^o)$	(100.00%)
7	17.1779	1^-	$2s_{1/2} 2p_{1/2}^2 2p_{3/2}^3$	(79.84%)	$2s\, 2p^5\ (^3P_1^o)$	(77.80%)
			$2s_{1/2} 2p_{1/2} 2p_{3/2}^4$	(20.16%)	$2s\, 2p^5\ (^1P_1^o)$	(22.00%)
8	19.6539	0	$2s_{1/2} 2p_{1/2} 2p_{3/2}^4$	(100.00%)	$2s\, 2p^5\ (^3P_0^o)$	(100.00%)
9	21.9780	1^-	$2s_{1/2} 2p_{1/2} 2p_{3/2}^4$	(79.84%)	$2s\, 2p^5\ (^1P_1^o)$	(77.80%)
			$2s_{1/2} 2p_{1/2}^2 2p_{3/2}^3$	(20.16%)	$2s\, 2p^5\ (^3P_1^o)$	(22.00%)
10	34.9459	0^+	$2p_{1/2}^2 2p_{3/2}^4$	(100.00%)	$2p^6\ (^1S_0^e)$	(100.00%)

Cross sections. The R-matrix boundary ($r = a$) was chosen at 1.5 Bohr radii. Twelve basis functions were obtained for each relativistic angular momentum quantum number $\kappa = \mp 1, \mp 2, \cdots, \mp 18$ by solving the appropriate radial Dirac equations (see Eq. (14.3)). The frozen core potential of the target was used in these equations and the constant b which appears in the R-matrix boundary condition (Eq. (14.6)) was taken as zero. The Dirac Hamiltonian matrix of the $(N + 1)$-electron system in the inner region was constructed for each of the J^Π symmetry/parity combinations $\frac{1}{2}^\pm, \ldots, \frac{29}{2}^\pm$. All possible channels were included for a given J^Π of the target-plus-electron system. The truncation error in the expression for the R-matrix due to the finite size of the basis set was mitigated by using the Buttle correction.

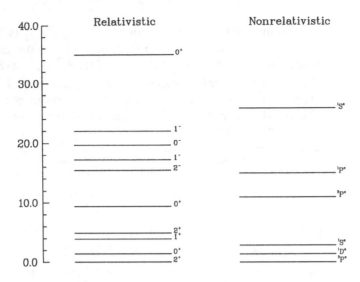

Fig. 14.1. Relativistic and nonrelativistic level schemes of Kr^{28+}; energies are given in rydbergs. The relativistic energies were calculated using GRASP^2 (present calculation) and nonrelativistic energies [568] using SUPERSTRUCTURE and CIV3.

We used a modified version of Seaton's program [569] to solve the nonrelativistic limiting form of the Dirac radial equations for the scattered electron in the asymptotic region. We expect that the error due to this approximation is small for the scattering energies considered. Seaton's program uses a perturbation approach to solve the coupled equations and calculate the cross sections.

Our calculated cross sections below incident energy 34.946 Ryd (the energy of the highest target level with respect to the ground state) are crowded with Rydberg resonances. In an effort to understand the resonances we focused on the resonances in the region from 21.978 to 34.946 Ryd which is least crowded. The dominant resonances in this region are the members of $1s^2_{1/2}2p^2_{1/2}2p^4_{3/2}n\ell$ series. To estimate the positions of the members of $n = 10$ series, we performed a GRASP^2 calculation [570] for the Kr XXVIII ion. These positions are compared with resonances obtained from the R-matrix calculation in Table 14.2. In the R-matrix calculation the positions were obtained by calculating the cross section for transition from level 1 to 9 in the region from 26.80 to 27.15 Ryd. The energy mesh was chosen so as to delinate the resonance structure with a high degree of resolution. The lowest ($\ell = s$) and the higher members ($\ell > g$) of the series are very narrow and it is difficult to resolve their structure in detail. The positions of the higher resonances are not presented in Table 14.2. Calculated cross sections for the electron-impact excitation from level 1

Table 14.2. Positions (in Ryd) of some $1s^2_{1/2}2p^2_{1/2}2p^4_{3/2}10\ell$ Rydberg resonances with respect to the Kr XXIX ground state. Positions obtained from the GRASP^2 calculation are tabulated in column 2 and those obtained from the Dirac R-matrix calculation are tabulated in column 3

Resonance	Position (Ryd)	
	GRASP^2	Dirac R-matrix
$1s^2_{1/2}2p^2_{1/2}2p^4_{3/2}10s_{1/2}$	26.827	26.817
$1s^2_{1/2}2p^2_{1/2}2p^4_{3/2}10p_{1/2}$	26.915	26.910
$1s^2_{1/2}2p^2_{1/2}2p^4_{3/2}10p_{3/2}$	26.941	26.936
$1s^2_{1/2}2p^2_{1/2}2p^4_{3/2}10d_{3/2}$	27.039	27.032
$1s^2_{1/2}2p^2_{1/2}2p^4_{3/2}10d_{5/2}$	27.046	27.039
$1s^2_{1/2}2p^2_{1/2}2p^4_{3/2}10f_{5/2}$	27.094	27.088
$1s^2_{1/2}2p^2_{1/2}2p^4_{3/2}10f_{7/2}$	27.097	27.091
$1s^2_{1/2}2p^2_{1/2}2p^4_{3/2}10g_{7/2}$	27.105	27.100
$1s^2_{1/2}2p^2_{1/2}2p^4_{3/2}10g_{9/2}$	27.107	27.101

to level 9 in the region from 26.80 to 27.15 Ryd is presented in Fig. 14.2. Our calculated cross sections for the scattering of electrons by Kr XXIX are in reasonable agreement with those of Bhatia *et al.* [571]

14.3.2 Hg

Target states. We performed a number of calculations using the GRASP^2 multiconfiguration Dirac-Fock (MCDF) program [565] to find a simple model for the Hg target. Three such models have been found whose level energies relative to the ground state are in good agreement with those observed. [572] We chose the smallest since the other models are too large for the computational facilities available.

The GRASP^2 program was used to perform the following calculation to obtain approximate wavefunctions of the Hg target. The $1s_{1/2}, \ldots,$ $6p_{1/2}$ and $6p_{3/2}$ orbitals were determined from an average level (AL) [566] calculation for the five jj-coupled levels arising from the relativistic generalization of the configurations $5d^{10}(6s^2 + 6s6p)$. (Here and elsewhere, closed subshell orbitals $1s_{1/2}, \ldots, 5p_{1/2}$ and $5p_{3/2}$ are omitted from the configuration descriptions.) These states are limited to the symmetry/parity combinations $J^\Pi = 0^\pm, 1^-, 2^-$. The orbitals determined in this manner were then held fixed for the remainder of the calculation. The

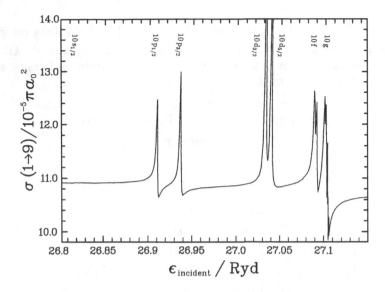

Fig. 14.2. Calculated cross sections for the electron-impact excitation of Kr^{28+} from level 1 to level 9 for incident electron kinetic energy from 26.80 to 27.15 Ryd. The Rydberg resonances have been identified as the members of $1s^2_{1/2}2p^2_{1/2}2p^4_{3/2}10\ell$ series: from left to right we have $10p_{1/2}$, $10p_{3/2}$, $10d_{3/2}$, $10d_{5/2}$, $10f$, $10g$.

orbitals $7p_{1/2}$ and $7p_{3/2}$ were obtained from an optimal level (OL) [566] calculation for the ground state. The jj-coupled CSFs (configuration state functions) with $J^\Pi = 0^\pm, 1^-, 2^-$ resulting from the configurations $5d^{10}(6s^2 + 6s6p + 6p^2 + 7p^2 + 6p7p + 6s7p)$ and $5d^9(6s^26p + 6s^27p)$ were included in this calculation.

The calculated orbitals were used to perform a CI calculation for the 98 jj-coupled levels with $J^\Pi = 0^\pm, 1^-, 2^-$ arising from the configurations $5d^{10}(6s^2 + 6s6p + 6p^2 + 7p^2 + 6p7p + 6s7p)$ and $5d^9(6s^26p + 6s^27p + 6p^3 + 7p^3 + 6s6p^2 + 6s7p^2 + 6s6p7p)$. Some of the core-polarization effects were included in these calculations. The lowest five levels obtained from this calculation were used to model the 80 electron target atom. Only the N-body Dirac–Coulomb Hamiltonian was used in the calculation.

The resulting level energies and jj-coupled compositions are presented in Table 14.3.2. The calculated energies and jj-coupled compositions of the levels arising from the configurations $6s^2$ and $6s6p$ (small model) are also given for comparison. Level energies of our target model which differ from those of the small model are in reasonable agreement with experiment. [572]

Level energies and jj-coupled compositions of the small model given in Table 14.3.2 were obtained from the first step of the $GRASP^2$ calculation described above. The additional CSFs were included in the final step

Table 14.3. The lowest five jj-coupled levels arising from the configurations $6s^x6p^y7p^z$, (where $x+y+z=2$) and $5d^96s^x6p^y7p^z$, (where $x+y+z=3$) of neutral Hg. Total angular momentum quantum numbers, J and parities, Π (+, even; − odd), are tabulated in column 2. We present the calculated energies (in eV) relative to the calculated ground state, $6s^2_{1/2}$ (94.00%) + $6p^2_{1/2}$ (1.20%) + $6p^2_{3/2}$ (1.10%), and jj-coupled compositions of the levels in columns 3, 4 and 5. The calculated energies (in eV) and jj-coupled compositions of the levels arising from the configurations $6s^2$ and $6s6p$, of Hg have also been given in columns 6, 7 and 8. Observed energies [572] are listed in column 9. The closed $1s$-$5d$ shells are omitted from the configuration descriptions. Contributions with weights less than 1.0% are not shown

| Level index | J^Π | Target model | | | Small model | | | Moore [572] |
		Energy	jj-coupled composition	Weights	Energy	jj-coupled composition	Weights	
1	0^+	0.00	$6s^2_{1/2}$	(94.00%)	0.00	$6s^2_{1/2}$	(100.00%)	0.00
			$6p^2_{1/2}$	(1.20%)				
			$6p^2_{3/2}$	(1.10%)				
2	0^-	4.32	$6s_{1/2}6p_{1/2}$	(98.70%)	3.20	$6s_{1/2}6p_{1/2}$	(100.00%)	4.67
3	1^-	4.55	$6s_{1/2}6p_{1/2}$	(75.20%)	3.45	$6s_{1/2}6p_{1/2}$	(74.30%)	4.89
			$6s_{1/2}6p_{3/2}$	(23.20%)		$6s_{1/2}6p_{3/2}$	(25.70%)	
4	2^-	5.06	$6s_{1/2}6p_{3/2}$	(97.50%)	3.97	$6s_{1/2}6p_{3/2}$	(100.00%)	5.46
5	1^-	6.81	$6s_{1/2}6p_{3/2}$	(72.50%)	6.09	$6s_{1/2}6p_{3/2}$	(74.30%)	6.70
			$6s_{1/2}6p_{1/2}$	(21.80%)		$6s_{1/2}6p_{1/2}$	(25.70%)	
			$6s_{1/2}7p_{1/2}$	(1.50%)				

of the calculation of the target model wave functions to account for interactions among valence electrons and for the core-polarization effects. The inclusion of these CSFs has changed the composition of all five states of Hg given in Table 14.3.2. The largest change is in the ground state. The improved representation of the ground state is responsible for most of the improvement of energies of the target model.

Resonances and cross sections. The R-matrix boundary ($r = a$) was chosen at 24.156 Bohr radii. Twenty basis functions were obtained for each relativistic angular momentum quantum number $\kappa = \mp1, \mp2, \cdots, \mp10$. The frozen core potential of the target was used and the constant b which appears in the R-matrix boundary condition (Eq. (14.6)) was taken as unity. The Dirac Hamiltonian matrix of the $(N + 1)$-electron system in the inner region was constructed for each of the J^{Π} symmetry/parity combinations $\frac{1}{2}^{\pm}, \ldots, \frac{15}{2}^{\pm}$. All possible channels were included for a given J^{Π} of the target-plus-electron system. The Buttle [522] correction was included.

The dynamics is essentially nonrelativistic in the asymptotic region, $r > a$, at the energies of interest and it is therefore possible to use standard nonrelativistic asymptotic packages to determine the wavefunction of the scattered electron. We have applied the VPM program [573] which implements the variable phase method of Calogero. [574] The nonrelativistic amplitude obtained from this code can be matched in the usual way to the Dirac large component amplitude obtained from the calculation in the inner region, $r < a$ with an error $O(\alpha^2)$. We expect the consequent error to be negligible in comparison with other errors of the model.

Our calculated partial and total cross sections show structures due to resonances.

Resonances. We present theoretical positions of the resonances in Table 14.3.2. In our Dirac R-matrix calculation the positions were obtained by calculating the cross sections for transitions from level 1 to levels 1–4 in the region from 0.0 to 8.0 eV. These positions are tabulated in column 3. Results of Scott *et al.* [575] and Heddle [576, 577] are given in columns 4 and 5 respectively. Results of three different R-matrix calculations by Bartschat and Burke [578] are tabulated in columns 6, 7 and 8. We also carried out a GRASP^2 calculationprovide further confirmation of our identification of resonances. These positions are tabulated in column 9 of Table 14.3.2.

In our calculation we used the Dirac Hamiltonian to define the dynamics of all electrons in the electron plus Hg system so that relativistic effects are included in both of the orbitals and in the interactions between all pairs of electrons. We have attempted to include what appeared to be the most significant CSFs according to our investigation to account for

Table 14.4. Positions (in eV) of some resonances of Hg⁻ ion with respect to the neutral Hg ground state. Total angular quantum numbers, J and parities, Π (+, even; − odd), are tabulated in column 2. We present the positions obtained from the Dirac R-matrix calculation and those obtained from the R-matrix calculation of Scott et al. [575] in columns 3 and 4. Calculated positions of Heddle [577] are tabulated in column 5. The positions obtained from three different R-matrix calculations of Bartschat and Burke [578] are presented in columns 6, 7 and 8. Estimated positions and possible LSJ-coupled compositions of the resonances obtained from the GRASP² calculation have been tabulated in columns 9 and 10

Resonance index	J^Π	Dirac R-matrix	Scott et al. [575]	Heddle [576, 577]	Bartschat and Burke [578] I	II	III	GRASP²	LSJ-coupled resonance
1	$1/2^-$	0.72						0.19	$6s^2(^1S)\,6p\ ^2P_{1/2}$
2	$3/2^-$	0.96						0.21	$6s^2(^1S)\,6p\ ^2P_{3/2}$
3	$1/2^+$		4.70	4.55	4.7	4.59	4.49	4.63	$6s(^2S)\,6p^2(^3P)\ ^4P_{1/2}$
4	$3/2^+$	4.58	4.70, 5.00	4.71	4.7	4.66	4.57	4.70	$6s(^2S)\,6p^2(^3P)\ ^4P_{3/2}$
5	$5/2^+$	4.81	4.90	4.94	4.9	4.86	4.76	5.0	$6s(^2S)\,6p^2(^3P)\ ^4P_{5/2}$
6	$3/2^+$	5.07	5.50	5.20	5.5	5.5	5.0	6.07	$6s(^2S)\,6p^2(^1D)\ ^2D_{3/2}$
7	$5/2^+$	5.46	5.50	5.51	5.5	5.5	5.4	6.20	$6s(^2S)\,6p^2(^1D)\ ^2D_{5/2}$
8	$1/2^+$	5.95	5.80(?)					6.86	$6s(^2S)\,6p^2(^3P)\ ^2P_{1/2}$
9	$1/2^+$	6.83	6.70(?)					7.26	$6s(^2S)\,6p^2(^1S)\ ^2S_{1/2}$

electron correlation and polarization effects. Our calculated positions are in reasonable agreement with those obtained from calculation III of Bartschat and Burke. [578]

Observed positions of the resonances are given in Table 14.3.2 indexed in accordance with Table 14.3.2. Observed positions of Kuyatt *et al.*, [580] Zapesochnyi and Shpenik, [581] Rockwood, [582] Düweke *et al.*, [583] Ottley and Kleinpoppen, [584] Burrow and Michejda, [585] Albert *et al.*, [586] Jost and Ohnemus [587] and Newman *et al.* [588] are given in columns 2–10. Results of the experiments which were performed before 1965 are not shown.

Cross sections. Figure 14.3 shows our results for the total cross section for elastic scattering of electrons by Hg in the energy range 0–4 eV along with a selection of results from previous calculations and from experiments. Walker [589, 592] was the first to carry out relativistic calculations of electron scattering from the Hg atom. His first set of calculations [589] dealt with elastic scattering in the range 3.5–500.0 eV. He used a Dirac-Fock static potential, including both direct and exchange contributions, but neglecting any correlation effects, the dynamical distortion of the atom by the scattering electron or coupling to inelastic channels. He computed differential and total cross sections at various energies as well as spin polarization parameters. The second set of calculations [590] added a polarization potential obtained from a polarized orbital method due to Temkin [593] to extend the calculations down to an energy of 0.5 eV. This polarization potential proved to be too large by a factor of two, and we have used the corrected results from a later paper1975 paper gives two sets of results: the first of these, the dashed curve in Fig. 14.3, included exchange contributions from all 80 electrons of the target; the second, dotted, curve, shows the effect of exchange interaction with the outermost shells ($n = 4, 5$ and 6) only. The experimental total cross sections are due to Jost and Ohnemus [587] together with earlier data by McCutchen [594] and Rockwood. [582]

Our calculation takes account of exchange effects with all target electrons in full inside the *R*-matrix boundary as well as the coupling to inelastic channels, and the CI calculation described in Sec. 14.3.1 represents some effect of polarization of the atom in terms of virtual excitations from the 6s and 5d shells. Our cross section has only a broad peak just below 1 eV due to the resonances 1 and 2 found at 0.72 and 0.96 eV in the calculated partial cross sections for the symmetry/parity combinations $J^\Pi = \frac{1}{2}^-$ and $\frac{3}{2}^-$. Walker's [592] calculation, the dashed curve, has two peaks: a sharp one at 0.2 eV due to $6s^2 6p\,^2P_{1/2}$ corresponding to our resonance 1, and a broad peak near 1 eV due to $6s^2 6p\,^2P_{3/2}$ corresponding to our resonance 2. Walker's second calculation, the dotted curve, has only

Table 14.5. Observed positions (in eV) of some resonances of Hg^- ion. See Table 14.4 for a full description of resonances designated by the indices 1–7 in column 1. Results of the experiments which were performed before 1965 are not shown

Resonance Index	Kuyatt et al. [580]	Zapesochnyi and Shepnik [581]	Rockwood [582]	Düweke et al. [583]	Ottley and Kleinpoppen [584]	Burrow and Michejda [585]	Albert et al. [586]	Jost and Ohnemus [587]	Newman et al. [588]
1			0.625			0.63±0.03		0.4	
3						4.55	4.55		
4	4.07			4.32		4.68	4.71		4.702
5	4.29	5.0		4.55	4.92	4.91	4.94		
6		5.3			5.23				4.9-5.4
7	4.89	5.6		5.15	5.50	5.50	5.51		5.59

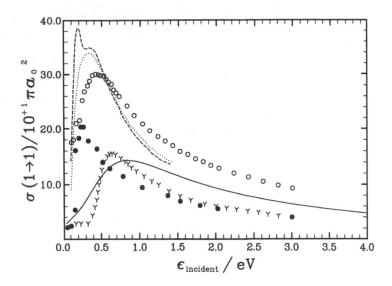

Fig. 14.3.   Calculated and observed total cross sections for elastic scattering of electrons by Hg in the region from 0 to 4 eV. The energy of the incident electron, $\varepsilon_{incident}$, is given in eV. The data are as follows: —, present calculation; – – –, total cross section obtained by Walker [592] including a polarization potential and exchange interactions with all the electrons; ..., total cross section obtained by Walker [592] including a polarization potential and exchange interactions only with the outermost three electrons shells ($n = 4, 5$ and 6); o, Observed values of Jost and Ohnemus [587]; •, Observed values of McCutchen [594]; Y, Observed values of Rockwood [582].

a single peak at 0.3 eV. Walker comments on the sensitivity of the cross section near threshold to the approximation used.

There is rather poor agreement between these different calculations below 1.5 eV. For reasons of economy, our target model does not have any virtual double excitations from the 5d shell. We did not have the resources to investigate the effect of these excitations in full in our calculation on scattering of electrons by neutral Hg, and it could be that the peak in our cross section at about 0.8 eV would have moved to lower energies, improving the agreement with experiment, had we been able to include such excitations.

Fig. 14.4 displays a comparison of our calculated total cross sections for elastic scattering of electrons by Hg in the region from 4 to 8 eV with those of the *R*-matrix calculation by Scott *et al.* [575] The inelastic total cross sections for excitation of electrons by Hg from level 1 to levels $2 - 5$ in the region from 4.0 to 8.0 eV and our calculated radiative transition data are given elsewhere. [579]

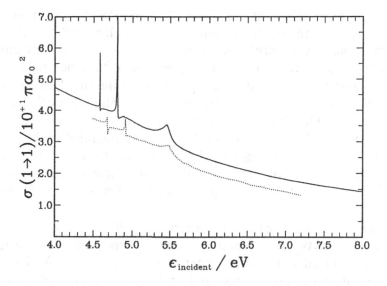

Fig. 14.4. Calculated total cross sections for elastic scattering of electrons by Hg in the region from 4 to 8 eV. The energy of the incident electron, $\varepsilon_{\text{incident}}$, is given in eV. The data are as follows: —, present calculation; ..., R-matrix calculation of Scott *et al.* [575].

14.3.3 Pb

Target states. The GRASP[2] program [565] was used to perform the following calculation to obtain approximate wavefunctions of the Pb target. The $1s_{1/2}, ..., 6p_{1/2}$ and $6p_{3/2}$ orbitals were determined from an average level (AL) [566] calculation for the five jj-coupled levels arising from the relativistic generalization of the configurations $6s^2 6p^2$. (Here and elsewhere, the closed subshell orbitals $1s_{1/2}, ..., 5d_{3/2}$ and $5d_{5/2}$ are omitted from the configuration descriptions.) These states are limited to the symmetry/parity combinations $J^{\Pi} = 0^+, 1^+, 2^+$. The orbitals determined in this manner were then held fixed for the remainder of the calculation. The orbitals $7p_{1/2}$ and $7p_{3/2}$ were obtained from an extended optimum level (EOL) [566] calculation for the lowest five states with $J^{\Pi} = 0^+, 1^+, 2^+$ arising from the relativistic generalization of the configurations $6s^2(6p^2 + 7p^2 + 6p7p)$.

The calculated orbitals were used to perform a CI calculation for the 68 jj-coupled levels with $J^{\Pi} = 0^+, 1^+, 2^+$ resulting from the configurations $6s^2(6p^2 + 7p^2 + 6p7p) + 6p^4 + 7p^4$. The lowest five levels obtained from this calculation were used to model the 82 electron target atom. Only the N-body Dirac-Coulomb Hamiltonian was used in the calculation.

We present the resulting level energies and jj-coupled compositions in

Table 14.3.3. The calculated energies and jj-coupled compositions of the levels arising from the configurations $6s^26p^3$ (small model) are also given for comparison. Level energies of our target model which differ from those of the small model are in reasonable agreement with experiment. [572]

Level energies and jj-coupled compositions of the small model given in Table 14.3.3 were obtained from the first step of the GRASP^2 calculation described above. The CSFs (configuration state functions) arising from the configurations $6s^2(7p^2 + 6p7p) + 6p^4 + 7p^4$ were included in the final step of the calculation of the target model to account for interactions among valence electrons and for the core-polarization effects. The inclusion of these CSFs has changed the composition of all five states of Pb given in Table 14.3.3.

Cross sections. The R-matrix boundary ($r = a$) was chosen at 24.156 Bohr radii. Twenty basis functions were obtained for each relativistic angular momentum quantum number $\kappa = \mp1, \mp2, \cdots, \mp10$. The frozen core potential of the target was used and the constant b which appears in the R-matrix boundary condition was taken as unity. [595] The Dirac Hamiltonian matrix of the $(N+1)$-electron system in the inner region was constructed for each of the J^Π symmetry/parity combinations $\frac{1}{2}^{\pm}, \ldots, \frac{15}{2}^{\pm}$. All possible channels were included for a given J^Π of the target-plus-electron system. The Buttle [522] correction was included. We used the VPM program [573] to solve the nonrelativistic limiting form of the Dirac radial equations for the scattered electron in the asymptotic region.

We present calculated positions of the resonances in Table 14.7. In our Dirac R-matrix calculation the positions were obtained by calculating the cross sections for transitions from level 1 to levels 1–3 in the region from 0.0 to 4.0 eV. We performed a CI calculation using GRASP^2 for the Pb$^-$ ion in order to confirm our identification of these resonances. The calculated LSJ-coupled compositions of the resonances are given in column 4.

Fig. 14.5 displays a comparison of our calculated total cross sections for elastic scattering of electrons by Pb in the region from 0 to 4.0 eV with those of the R-matrix calculation of Bartschat. [595] Our calculated cross section has two peaks. The broad peak near 0.2 eV is due to the resonance 1 and the other peak at 0.81 eV is due to the resonance 2. The resonance 3 which is found at 1.1 eV in our calculated partial cross section for symmetry/parity combinations $J^\Pi = \frac{5}{2}^-$ is too weak to be seen in the total cross section for elastic scattering. The resonances in the calculated cross section (dotted curve) of Bartschat [595] have been identified as $6s^26p^3$ $^4S_{3/2}$ at 0.7 eV (our resonance 1 at 0.22 eV) and two degenerate resonances, $6s^26p^3$ $^2D_{3/2}$ (our resonance 2 at 0.81 eV) and $6s^26p^3$ $^2D_{5/2}$, at 1.3 eV (our resonance 3 at 1.11 eV).

Table 14.6. The lowest five jj-coupled levels arising from the configurations $6s^2(5p^2 + 7p^2 + 6p7p)$, $6p^4$ and $7p^4$, of neutral Pb. Total angular momentum quantum numbers, J and parities, Π (+, even; −, odd), are tabulated in column 2. We present the calculated energies (in eV) relative to the calculated ground state, $6s^2_{1/2}\,6p^2_{1/2}$ (93.5%) + $6s^2_{1/2}\,6p^2_{3/2}$ (5.3%) + $6p^2_{1/2}\,6p^2_{3/2}$ (0.6%), and jj-coupled compositions of the levels in columns 3, 4 and 5. The calculated energies (in eV) and jj-coupled compositions of the levels arising from the configurations $6s^2\,6p^2$, of Pb have also been given in columns 6, 7 and 8. Observed energies [572] are listed in column 9. The closed $1s$–$5d$ subshells are omitted from the configuration descriptions. Contributions with weights less than 0.5% are not shown

Level index	J^Π	Target model			Small model			Moore [572]
		Energy	jj-coupled composition	Weights	Energy	jj-coupled composition	Weights	
1	0^+	0.000	$6s^2_{1/2}6p^2_{1/2}$ $6s^2_{1/2}6p^2_{3/2}$ $6p^2_{1/2}6p^2_{3/2}$	(93.5%) (5.3%) (0.6%)	0.000	$6s^2_{1/2}6p^2_{1/2}$ $6s^2_{1/2}6p^2_{3/2}$	(92.5%) (7.5%)	0.000
2	1^+	0.923	$6s^2_{1/2}6p_{1/2}6p_{3/2}$ $6p_{1/2}6p^3_{3/2}$	(99.2%) (0.5%)	0.860	$6s^2_{1/2}6p_{1/2}6p_{3/2}$	(100.0%)	0.969
3	2^+	1.366	$6s^2_{1/2}6p_{1/2}6p_{3/2}$ $6s^2_{1/2}6p^2_{3/2}$	(88.8%) (10.2%)	1.331	$6s^2_{1/2}6p_{1/2}6p_{3/2}$ $6s^2_{1/2}6p^2_{3/2}$	(90.2%) (9.8%)	1.320
4	2^+	2.703	$6s^2_{1/2}6p^2_{3/2}$ $6s^2_{1/2}6p_{1/2}6p_{3/2}$	(88.8%) (10.2%)	2.750	$6s^2_{1/2}6p^2_{3/2}$ $6s^2_{1/2}6p_{1/2}6p_{3/2}$	(90.2%) (9.8%)	2.660
5	0^+	3.682	$6s^2_{1/2}6p^2_{3/2}$ $6s^2_{1/2}6p^2_{1/2}$ $6p^2_{1/2}6p^2_{3/2}$	(92.0%) (4.7%) (1.6%)	4.096	$6s^2_{1/2}6p^2_{3/2}$ $6s^2_{1/2}6p^2_{1/2}$	(92.5%) (7.5%)	3.652

Table 14.7. Positions (in eV) of the $6s^2 6p^3$ resonances of Pb$^-$ ion with respect to the neutral Pb ground state. Total angular quantum numbers, J and parities, Π (+, even; − odd), are tabulated in column 2. We present the positions obtained from the Dirac R-matrix calculation in columns 3. Estimated positions and possible LSJ-coupled compositions of the resonances obtained from the GRASP^2 calculation have been tabulated in columns 6 and 7

Resonance index	J^Π	Dirac R-matrix	GRASP^2	LSJ-coupled resonance
1	$3/2^-$	0.22	0.28	$6s^2(^1S)6p^3\ (^4S_{3/2})$
2	$3/2^-$	0.81	1.31	$6s^2(^1S)6p^3\ (^2D_{3/2})$
3	$5/2^-$	1.11	1.68	$6s^2(^1S)6p^3\ (^2D_{5/2})$
4	$1/2^-$	1.47	2.25	$6s^2(^1S)6p^3\ (^2P_{1/2})$
5	$3/2^-$	2.50	2.92	$6s^2(^1S)6p^3\ (^2P_{3/2})$

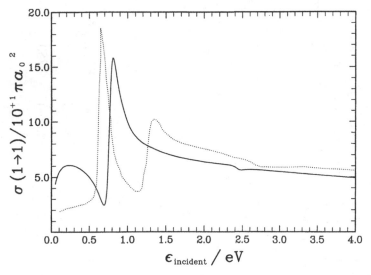

Fig. 14.5. Calculated total cross sections for elastic scattering of electrons by Pb in the region from 0 to 4 eV. The energy of the incident electron, $\varepsilon_{\text{incident}}$, is given in eV. The data are as follows: —, present calculation; ..., R-matrix calculation of Bartschat [595].

Our calculated total cross sections for the impact excitation of electrons by Pb from level 1 to levels 2–3 in the region from 0 to 4 eV are presented elsewhere. [596]

14.3.4 Cs

The GRASP^2 program [565] was used to obtain approximate wavefunctions of the Cs target. The jj-coupled levels with $J^\Pi = \frac{1}{2}^+, \frac{1}{2}^-, \frac{3}{2}^+, \frac{3}{2}^-, \frac{5}{2}^+,$

Table 14.8. Levels arising from the configurations $6s_{1/2}$, $6p_{1/2}$, $6p_{3/2}$, $5d_{3/2}$ and $5d_{5/2}$ of Cs. Tabulated are the total angular momentum quantum numbers, J, and parities, Π (+, even; − odd), dominant configuration state functions (CSFs), calculated energies (eV) relative to the calculated ground state, $J^{\Pi} = 1/2^{+}6s_{1/2}$ and observed energies (eV) (Moore [572]). The closed subshells, $1s_{1/2}^{2}, \ldots, 5p_{1/2}^{2}$ and $5p_{3/2}^{4}$ are omitted from the description of CSFs

Level index	J^{Π}	Dominant CSF(s)	Present calculation	Moore [572]
1	$1/2^{+}$	$6s_{1/2}$	0	0
2	$1/2^{-}$	$6p_{1/2}$	1.187	1.385
3	$3/2^{-}$	$6p_{3/2}$	1.240	1.454
4	$3/2^{+}$	$5d_{3/2}$	1.756	1.797
5	$5/2^{+}$	$5d_{5/2}$	1.754	1.809

resulting from the configurations $5p^{6}(6s + 6p + 5d + 7s) + 5p^{5}(6s6p + 6p7s + 6s5d + 6s7s + 6p5d)$ were included. The lowest five levels obtained from his calculation were used to model the 55 electron target atom. Only the N-body Dirac-Coulomb Hamiltonian was used in the calculation.

We present the resulting level energies and jj-coupled compositions in Table 14.8.

The R-matrix boundary ($r = a$) was chosen at 33.00 Bohr radii. Twenty basis functions were obtained for each relativistic angular momentum quantum number $\kappa = \mp 1, \mp 2, \cdots, \mp 12$. The frozen core potential of the target was used and the constant b which appears in the R-matrix boundary condition was taken as unity. The Dirac Hamiltonian matrix of the $(N + 1)$-electron system in the inner region was constructed for each of the J^{Π} symmetry/parity combinations $0^{\pm}, \ldots, 9^{\pm}$. All possible channels were included for a given J^{Π} of the target-plus-electron system. The Buttle [522] correction was included. We used the VPM program [573] to solve the nonrelativistic limiting form of the Dirac radial equations for the scattered electron in the asymptotic region.

Fig. 14.6 displays our calculated partial cross sections with symmetry/parity combinations $J^{\Pi} = 3^{-}$ and $J^{\Pi} = 4^{-}$ for elastic scattering. Two peaks are due to the LSJ-coupled $6s5d(^{3}F_{3,4})$ resonances. This work is still in progress.

14.4 Conclusions

We have used the revised JJMTRX Dirac R-matrix package to calculate cross sections for the scattering of electrons by atomic targets. In our

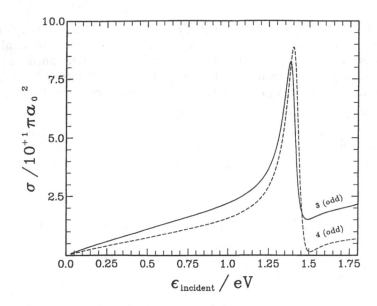

Fig. 14.6. Calculated partial cross sections with symmetry/parity combinations $J^{\Pi} = 3^{-}, 4^{-}$ for elastic scattering of electrons by Cs in the region from 0 to 1.8 eV. The energy of the incident electron, $\varepsilon_{\text{incident}}$, is given in eV.

calculations we used the Dirac Hamiltonian to define the dynamics of all electrons in the electron plus target system so that relativistic effects are included in both the orbitals and in the interactions between all pairs of electrons. We have computed the profiles of, and identified, resonances in the cross sections for the scattering of electrons by Hg, Pb and Cs. Our calculated positions of the resonances for the scattering of electrons by Hg are in good agreement with recent experiments. Our calculations of resonance structures and cross sections for the electron-impact excitation of neutral Pb and Cs need to be compared with experiment.

We are planning further investigations of the above systems, including differential cross sections and spin-dependent effects.

15

Electron-ion collisions using close-coupling and distorted-wave theory

D. C. Griffin and M. S. Pindzola

15.1   Introduction

In this chapter, we consider various theoretical methods for the calculation of electron-impact excitation and ionization in positively charged ions. First, we compare the distorted-wave and close-coupling approximations for the calculation of the non-resonant contributions to electron-impact excitation. The close-coupling method takes into account coupling effects to all orders of perturbation theory for those states included in the close-coupling expansion, while the distorted-wave method treats coupling effects by first-order perturbation theory. By comparing calculations using these two approximations along an isoelectronic sequence, we study continuum coupling as a function of ionization stage.

In many cases, excitation cross sections are dominated by strong resonance features. These occur when an electron in the continuum of an N-electron ion coincides in energy with a compound doubly-excited state of the $(N + 1)$-electron ion. The continuum electron can be captured into the compound state, and if this state then undergoes autoionization to an excited state of the N-electron ion, we have a resonance in the excitation cross section. Such resonances are treated in a natural way within the close-coupling method. However, using perturbation theory, it is also possible to include them within a distorted-wave formalism. We focus on the independent-processes approximation, in which interference between the individual resonances and between the resonances and the non-resonant background is ignored. In order to study the validity of this approximation, we consider transitions for which the resonant contributions are significant and compare close-coupling calculations with independent processes calculations along an isoelectronic sequence.

There are relatively few experimental measurements of excitation cross sections for ionized species. However, there exist a large number of

349

crossed-beam measurements of electron-impact ionization for which indirect contributions dominate the total ionization cross section. The
most significant contributions come from inner-shell excitation followed
by autoionization or dielectronic-capture followed by sequential double
autoionization. If one ignores the interference between direct ionization
and the indirect contributions, both the close-coupling and the distorted-
wave approximations can be employed to calculate the indirect processes.
In low-charge state ions, where coupling and interference effects are important, the close-coupling approximation seems the more appropriate.
However, it is often very difficult to include a sufficient number of states
within the close-coupling expansion to assure the accuracy of such calculations.

For highly ionized species, where coupling and interference effects are
greatly reduced, the independent-processes approximation appears to provide an accurate approach for including these contributions; it has the
added feature that radiation damping, which can become quite important in highly charged ions, can easily be included within the formalism.
In this chapter, we consider comparisons of such calculations with each
other and with experimental measurements, in order to illustrate these
important points.

15.2   Electron-impact excitation

15.2.1   Theoretical methods

We shall consider two methods for the calculation of electron-impact
excitation; the first is the close-coupling approximation, and the second is
based on the distorted-wave approximation. [597, 598] In the close-coupling
approximation, one first expands the wave function for the $(N+1)$–electron
system in the form:

$$\Psi = \mathbf{A} \sum_i \theta_i(\mathbf{x}_i, \cdots, \mathbf{x}_N, \hat{\mathbf{r}}_{N+1}, \sigma_{N+1}) \frac{1}{r_{N+1}} F_i(r_{N+1})$$
$$+ \sum_j c_j \phi_j(\mathbf{x}_i, \cdots, \mathbf{x}_{N+1}) , \tag{15.1}$$

where $\mathbf{A}$ is an operator that antisymmetrizes the scattered-electron coordinate with the N–electron coordinates; θ_i are channel functions formed
by coupling a target wavefunction with the angular and spin parts of
the scattered-electron wavefunction, with a radial function $F_i(r_{N+1})$; $\mathbf{x}_n = \mathbf{r}_n \sigma_n$, represents the space and spin coordinates of the nth electron. The
functions ϕ_j require a more complete explanation. They must be included
in the expansion when the radial wavefunctions for the continuum orbitals are forced to be orthogonal to radial wavefunctions for the bound
orbitals with the same angular momentum; by forcing this orthogonality,

closed channels, for which the scattered-electron wavefunction has the radial character of those particular bound orbitals, are projected out of the close-coupling expansion. Therefore, if the closed-channel portion of the close-coupling expansion is to be complete, these bound-states must be added back in. One can also add additional $(N + 1)$-electron bound states in order to better represent correlation within the $(N + 1)$-electron system.

The close-coupling equations are derived by employing the wavefunction of Eq. (15.1) and applying the variational principle. Although most formulations of the close-coupling approximation force complete orthogonality between the continuum and bound radial orbitals, and therefore include the $(N + 1)$-electron bound states in the close-coupling expansion, for simplicity, we will consider the form of these equations when this condition is not imposed. This form will also allow us to more clearly outline the development of the distorted-wave approximation for the calculation of non-resonant excitation cross sections. In this case, the close-coupling equation is as follows:

$$\left[-\frac{1}{2}\frac{d^2}{dr^2} + \frac{\ell_i(\ell_i + 1)}{2r^2} - \frac{Z}{r} + V_{ii} - \frac{k_i^2}{2} \right] F_{ii'}(r)$$
$$+ \sum_{j \neq i} V_{ij} F_{ji'}(r) = 0 . \qquad (15.2)$$

For atomic systems with low Z, the conserved quantum numbers are $LS\Pi$, where L is the total orbital angular momentum, S is the total spin, and Π is the parity of the $(N + 1)$-electron system. The channel index i then denotes the collection of quantum numbers $\gamma_i L_i S_i k_i \ell_i LS\Pi$, where γ_i represents all additional quantum numbers that are needed to completely characterize a term of the N-electron bound system with total orbital angular momentum L_i and total spin S_i; k_i and ℓ_i are the linear momentum and orbital angular momentum respectively of the continuum electron. The second index, i', on the radial wavefunction represents the channel index of the incident channel. The potential operators, V_{ij}, contain direct and exchange electrostatic terms. With the above equation, for which orthogonality between the bound and continuum orbitals is not forced, the potential operator also includes exchange-overlap terms, which arise because of the non-zero overlap between bound and continuum orbitals with the same orbital angular momentum.

Elements of the K-matrix are defined in terms of the asymptotic form of the radial functions, $F_{ii'}(r)$, for all open channels as

$$\lim_{r \to \infty} F_{ii'} \sim \frac{1}{\sqrt{k_i}} \left[\delta_{ii'} \sin(k_i x) + K_{ii'} \cos(k_i x) \right] , \qquad (15.3)$$

where $F_{ii'}(r) \to 0$ as $r \to 0$ and where

$$k_i x = k_i r - \frac{\ell_i \pi}{2} + \left(\frac{q}{k_i}\right) \ln(2k_i r) + \arg \Gamma \left[\ell_i + 1 - i\left(\frac{q}{k_i}\right)\right], \tag{15.4}$$

is the asymptotic phase of the regular Coulomb function.

In atomic units, the excitation cross section for the transition $\gamma_i L_i S_i \to \gamma_f L_f S_f$ is given by

$$\sigma_{if} = \frac{\pi}{k_i^2 2(2L_i + 1)(2S_i + 1)} \sum_{LS\ell_i \ell_f} (2L + 1)(2S + 1)|T_{fi}|^2, \tag{15.5}$$

where the transition matrix T is related to the K-matrix by the equation

$$\underline{T} = -\frac{2i\underline{K}}{1 - i\underline{K}}. \tag{15.6}$$

In this chapter, we will present the results of close-coupling calculations performed using two approaches. In the first approach, the close-coupling equations are solved using finite difference formulae, as in the University College program, IMPACT. [599] The second approach is to employ the R-matrix method, [518] first applied to atomic scattering by Burke and collaborators and described by Professor Burke in chapter thirteen. The R-matrix method is especially advantageous when a fine energy mesh must be employed in order to describe a rapidly varying cross section in the region of complex resonances. Some of the close-coupling calculations described here were performed using a modified version of the IMPACT program, while others employed the version of the RMATRX program developed for an opacity project. [600]

We now consider the distorted-wave approximation for direct, non-resonant excitation. We drop all potential terms V_{ij} (with $i \neq j$) in Eq. (15.2), which couple the various channels, and solve the differential equation

$$\left[-\frac{1}{2}\frac{d^2}{dr^2} + \frac{\ell_i(\ell_i + 1)}{2r^2} - \frac{Z}{r} + V_{ii} - \frac{k_i^2}{2}\right] f_i(r) = 0, \tag{15.7}$$

where the asymptotic form of the radial distorted-wave function $f_i(r)$ is given by

$$\lim_{r \to \infty} f_i(r) \sim \sin(k_i x + \delta_i), \tag{15.8}$$

and δ_i is the distorted-wave phase shift.

Now we can write the asymptotic form of the solutions to the close-coupling equations in terms of the distorted-wave phase shift by defining the elements of the ρ-matrix by the equation

$$\lim_{r \to \infty} F_{ii'} \sim \frac{1}{\sqrt{k_i}} \left[\delta_{ii'} \sin(k_i x + \delta_i) + \rho_{ii'} \cos(k_i x + \delta_i)\right]. \tag{15.9}$$

The ρ-matrix elements are derived using first-order perturbation theory. One first multiplies Eq. (15.2) from the left by $f_i(r)$ and integrates the resulting equation from 0 to ∞. This is integrated by parts using the asymptotic form of $F_{ii'}(r)$ given in Eq. (15.9) and $f_i(r)$ given in Eq. (15.8), as well as the distorted-wave differential equation given in Eq. (15.7). The off-diagonal elements of the ρ-matrix are found to be given by the expression

$$\rho_{fi} = -\frac{2}{\sqrt{k_i k_f}} \int_0^\infty f_f(r) V_{fi}(r) f_i(r) dr , \qquad (15.10)$$

and the diagonal elements of ρ are zero.

As in Eq. (15.6), one can then relate a transition matrix τ to the ρ-matrix by the equation

$$\underline{\tau} = -\frac{2i\underline{\rho}}{(1 - i\underline{\rho})} . \qquad (15.11)$$

We can then employ elements of the τ-matrix, in place of T-matrix elements, in Eq. (15.5) to calculate the unitarized distorted-wave cross section for the transition $\gamma_i L_i S_i \rightarrow \gamma_f L_f S_f$. In this form, the distorted-wave approximation conserves probability and includes a limited amount of coupling. In many cases, it is useful to compare K-matrix elements and T-matrix elements calculated from the distorted-wave approximation with those calculated in the close-coupling approximation. This can be accomplished by using the transformation [601]

$$\underline{K} = \frac{\sin(\delta)\underline{1} + \cos(\delta)\underline{\rho}}{\cos(\delta)\underline{1} - \sin(\delta)\underline{\rho}} . \qquad (15.12)$$

The elements of the T-matrix can then be determined from Eq. (15.6) or from the transformation [601]

$$\underline{T} = e^{i\delta} \underline{\tau} e^{-i\delta} . \qquad (15.13)$$

In Eqs. (15.12) and (15.13), δ is a diagonal matrix with elements δ_i.

When the elements of the ρ-matrix are small, then

$$\underline{\tau} \approx -2i\underline{\rho} , \qquad (15.14)$$

and we can determine a non-unitarized distorted-wave direct (non-resonant) excitation cross section for the transition $\gamma_i L_i S_i \rightarrow \gamma_f L_f S_f$ from the equation

$$\sigma_{if} = \frac{4\pi}{k_i^2 2(2L_i + 1)(2S_i + 1)} \sum_{LS\ell_i\ell_f} (2L + 1)(2S + 1) |\rho_{fi}|^2 . \qquad (15.15)$$

However, the non-unitarized distorted-wave approximation does not conserve probability.

We now consider the inclusion of the resonance contributions using the distorted-wave approximation and perturbation theory. We first define the $(N + 1)$-electron wavefunction for an open channel, i, by the equation

$$\Phi_i = A\theta_i(\mathbf{x}_i, \cdots, \mathbf{x}_N, \hat{\mathbf{r}}_{N+1}, \sigma_{N+1})\frac{1}{r_{N+1}}f_i(r_{N+1}), \qquad (15.16)$$

while for a closed channel, m, it is given by

$$\Phi_i = A\theta_m(\mathbf{x}_i, \cdots, \mathbf{x}_N, \hat{\mathbf{r}}_{N+1}, \sigma_{N+1})\frac{1}{r_{N+1}}P_m(r_{N+1}), \qquad (15.17)$$

where $P_m(r) \to 0$ as $r \to \infty$. The non-unitarized distorted-wave excitation cross section for the transition $\gamma_i L_i S_i \to \gamma_f L_f S_f$, which includes the lowest-order contributions from resonances arising from dielectronic capture followed by autoionization may be given by the Breit-Wigner expression:

$$\sigma_{if} = \frac{16\pi}{k_i^3 k_f 2(2L_i + 1)(2S_i + 1)}\sum_{LS\ell_i\ell_f}(2L + 1)(2S + 1)$$

$$\times \left|\langle\Phi_f \mid V \mid \Phi_i\rangle + \sum_m \frac{\langle\Phi_f \mid V \mid \Phi_m\rangle\langle\Phi_m \mid V \mid \Phi_i\rangle}{\varepsilon - \varepsilon_m + i\Gamma_m/2}\right|^2, \qquad (15.18)$$

where $\varepsilon = k_i^2/2$ is the electron energy, ε_m is the energy of a resonance state m, with respect to the energy of the initial-state ion energy; and Φ_m is the wavefunction for the resonance state given by Eq. (15.17). Γ_m is the total width due to all radiative and autoionizing transitions:

$$\Gamma_m = \sum_n \mathscr{A}_a(m \to n) + \sum_k \mathscr{A}_r(m \to k); \qquad (15.19)$$

where the autoionizing rate $\mathscr{A}_a(m \to n)$, from the resonance state m to a final state n of the N–electron ion, is given by

$$\mathscr{A}_a(m \to n) = \frac{4}{k_n}\sum_{\ell_n}|\langle\Phi_n \mid V \mid \Phi_m\rangle|^2; \qquad (15.20)$$

and the radiative rate $\mathscr{A}_r(m \to k)$, from the resonance state m to the final bound state k of the $(N + 1)$–electron ion, is given by

$$\mathscr{A}_r(m \to k) = \frac{4\omega^3}{3c^3}\left|\left\langle\Phi_k \mid \sum_{s=1}^{N+1}\mathbf{r}_s \mid \Phi_m\right\rangle\right|^2. \qquad (15.21)$$

With the inclusion of the radiative rate within the width Γ_m, one accounts for the effects of radiation damping, which can become important in highly charged ions.

If one ignores all interference terms in Eq. (15.18), we obtain the independent-processes approximation for the excitation cross section,

$$\sigma_{if} = \sigma_{if}^{NR} + \sum_m \sigma_{im}^{DC} \frac{\mathscr{A}_a(m \to f)}{\Gamma_m}, \qquad (15.22)$$

where the first term is the non-resonant excitation cross section, given by Eq. (15.15); and

$$\sigma_{im}^{DC} = \frac{8\pi^2 (2L + 1)(2S + 1)}{2k_i^3 (2L_i + 1)(2S_i + 1)} \sum_{\mathscr{L}_i} |\langle \Phi_m | V | \Phi_i \rangle|^2 \mathscr{L}_m(\varepsilon) \qquad (15.23)$$

is the cross section for dielectronic-capture to the state m, with total angular momentum L and total spin S; and

$$\mathscr{L}_m(\varepsilon) = \frac{\Gamma_m/2\pi}{(\varepsilon - \varepsilon_m)^2 + \Gamma_m^2/4} \qquad (15.24)$$

is the Lorentz profile. In many cases, when the detailed resonance shapes are not important, one simply integrates Eq. (15.23) over a small energy interval $\Delta\varepsilon$ which is large compared to the largest resonance width. One then obtains an expression identical to Eq. (15.23), except that $\mathscr{L}_m(\varepsilon)$ is replaced by $1/(\Delta\varepsilon)$.

15.2.2 Non-resonant excitation in K-like ions

In order to illustrate the effects of continuum coupling in direct (non-resonant) excitation cross sections as a function of ionization stage, we have compared three-state close-coupling calculations with distorted-wave calculations for the K-like ions Ca^+, Sc^{2+}, Ti^{3+}, Cr^{5+}, and Fe^{7+}. [602] The close-coupling calculations were performed using a modified version of the program IMPACT. [599] They were not intended to provide accurate cross sections, since more complete calculations with more states are required to obtain accurate results, that include important resonance contributions. [603] However, these results nicely illustrate the way in which coupling effects, not included in the distorted-wave approximation, decrease as a function of ionization stage.

The results are summarized in Fig. 15.1, where we plot the ratio of the non-unitarized distorted-wave cross sections to the three-state close-coupling cross sections for the $4s \to 4p$ and $3d \to 4s$ transitions as a function of electron energy, in threshold units. These results are somewhat complicated by the fact that $4s$ is the ground state for Ca^+, while $3d$ is the ground state for all other ions. The $4s \to 4p$ dipole transition shows a uniform convergence to 1.0 as a function of both energy and charge state. However, this ratio for the weaker $3d \to 4s$ quadrupole transition changes rapidly at low energies due to strong coupling effects, and only begins to converge to 1.0 at high energies. For all these ions, the unitarized

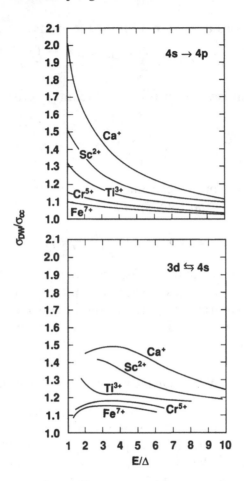

Fig. 15.1. Non-unitarized distorted-wave to close-coupling cross-section ratios for the $4s \to 4p$ and $3d \rightleftharpoons 4s$ transitions along the K-isoelectronic sequence (see Ref. [602]). The energy is in threshold units E/Δ, where Δ is the threshold energy of the transition and E is the electron energy.

distorted-wave cross sections are in better agreement with the three-state close-coupling results, especially for the $4s \to 4p$ transition, indicating the improvement provided by the unitarization step.

These results support the general assumption that coupling of states through the presence of the continuum electron decreases as a function of ionization stage. Thus the non-unitarized distorted-wave method provides a reasonably accurate method of obtaining the non-resonant background cross section for more highly charged ions. This is less true for non-dipole transitions; however, the non-resonant cross sections for these excitations are normally small and the total cross sections are often dominated by the resonant contributions.

An additional point regarding the effects of continuum coupling on non-resonant excitation is important. Angular differential cross sections are more sensitive to the details of a scattering calculation, than total cross sections. For example, calculations in Li^+ show that the differential cross sections determined from the close-coupling method, the unitarized distorted-wave method, and the non-unitarized distorted-wave method can be quite different, even when the total cross sections are in reasonably close agreement. [604] However, as illustrated by calculations in Ar^{7+}, these differences are found again to fall off with electron energy and ionization stage. [605, 606, 607]

15.2.3 *Electron-impact excitation including resonances for* Na-*like ions*

Although the effects of continuum coupling are not pronounced for ions with intermediate and high charge states, the effects of resonances are often dominant for all stages of ionization. It is important to determine the accuracy of the independent-processes approximation, using distorted-waves, in representing these resonant contributions. A series of detailed close-coupling and independent-processes calculations of the $3s \rightarrow 3p$ and $3s \rightarrow 3d$ transitions in the Na-like ions Si^{3+}, Ar^{7+}, and Ti^{11+} was performed, in order to try to obtain a better understanding of the validity of the independent-processes approximation. [608] Seven-state close-coupling calculations ($3s$, $3p$, $3d$, $4s$, $4p$, $4d$, and $4f$) were made using the R-matrix programs developed for an opacity project. [600] In the independent-processes calculations, we included resonances of the form $3dn\ell$ and $4\ell n\ell'$ for the $3s \rightarrow 3p$ transition and of the form $4\ell n\ell'$ for the $3s \rightarrow 3d$ transition. For both sets of calculations, we used the same Hartree-Fock bound-state orbitals. These calculations were performed in LS coupling; effects of intermediate coupling and radiation damping were investigated using the independent-processes approximation, and were found to be negligible for these transitions in the ions considered.

The results of these calculations for the $3s \rightarrow 3p$ transition in Si^{3+} and Ti^{11+} are shown in Figs. 15.2 and 15.3, respectively. For Si^{3+}, there are some noticeable differences between the close-coupling and independent-processes results. For example, there is a large window resonance in the close-coupling results at 11 eV, which is apparently due to interference between the background cross section and the $3d4d\,^3G$ resonance. Similar features are repeated at higher energies. As can be seen from Fig. 15.3, the agreement is much better in Ti^{11+} in terms of both the background and resonance contributions. Improvement in the agreement between the two approximations as a function of ionization stage is also observed for the $3s \rightarrow 3d$ transition. These results indicate that the independent-processes approximation should provide an efficient and accurate method

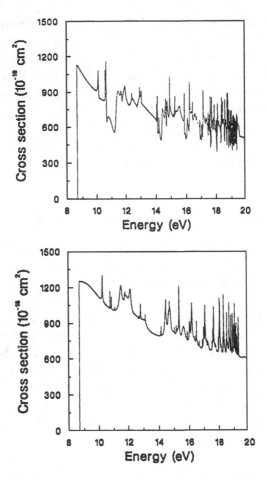

Fig. 15.2. Excitation cross section for the $3s \to 3p$ excitation in Si^{3+}. Upper graph, seven-state close-coupling approximation; lower graph, independent-processes approximation (see Ref. [608]).

of incorporating the details of the resonance contributions for intermediate and high charge states. Even in lower charge-state ions, the independent-processes approximation may be sufficiently accurate for the determination of rate coefficients. However, additional studies of this type are required to confirm these conclusions, and we are now in the process of performing similar calculations in Mg-like ions.

Although very few measurements of electron-impact excitation exist for multiply-charged ions, measurements of the $3s \to 3p$ transition in Si^{3+}, and more recently in Ar^{7+}, have been completed in the threshold region using a merged-beams, energy-loss technique. [609, 610] The experimental measurements [609] are shown in comparison to our close-coupling cal-

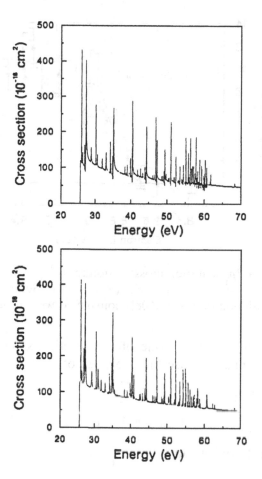

Fig. 15.3. Excitation cross section for the $3s \rightarrow 3p$ excitation in Ti^{11+}. Upper graph, seven-state close-coupling approximation; lower graph, independent-processes approximation (see Ref. [608]).

culations for Si^{3+} in Fig. 15.4. As can be seen, the agreement is quite good; however, the measurements were not performed at a high enough energy to include the important resonance contributions. High resolution measurements for such ions in the resonance region are urgently needed in order to test theory.

15.3 Electron-impact ionization

15.3.1 Theoretical methods

The numerous reaction pathways found in the electron-impact ionization of atomic ions may be grouped by the highest-order of perturbation

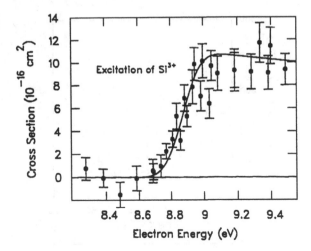

Fig. 15.4. The $3s \to 3p$ excitation cross section for Si^{3+} in the threshold region. Experimental points from Wahlin *et al.* (Ref. [609]). Solid curve, seven-state close-coupling calculation (see Ref. [608]) convoluted with a 0.2 eV Gaussian.

theory needed to describe any one of the reaction steps. In first order, leading to single ionization, are the processes of

- direct ionization

$$e^- + A^{q+} \to A^{(q+1)+} + e^- + e^- \; ; \tag{15.25}$$

- excitation-autoionization

$$\begin{aligned} e^- + A^{q+} \to \quad & [A^{q+}] \quad + e^- \\ & \hookrightarrow \quad A^{(q+1)+} + e^- \; ; \end{aligned} \tag{15.26}$$

- dielectronic capture followed by sequential double autoionization

$$\begin{aligned} e^- + A^{q+} \to [A^{(q-1)+}] \\ & \hookrightarrow \quad [A^{q+}] \quad + e^- \\ & \hookrightarrow \quad A^{(q+1)+} + e^- . \end{aligned} \tag{15.27}$$

The bracket indicates a resonance state, while q is the charge on atom A. All of these first-order processes can make substantial contributions to the total single-ionization cross section and have been observed experimentally.

Among the many second-order reaction pathways, leading to single ionization, is the process of

- dielectronic capture followed by simultaneous double autoionization

$$e^- + A^{q+} \rightarrow [A^{(q-1)+}]$$
$$\hookrightarrow A^{(q+1)+} + e^- + e^- . \qquad (15.28)$$

This second-order process has also been observed experimentally.

A mixture of first- and second-order reaction pathways leads to double ionization. The direct knockout of two electrons is a correlated second-order process, while inner-shell ionization followed by autoionization is a sequence of first-order processes. The theoretical study of electron-ion scattering leading to single, double, and even triple ionization affords the opportunity to explore many different facets of the many-body problem in atomic physics.

In LS coupling, the direct-ionization cross section from an N–electron initial term $(\gamma_i L_i S_i)$ to an $(N - 1)$–electron final term $(\gamma_f L_f S_f)$ is given by [611, 612]

$$\sigma = \sum_{\ell_i, \ell_e, \ell_f} \sum_{L'S'} \sum_{LS} \int_0^{\frac{E_{\max}}{2}} \frac{d\sigma}{d\varepsilon_e} d\varepsilon_e , \qquad (15.29)$$

and the differential cross section with ejected energy, ε_e, is given by

$$\frac{d\sigma}{d\varepsilon_e} = \frac{16}{k_i^3 k_e k_f} \frac{(2L+1)(2S+1)}{(2L_i+1)(2S_i+1)} |\langle \Phi_f | V | \Phi_i \rangle|^2 . \qquad (15.30)$$

The initial-state wavefunction, Φ_i, carries quantum numbers $(\gamma_i L_i S_i \ell_i L S)$, and that of the final state, Φ_f, carries the quantum numbers $(\gamma_f L_f S_f \ell_e L'S' \ell_f L S)$. The expression (15.30) is first-order in the perturbation

$$V = \sum_{i<j=1}^{N+1} \frac{1}{r_{ij}} - \sum_{i=1}^{N} V_{N-1}(r_i) - V_N(r_{N+1}) , \qquad (15.31)$$

which is simply the difference between the "exact" electrostatic interaction between electrons and their central-field potentials. The triads (ℓ_i, ℓ_e, ℓ_f) and (k_i, k_e, k_f) are the angular- and linear-momentum quantum numbers for the initial, ejected, and final continuum electrons. Energy conservation is given by

$$E_{\max} = -I_p + \frac{k_i^2}{2} = \frac{k_e^2}{2} + \frac{k_f^2}{2} , \qquad (15.32)$$

where I_p is the N–electron target ionization potential.

The differential cross section with ejected energy, including direct-ionization and excitation-autoionization contributions, may be written in the Breit-Wigner form as

$$\frac{d\sigma}{d\varepsilon_e} = \frac{16}{k_i^3 k_e k_f} \frac{(2L+1)(2S+1)}{(2L_i+1)(2S_i+1)}$$

$$\times \left| \langle \Phi_f \mid V \mid \Phi_i \rangle + \sum_n \frac{\langle \Phi_f \mid V \mid \Phi_n \rangle \langle \Phi_n \mid V \mid \Phi_i \rangle}{\varepsilon_f - \varepsilon_n + i\Gamma_n'/2} \right. \tag{15.33}$$

$$\left. + \sum_n \frac{\langle \Phi_f \mid V \mid \Phi_n \rangle \langle \Phi_n \mid V \mid \Phi_i \rangle}{\varepsilon_e - \varepsilon_n + i\Gamma_n'/2} \right|^2 .$$

The resonant-state wavefunction, Φ_n, carries the quantum numbers $(\gamma_n L_n S_n \ell L S)$, the resonant energy, ε_n, is with respect to the final-state ion energy, and the quantity Γ_n' is the sum over all possible autoionizing and radiative rates for the decay of each resonant state. As discussed by Moores and Reed, [613] the energy dependence of a resonant contribution to $d\sigma/d\varepsilon_e$ is unusual. For low energies, such that $E_{max} < \varepsilon_n$, no resonance structure appears and only the first term of Eq. (15.33) makes a contribution. For slightly higher energies, such that $E_{max}/2 < \varepsilon_n < E_{max}$, the second term of Eq. (15.33) contributes a resonance structure to $d\sigma/d\varepsilon_e$. In this energy range the resonance appears to move, since it is a function of ε_f and not ε_e. Finally at higher energies, such that $\varepsilon_n < E_{max}/2$, the third term of Eq. (15.33) contributes a resonance structure to $d\sigma/d\varepsilon_e$. For all higher energies the resonance remains fixed. At present no electron-ion collision experiments have been performed which measure resonance structures in the differential cross section with ejected energy.

If one ignores all interference terms in Eq. (15.33), we obtain the independent-processes approximation for the total ionization cross section:

$$\sigma = \sigma_{DI} + \sum_n \sigma_{in}^{NR} \frac{\mathscr{A}_a(n \to f)}{\Gamma_n'} , \tag{15.34}$$

where σ_{DI} is the direct-ionization cross section of Eqs. (15.29)-(15.30) and σ_{in}^{NR} is the non-resonant excitation cross section given by Eq. (15.15). To further include dielectronic-capture followed by sequential double autoionization contributions, within the independent-processes approximation, we simply replace σ_{in}^{NR} in Eq. (15.34) by σ_{in} of Eq. (15.22). The resulting expression for the total-ionization cross section may be written as

$$\sigma = \sigma_{DI} + \sum_n \sigma_{in}^{NR} \frac{\mathscr{A}_a(n \to f)}{\Gamma_n'} + \sum_n \sum_m \sigma_{im}^{DC} \frac{\mathscr{A}_a(m \to n)}{\Gamma_m} \frac{\mathscr{A}_a(n \to f)}{\Gamma_n'} ,$$

$$\tag{15.35}$$

which clearly reflects the three (independent) processes of Eqs. (15.25)-(15.27).

In the independent-processes approximation, we may also add a contribution from dielectronic capture followed by simultaneous double autoionization,

$$\sum_m \sigma_{im}^{DC} \frac{\mathscr{A}_{da}(m \to f)}{\Gamma_m''} , \qquad (15.36)$$

where the simultaneous double autoionization rate is given by

$$\mathscr{A}_{da}(m \to f) = \frac{8}{\pi} \int_0^{k_{max}} \frac{dk_e}{k_f} \left| \sum_p \int_{k_p} \frac{\langle \Phi_f \mid V \mid \Phi_p \rangle \langle \Phi_p \mid V \mid \Phi_m \rangle}{\varepsilon_m - \varepsilon_p - k_p^2/2} \right|^2 , \qquad (15.37)$$

and Γ_m'' is the sum over all possible autoionizing and radiative decay rates, including the simultaneous double autoionizing decay rate.

15.3.2 Direct ionization of F^{2+} and W^+

In Fig. 15.5 we compare theory and experiment for the single ionization of

F^{2+}. The solid curve is the contribution from the direct ionization of the $2s$ and $2p$ subshells calculated using the first-order expression of Eqs. (15.29),(15.30). As shown there is good agreement over a wide energy range with the crossed-beams experiment of Mueller *et al.* [614] In Fig. 15.6 we compare theory and experiment for the single ionization of W^+. The solid curve is the contribution from the direct ionization of the $5d$ and $6s$ subshells calculated [615] using Eqs. (15.29)-(15.30). As shown the crossed-beams measurements of Montague and Harrison [616] are approximately 40% below the theory at the peak of the data. For both F^{2+} and W^+ we found little contribution from the resonance processes of Eqs. (15.26) and (15.27). The large differences found between theory and experiment for W^+ may then be attributed to the inadequacy of first-order theory for the direct ionization process. Many other examples of significant differences between theory and experiment for the direct ionization process may be given, especially for low-charged heavy atomic ions.

Efforts have been made to go beyond the first-order theory of Eqs. (15.29),(15.30) for the calculation of the direct ionization cross section. Pan and Kelly [475] have examined the relative strengths of the numerous second-order contributions in a calculation for the direct ionization cross section of Helium. A limited amount of initial and final state correlation has also been included in the study of inner-shell direct ionization of heavy atomic ions. [617, 618] Certainly more work is needed before accurate cross section calculations can be made for the direct ionization of heavy atomic ions.

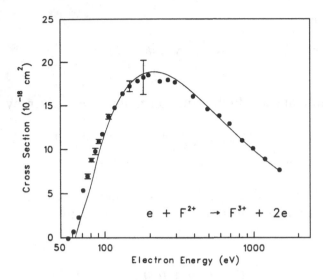

Fig. 15.5. Electron-impact ionization of F^{2+}. Experimental points from Mueller *et al.* (Ref. [614]). Solid curve is a distorted-wave calculation including only direct ionization.

15.3.3 *Excitation-autoionization of* Xe^{6+} *and* Fe^{15+}

In Fig. 15.7 we compare theory and experiment for the single ionization of Xe^{6+}. The dashed curve is the contribution from the direct ionization of the $5s$ subshell, while the solid curve includes additional excitation-autoionization contributions. [619] As shown there is good agreement over a wide energy range with the crossed-beams experiment of Gregory and Crandall. [620] As predicted by Eq. (15.34) the total ionization cross section steps up every time a resonance state becomes energetically accessible. The large step at 101 eV is due to the $4d^{10}5s^2\ {}^1S \rightarrow 4d^95s^24f\ {}^1P$ excitation followed by autoionization. Further contributions are obtained from excitation to autoionizing states of the $4d^95s^25d$ and $4d^95s^25f$ configurations. Since the branching ratio for autoionization, $\mathscr{A}_a(n \rightarrow f)/\Gamma'_n$, is approximately one for most of the resonant states of Xe^{6+}, the comparison with experiment becomes a direct test of our ability to calculate excitation cross sections. For Xe^{6+} the first-order perturbation theory expression of Eq. (15.15) is found to be perfectly adequate.

In Fig. 15.8 we give an energy level diagram for the single ionization of Fe^{15+}. The main excitation-autoionization contributions involve the excitation of the $2p^53s3\ell$ configurations, although contributions from the $2p^53s4\ell$, $2p^53s5\ell$, $2s2p^63s3\ell$, and $2s2p^63s4\ell$ configurations are non-negligible. We choose Fe^{15+} to illustrate the importance of a proper calculation for the branching ratio for autoionization in more highly

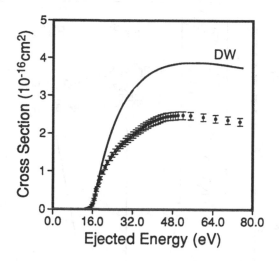

Fig. 15.6. Electron-impact ionization of W^+. Experimental points from Montague and Harrison [616]. Solid curve is a distorted-wave calculation including only direct ionization (see [615]).

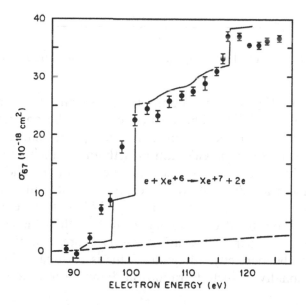

Fig. 15.7. Electron-impact ionization of Xe^{6+}. Experimental points from Gregory and Crandall [620]. Solid curve is a distorted-wave calculation including direct ionization and excitation-autoionization contributions (see [619]); dashed curve is direct ionization only.

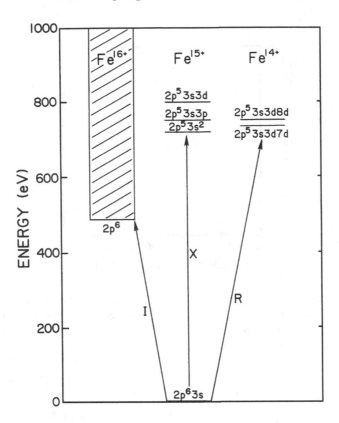

Fig. 15.8.   Energy level diagram for electron ionization of Fe^{15+}.

charged ions. In Fig. 15.9, we compare theory and experiment for the single ionization of Fe^{15+}. The dashed curve is the contribution from the direct ionization of the $3s$ subshell, the upper, chained, curve is a total ionization cross section calculation with unit branching ratios for the excitation-autoionization contributions, while the lower, solid, curve properly includes radiation damping of the excitation-autoionization contributions (Ref. [621]). The solid curve is in slightly better agreement with the crossed-beams experiment of Gregory et al. [622] than the chained curve. For Fe^{15+} the first-order perturbation theory expression of Eq. (15.15) is again found to be perfectly adequate. This is fortunate since it is difficult computationally to include radiation damping in a non-perturbative approach.

Also shown in Fig. 15.8 are the configurations $2p^53s3dnd$ of Fe^{14+}, whose states are formed through dielectronic capture of the incident electron by Fe^{15+}. They, along with hundreds of other possible configurations, may contribute to the ionization of Fe^{15+} by the further process of sequential double autoionization (see Eq. 15.27). Contributions from dielectronic

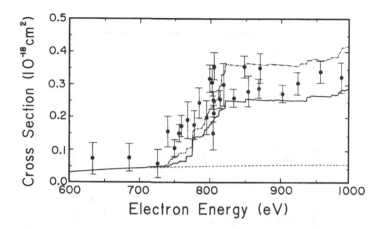

Electron Energy (eV)

Fig. 15.9. Electron-impact ionization of Fe^{15+}. Experimental points from Gregory *et al.* [622]. Solid curve is a distorted-wave calculation including direct ionization and excitation-autoionization contributions (see [621]); chained curve is a total cross section with unit branching ratios; dashed curve is direct ionization only.

capture resonances to the total ionization cross section of Fe^{15+} have been calculated by several groups. [623, 624, 625] The independent processes approximation calculations by Chen *et al.* [625] find that the dielectronic capture resonances contribute about 30% to the average total ionization cross section in the energy range from 720 eV to 940 eV. Tests of our ability to calculate the detailed resonance structure for Fe^{15+} await further improvements in the experimental observations.

15.3.4 Dielectronic-capture resonances in C^{3+}

In Fig. 15.10 we give an energy-level diagram for the single ionization of C^{3+}. The main excitation-autoionization contributions involve the excitation of the $1s2s2\ell$ configurations with almost unit branching ratios of almost unity. In Fig. 15.11 the crossed-beams measurements of Müller *et al.* [626] are compared with close-coupling calculations [627] for the $1s^2 2s \rightarrow 1s2s2\ell$ excitations added to a scaled distorted-wave calculation [611] for the direct $2s$ ionization cross section. The individual steps in the excitation cross section predicted by theory are beginning to be resolved by the experiment. Further close-coupling [628] and distorted-wave [629] calculations attribute the largest resonance features in the experimental cross section above 310 eV as due to dielectronic capture to the $1s2s3\ell3\ell'$ configurations which then undergo a sequential double autoionization process.

Also shown in Fig. 15.10 are the configurations $1s2s^2 2p$ and $1s2s2p^2$ of

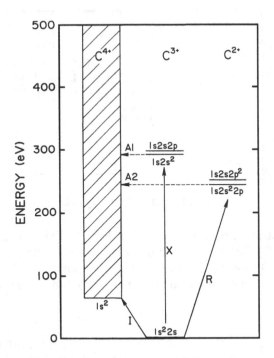

Fig. 15.10.   Energy level diagram for electron ionization of C^{3+}.

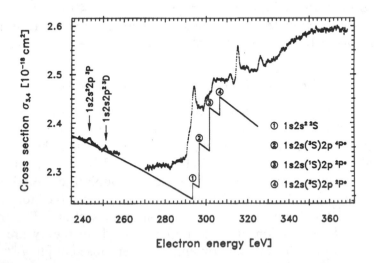

Fig. 15.11.   Electron-impact ionization of C^{3+}. Experimental points from Müller *et al.* [626]. Solid curve is a scaled distorted-wave calculation for direct ionization (see Ref. [611]) plus a close-coupling calculation for excitation-autoionization contributions.

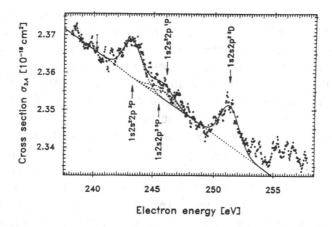

Fig. 15.12. Electron-impact ionization of C^{3+}. Experimental points from Müller *et al.* [626]. Dashed curve is a scaled distorted-wave calculation for direct ionization (see [611]). Solid curve is an experimental fit to the scan data.

C^{2+}, whose states are formed through dielectronic capture of the incident electron by C^{3+}. They, due to their energy position, may only contribute to the ionization of C^{3+} by the further process of simultaneous double autoionization (see Eq. 15.28). In Fig. 15.12 the crossed-beams measurements of Müller *et al.* [626] are shown on an expanded energy scale to show resonance structure which may be attributed to this exotic process. The experimentally determined resonance cross section through the $1s2s^22p\ {}^3P$ multiplet of C^{2+} is in good agreement with perturbation-theory calculations [630] made using Eqs. (15.36), (15.37). The key element of the theory is the determination of the double Auger rate of Eq. (15.37), which was first calculated using many-body perturbation theory for the $2s^22p\ {}^2P$ multiplet in Li by Simons and Kelly. [287] These particular experiments enable us to check our understanding of resonance-decay processes which proceed only through non-central-field electron correlation.

15.4 Summary

The independent-processes approximation using distorted waves provides an efficient means of calculating electron-impact excitation and ionization cross sections in highly charged ions, where continuum-coupling effects and interference effects are relatively small. It also provides an effective means of incorporating radiation damping, which can be important in such ions. However, additional calculations of the type reviewed in this chapter, along with high-resolution measurements, are needed to further test the validity of these perturbation-theory approaches.

The most significant challenge facing electron-ion scattering theory is the determination of accurate excitation and ionization cross sections for heavy atomic ions in relatively low charge states. In such ions, the bound-state atomic structure is quite complex and must incorporate relativistic effects within intermediate-coupling calculations of individual levels. In addition, continuum coupling effects and interference between various resonant processes can become very important. However, it is very difficult in such complex systems to include a sufficient number of states within non-perturbative close-coupling calculations. This is especially true for the inner-shell excitation-autoionization contributions to ionization, where a large number of states for both singly excited and doubly excited configurations must be included. In addition, in such complex ions, first order theory, using distorted waves, does not provide accurate cross sections for direct ionization. It is clear that a great deal of additional research is needed in order to find an efficient means of dealing with these processes in complex heavy atomic ions.

Appendix: Units and notation

Introduction

The purpose of the appendix is to acquaint the reader with the common notation used throughout this book, as well as to point out where there are differences.

For example, there occur two conventions of diagramatic notation: vertical particle-hole lines (Chapters 1, 2, 4, 6, 11); and left-right particle-hole lines (Chapter 8). There is also the use of atomic units (Hartree units), Rydberg units of energy, and the occasional mention of electronvolts. Finally, there are two phase conventions that are used in relativistic quantum mechanics. One convention is utilized in Chapters 2 and 3, [631] and the other is utilized in Chapter 14. [632]

An effort was made to coalesce the majority of notation in this book regarding the physical variables, while the phase conventions and other notation were left unaltered in order to reflect the range of notations currently used.

Units

Atomic units and conversion factors [633] are as follows (in SI).

$$
\begin{aligned}
e &= \text{charge on the electron} \\
&= 1 \text{ a.u.} \\
&= 1.602177 \times 10^{-19} \text{ C}; \\
m_e &= \text{mass of the electron} \\
&= 1 \text{ a.u.} \\
&= 9.109390 \times 10^{-31} \text{ kg};
\end{aligned}
$$

$$\hbar \ = \ \text{Planck's constant}/2\pi$$
$$= \ 1 \ \text{a.u.}$$
$$= \ 1.054573 \times 10^{-34} \ \text{J s} \ ;$$
$$4\pi\epsilon_0 \ = \ \text{unit of permittivity}$$
$$= \ 1 \ \text{a.u.}$$
$$= \ 1.112650 \times 10^{-10} \ \text{C}^2/(\text{N m}^2) \ ;$$
$$\text{unit of energy} \ = \ 1 \ \text{a.u.}$$
$$= \ 2 \times 13.60570 \text{eV}$$
$$= \ 2 \times \text{rydbergs (Ryd)}$$
$$= \ 4.359748 \times 10^{-18} \ \text{J};$$
$$a_0 \ = \ \text{Bohr radius}$$
$$= \ 1 \ \text{a.u.}$$
$$= \ 5.291772 \times 10^{-11} \ \text{m} \ ;$$
$$\text{sr} \ = \ \text{designation for steradians};$$
$$1 \ \text{Mb} \ = \ 1 \ \text{megabarn}$$
$$= \ 1 \times 10^{-22} \ \text{m}^2 \ .$$

Notation

Some of the common notation which is used among the chapters is as follows. Also included, for some of the variables, is the equation number where it is first used.

$\mathbf{A}$ = antisymmetrization operator, Eq. (1.25);

$\mathbf{A}(t)$ = electromagnetic vector potential, Eq. (13.28);

A = designation for an arbitrary atom;

a_0 = designation for bohr radius;

a_μ = designation for the bohr radius of muonic hydrogen, Eq. (8.54);

$a^\dagger, a$ = creation, annihilation operator, Eq. (1.76);

$\boldsymbol{\alpha}$ = relativistic spin matrix, Eq. (2.2);

α_{pol} = dipole polarizability, Eq. (4.38);

α = $\dfrac{e^2}{4\pi\epsilon_0 \hbar c}$, fine structure constant;

β = relativistic spin matrix, Eq. (2.2);

$\beta(\omega)$ = beta asymmetry parameter, Eq. (7.1);

β_{pol} = vector polarizability, Eq. (3.66);

β_μ = inverse of the muon mass: $\approx 1/205$ a.u., Eq. (8.54);

C_{LM} = $\sqrt{\dfrac{4\pi}{2L+1}}\,Y_{LM}$, Eq. (2.22);

c = speed of light, Eq. (2.2);

γ_i = unspecified angular momentum quantum number, Eq. (4.65);

γ = Redmond phase, Eq. (9.6);

γ^μ = relativistic gamma matrices, Eq. (3.1), for $\mu = 0, 1, 2, 3$, and $i = 1, 2, 3$, $\gamma^0 = \beta$, $\gamma^i = \beta\alpha_i$;

γ_5 = $i\gamma^0\gamma^1\gamma^2\gamma^3$, Eq. (3.1);

Γ_i = width of resonance, Eq. (5.47);

$\widehat{\Gamma}$ = RPAE operator, Eq. (8.50);

Γ = photon momentum, Eq. (2.46);

$\mathbf{D}$ = dipole operator, Eq. (3.7);

D = energy denominator, Eq. (1.92);

d_{ij} = dipole matrix element, Eq. (3.13);

δ_ℓ = non-coulomb phase shift associated with angular momentum ℓ, Eq. (4.9);

δ_ℓ^C = coulomb phase shift associated with angular momentum ℓ, Eq. (4.9);

$\delta(x)$ = Dirac delta function, Eq. (4.6);

$e^{-,+}$ = designation for the electron, positron;

$e(\mathbf{x})$ = four-component spinor for the electron, Eq. (3.1);

e = base of natural logarithms $= 2.71828\cdots$;

E_i = total energy of state i ;

ε_i = single particle energy of orbital ϕ_i ;

$\hat{\epsilon}$ = polarization vector, Eq. (13.29);

ϵ_0 = permittivity of vacuum ;

F_{ij} = radial close-coupling matrix, Eq. (13.20);

$f(E, \theta)$ = scattering amplitude, Eq. (8.22);

g_{ijkl} = two-particle matrix element, Eq. (1.79), or coulomb matrix element, Eq. (2.37);

$\tilde{g}_{ijkl}$ = antisymmetrized coulomb matrix element, Eq. (2.69);

$\eta \quad = \quad$ arbitrary small quantity ≥ 0;

$H \quad = \quad$ many-particle Hamiltonian;

$h_i \quad = \quad$ single-particle Hamiltonian for orbital i;

$I_i \quad = \quad$ ionization threshold for state i, $I_i \geq 0$, Eq. (4.6);

$J_i \quad = \quad$ total $L + S$ quantum number of state i,
$\qquad\qquad\quad$ LSJ-coupling);

$K \quad = \quad -2\mathbf{L}\cdot\mathbf{S} - 1$, Eq. (2.8);

$\underline{K} \quad = \quad K$-matrix, Eq. (13.21) ;

$\mathbf{k}_i \quad = \quad$ linear momentum of orbital i;

$L_i \quad = \quad$ total orbital angular momentum quantum number of state i;

$\ell_i \quad = \quad$ orbital angular momentum quantum number of orbital i;

$\Lambda_{++} \quad = \quad$ positive energy projection operator, Eq. (1.19);

$M_i \quad = \quad$ total magnetic quantum number of state i;

$m_\ell \quad = \quad$ single-particle magnetic quantum number of orbital i;

$m_\mu \quad = \quad$ mass of muon: ≈ 205 a.u., Eq. (8.54);

$\mathbf{P} \quad = \quad$ projection operator over occupied orbitals, Eq. (1.29);

$\mathbf{p} \quad = \quad$ linear momentum, Eq. (2.2);

$P_i(r)/r \quad = \quad$ radial part of orbital i, Eq. (5.5)
$\qquad\qquad\quad$ or large component of relativistic orbital, Eq. (2.12);

$P_j(\cos(\theta)) \quad = \quad$ Legendre function of $\cos(\theta)$ of order j;

$\widehat{P}_{ij} \quad = \quad$ parity operator between orbitals i and j, Eq. (10.4);

$\mathsf{P} \quad = \quad$ principal value integration, Eq. (4.45);

$\Pi \quad = \quad$ parity quantum number, Eq. (7.5);

$\Pi(\ell_1, \ell_2, \ell_3) \quad = \quad$ parity factor, Eq. (2.25);

$\pi \quad = \quad 3.14159\cdots$;

$\mathbf{Q} \quad = \quad$ complementary operator to $\mathbf{P}$: $1 - \mathbf{P}$, Eq. (1.30);

$Q_i(r)/r \quad = \quad$ radial part of small component of relativistic orbital,
$\qquad\qquad\quad$ Eq. (2.12);

$q \quad = \quad$ residual charge with which continuum electron
$\qquad\qquad\quad$ interacts, Eq. (4.9);

$q(\mathbf{x}) \quad = \quad$ four-component spinor for the quark, Eq. (3.1);

$\mathbf{r} \quad = \quad$ space vector;

$\mathbf{S} \quad = \quad$ four component spin matrix, Eq. (2.6);

$S_i \quad = \quad$ total spin angular momentum quantum number of state i;

$\underline{S} \quad = \quad S$-matrix, Eq. (13.21);

$\Sigma(\varepsilon)$ = self-energy operator, Eq. (3.11);

$\boldsymbol{\sigma}$ = Pauli spin matrix, Eq. (2.3);

σ = cross section, Eq. (4.3);

σ_i = spin designation of orbital i, Eq. (4.54);

$\underline{T}$ = T-matrix, Eq. (9.2);

Φ = approximate or initial many-particle function;

ϕ = approximate or initial single-particle function;

χ_μ = two-component spinor, Eq. (2.7);

$Y_{\ell m}$ = spherical harmonic, Eq. (2.7);

Ψ = correlated many-particle function;

ψ = correlated single-particle function;

Ω = solid angle;

Ω_{ij} = collision strength, Eq. (13.23);

$\Omega_{j\ell m}(\hat{\mathbf{r}})$ = spherical spinor, Eq. (2.7);

$\boldsymbol{\Omega}$ = wave operator, Eq. (1.37);

ω = photon energy;

Z = nuclear charge;

Z_{op} = dipole operator, Eq. (3.12).

References

[1]   R. P. Feynman, *Phys. Rev.* **76**, 749 (1949).

[2]   R. P. Feynman, *Phys. Rev.* **76**, 769 (1949).

[3]   F. J. Dyson, *Phys. Rev.* **75**, 486 (1949).

[4]   F. J. Dyson, *Phys. Rev.* **75**, 1736 (1949).

[5]   M. Goeppert Mayer, *Phys. Rev.* **75**, 1969 (1949).

[6]   O. Haxel, J. H. D. Jensen, and H. E. Suess, *Phys. Rev.* **75**, 1766 (1949).

[7]   M. Goeppert Mayer and J. H. D. Jensen, *Elementary Theory of Nuclear Shell Structure* (Wiley, New York, 1955).

[8]   R. J. Eden and N. C. Francis, *Phys. Rev.* **97**, 1366 (1955).

[9]   K. A. Brueckner and C. A. Levinson, *Phys. Rev.* **97**, 1344 (1955).

[10]   B. A. Lippmann and J. Schwinger, *Phys. Rev.* **79**, 469 (1950).

[11]   K. M. Watson, *Phys. Rev.* **89**, 575 (1953).

[12]   N. C. Francis and K. M. Watson, *Phys. Rev.* **92**, 291 (1953).

[13]   H. A. Bethe, *Phys. Rev.* **103**, 1353 (1956).

[14]   H. A. Bethe and J. Goldstone, *Proc. R. Soc. London, Ser. A* **238**, 551 (1957).

[15]   K. A. Brueckner and J. L. Gammel, *Phys. Rev.* **109**, 1023 (1958).

[16]   K. A. Brueckner, *Phys. Rev.* **100**, 36 (1955).

[17]   J. Goldstone, *Proc. R. Soc. London, Ser. A* **239**, 267 (1957).

[18]   K. A. Brueckner, J. L. Gammel, and H. Weitzner, *Phys. Rev.* **110**, 431 (1958).

[19]   K. A. Brueckner and D. T. Goldman, *Phys. Rev.* **116**, 424 (1959).

[20]   B. R. Barrett and M. W. Kirson, in *Advances in Nuclear Physics,* edited by M. Baranger and E. Vogt (Plenum, New York, 1973) Vol. 6, p. 219.

[21]   P. J. Ellis and E. Osnes, *Rev. Mod. Phys.* **49**, 777 (1977).

[22] H. P. Kelly, *Phys. Rev.* **131**, 684 (1963).

[23] H. P. Kelly and A. M. Sessler, *Phys. Rev.* **132**, 2091 (1963).

[24] H. P. Kelly, *Phys. Rev.* **134**, A1450 (1964).

[25] H. P. Kelly, *Phys. Rev.* **136**, B896 (1964).

[26] G. C. Wick, *Phys. Rev.* **80**, 268 (1950).

[27] L. M. Frantz and R. L. Mills, *Nucl. Phys.* **15**, 16 (1960).

[28] A. P. Yutsis, I. B. Levinson, and V. V. Vanagas, *Mathematical Apparatus of the Theory of Angular Momentum* (Israel) Program for Scientific Translations, Jerusalem, 1962).

[29] I. Lindgren and J. Morrison, *Atomic Many-Body Theory*, 2nd ed. (Springer-Verlag, Berlin, 1986).

[30] S. Salomonson and P. Öster, *Phys. Rev. A* **41**, 4670 (1990).

[31] C. Froese Fischer, *J. Phys. B* **26**, 855 (1993).

[32] W. Brenig, *Nucl. Phys.* **4**, 363 (1957).

[33] R. K. Nesbet, *Phys. Rev.* **155**, 51 (1967); **155**, 56 (1967); **175**, 2 (1968).

[34] *Advances in Chemical Physics*, edited by R. Lefebvre and C. Moser (Interscience, London, 1969), Vol. 14.

[35] O. Sinanoğlu, *J. Chem. Phys.* **36**, 706 (1962).

[36] W. Meyer, *J. Chem. Phys.* **58**, 1017 (1973); **64**, 2901 (1976).

[37] W. Kutzelnigg, in *Methods of Electronic Structure Theory*, edited by H. F. Schaefer III (Plenum Press, New York, 1977) pp. 129-188.

[38] H. P. Kelly, *Phys. Rev.* **173**, 142 (1968).

[39] H. P. Kelly, *Phys. Rev.* **180**, 55 (1969).

[40] H. P. Kelly and A. Ron, *Phys. Rev. A* **2**, 1261 (1970).

[41] C. Bloch and J. Horowitz, *Nucl. Phys.* **8**, 91 (1958).

[42] H. B. Brandow, *Rev. Mod. Phys.* **39**, 771 (1967); *Adv. Quantum Chem.* **10**, 187 (1977).

[43] P. G. H. Sandars, in *La Structure Hyperfine Magnétique des Atomes et des Molécules* (Centre National de la Recherche Scientifique, Paris, 1967).

[44] C. Bloch, *Nucl. Phys.* **6**, 329 (1958); **7**, 451 (1958).

[45] C. Møller, *K. Dan. Vidensk. Selsk. Mat. Fys. Medd.* **23** No. 1 (1945); **22** No. 19 (1946).

[46] J. des Cloizeaux, *Nucl. Phys.* **20**, 321 (1960).

[47] I. Lindgren, *J. Phys. B* **7**, 2441 (1974).

[48] V. Kvasnička, *Czech. J. Phys. B* **24**, 605 (1974); in *Advances in Chemical Physics*, edited by I. Prigogne and S. A. Rice (Interscience, New York, 1977), Vol. 36, p. 345.

[49]   S. Garpman, I. Lindgren, J. Lindgren, and J. Morrison, *Z. Phys. A* **276**, 167 (1976); I. Lindgren, J. Lindgren, and A.-M. Mårtensson, *ibid.* **279**, 113 (1976).

[50]   A.-M. Mårtensson, *J. Phys. B* **12**, 3995 (1979).

[51]   I. Lindgren, *Phys. Rev. A* **31**, 1273 (1985).

[52]   S. Salomonson and P. Öster, *Phys. Rev. A* **40**, 5559 (1989).

[53]   J. Hubbard, *Proc. Roy. Soc. London, Ser. A* **240**, 539 (1957).

[54]   F. Coester, *Nucl. Phys.* **7**, 421 (1958); F. Coester and H. Kümmel, *Nucl. Phys.* **17**, 477 (1960).

[55]   H. Kümmel, K. H. Lührmann, and J. G. Zabolitzky, *Physics Rep.* **36**, 1 (1978).

[56]   J. Čížek, *J. Chem. Phys.* **45**, 4256 (1966).

[57]   *Proceedings of the Symposium on Coupled-Cluster Theory in Atomic Physics and Quantum Chemistry, Theor. Chim. Acta* **80**, pp. 71-507 (1991).

[58]   H. Primas, in *Modern Quantum Chemistry, Part II: Interactions*, edited by O. Sinanoğlu (Academic Press, New York, 1965) pp. 45-74.

[59]   J. A. Pople, R. Krishnan, H. B. Schlegel, and J. S. Binkley, *Int. J. Quantum Chem.* **14**, 545 (1978).

[60]   D. Mukherjee, R. K. Moitra, and A. Mukhopadhyay, *Mol. Phys.* **33**, 955 (1977).

[61]   R. Offermann, W. Ey, and H. Kümmel, *Nucl. Phys. A* **273**, 399 (1976); R. Offermann, *ibid. A* **273**, 368 (1976).

[62]   J. Paldus, J. Čížek, M. Saute, and L. Laforgue, *Phys. Rev. A* **17**, 805 (1978).

[63]   I. Lindgren, *Int. J. Quantum Chem.* Quantum Chemistry Symposium **12**, 33 (1978).

[64]   W. Ey, *Nucl. Phys. A* **296**, 189 (1978).

[65]   G. E. Brown and D. G. Ravenhall, *Proc. Roy. Soc. London, Ser. A* **208**, 552 (1951).

[66]   J. Sucher, *Phys. Rev. A* **22**, 348 (1980).

[67]   W. R. Johnson, S. A. Blundell, and J. Sapirstein, *Phys. Rev. A* **37**, 307 (1988).

[68]   W. R. Johnson, S. A. Blundell, and J. Sapirstein, *Phys. Rev. A* **37**, 2764 (1988).

[69]   W. R. Johnson, S. A. Blundell, and J. Sapirstein, *Phys. Rev. A* **38**, 2699 (1988).

[70]   W. R. Johnson, S. A. Blundell, and J. Sapirstein, *Phys. Rev. A* **42**, 1087 (1990).

[71] H. M. Quiney, I. P. Grant, and S. Wilson, *J. Phys. B* **23**, L271 (1990); S. Wilson, in *The Effects of Relativity in Atoms, Molecules and the Solid State*, edited by S. Wilson, I. P. Grant, and B. L. Gyorffy (Plenum Press, New York, 1991), p. 217.

[72] E. Lindroth, *Phys. Rev. A* **37**, 316 (1988); S. Salomonson and P. Öster, *Phys. Rev. A* **40**, 5548 (1989); E. Lindroth and S. Salomonson, *Phys. Rev. A* **41**, 4659 (1990); E. Lindroth, H. Persson, S. Salomonson, and A.-M. Mårtensson-Pendrill, *Phys. Rev. A* **45**, 1493 (1992).

[73] E. Ilyabaev and U. Kaldor, *Phys. Rev. A* **47**, 137 (1993).

[74] J. Sucher, *J. Phys. B* **21**, L585 (1988).

[75] I. Lindgren, *J. Phys. B* **23**, 1085 (1990).

[76] See, for instance, *Proceedings of the Nobel Symposium 85, Heavy-Ion Spectroscopy and QED Effects in Atomic Systems*, eds. I. Lindgren, I. Martinson, and R. Schuch, *Physica Scripta* **T46** (1993).

[77] L. Labzowsky, V. Karasiev, I. Lindgren, H. Persson, and S. Salomonson, *Physica Scripta* **T46**, 150 (1993).

[78] V. A. Dzuba, V. V. Flambaum, P. G. Silverstrov, and O. P. Sushkov, *J. Phys. B* **16**, 715 (1983).

[79] W. R. Johnson and J. Sapirstein, *Phys. Rev. Lett.* **57**, 1126 (1986).

[80] Z. W. Liu and H. P. Kelly, *Theor. Chim. Acta* **80**, 307 (1991).

[81] Z. W. Liu and H. P. Kelly, *Phys. Rev. A* **45**, R4211 (1992).

[82] G. Hardekopf and J. Sucher, *Phys. Rev. A* **30**, 703 (1984).

[83] M. H. Mittleman, *Phys. Rev. A* **4**, 893 (1971); *Phys. Rev. A* **5**, 2395 (1972); *Phys. Rev. A* **24**, 1167 (1981).

[84] H. Araki, *Prog. Theor. Phys.* **17**, 619 (1957); P. K. Kabir and E. E. Salpeter, *Phys. Rev.* **108**, 1256 (1957); J. Sucher, *Phys. Rev.* **109**, 1010 (1958); M. Douglas and N. M. Kroll, *Ann. Phys. (NY)* **82**, 89 (1974).

[85] K. T. Cheng, W. R. Johnson, and J. Sapirstein, *Phys. Rev. Lett.* **66**, 2960 (1991).

[86] P. Indelicato and P. J. Mohr, *Theor. Chim. Acta* **80**, 207 (1991).

[87] J. P. Desclaux, *Comput. Phys. Commun.* **9**, 31 (1975).

[88] I. P. Grant, B. J. McKenzie, P. H. Norrington, D. F. Mayers, and N. C. Pyper, *Comput. Phys. Commun.* **21**, 207 (1980).

[89] W. R. Johnson and G. Soff, *At. Data and Nucl. Data Tables*, **33**, 405 (1985).

[90] C. W. de Jager, H. de Vries, and C. de Vries, *At. Data and Nucl. Data Tables*, **14**, 479 (1974); R. Engfer, H. Schneuwly, J. L. Vuilleumier, H. K. Walter, and A. Zehnder, *At. Data and Nucl. Data Tables*, **14**, 509 (1974).

[91] C. E. Moore, in *Atomic Energy Levels as Derived From the Analyses of Optical Spectra*, Natl. Bur. Stand. Ref. Data. Ser., (US) Circ. No. 35 (US GPO, Washington, DC 1971), Vol. II.

[92] G. E. Brown, *Many-Body Problems* (North-Holland, Amsterdam, 1972).

[93] S. A. Blundell, W. R. Johnson, Z. W. Liu, and J. Sapirstein, *Phys. Rev. A* **39**, 3768 (1989).

[94] H. A. Bethe and E. E. Salpeter, *Quantum Mechanics of One- and Two-Electron Atoms* (Plenum, New York, 1977).

[95] J. A. Gaunt, *Proc. Roy. Soc. London, Ser. A* **122**, 513 (1929).

[96] G. Breit, *Phys. Rev.* **34**, 553 (1929); **36**, 383 (1930); **39**, 616 (1932).

[97] J. B. Mann and W. R. Johnson, *Phys. Rev. A* **4**, 41 (1971).

[98] J. Sucher in *Relativistic, Quantum Electrodynamic, and Weak Interaction Effects in Atoms*, A.I.P. Conference Proceedings **189**, eds. W. R. Johnson, P. Mohr, and J. Sucher (American Institute of Physics, New York, 1989), p. 28.

[99] A. Dalgarno and J. T. Lewis, *Proc. Roy. Soc. London, Ser. A* **233**, 70 (1955); V. McKoy and N. W. Winter, *J. Chem. Phys.* **48**, 5514 (1968).

[100] J. J. Sakurai, *Advanced Quantum Mechanics* (Addison-Wesley, Reading MA, 1967).

[101] A. Chodos, R. L. Jaffe, K. Johnson, C. B. Thorn, and V. W. Weisskopf, *Phys. Rev. D* **9**, 3471 (1974).

[102] C. deBoor, *A Practical Guide to Splines* (Springer-Verlag, New York, 1978).

[103] S. A. Blundell, D. S. Guo, W. R. Johnson, and J. Sapirstein, *At. Data Nucl. Data Tables* **37**, 103 (1987).

[104] J. F. Seely, C. M. Brown, and U. Feldman, *At. Data Nucl. Data Tables* **43**, 145 (1989).

[105] K. T. Cheng, W. R. Johnson and J. Sapirstein, *Phys. Rev. Lett.* **66**, 2960 (1991).

[106] Ya. B. Zel'dovich, *Zh. Eksp. Teor. Fiz.* **36**, 964 (1959) *Sov. Phys. JETP* **9**, 682 (1959).

[107] M. A. Bouchiat and C. C. Bouchiat, *J. Phys. (Paris)* **35**, 899 (1974).

[108] M. A. Bouchiat *et al.*, *J. Phys. (Paris)* **47**, 1709 (1986).

[109] M. C. Noecker, B. P. Masterson, and C. E. Wieman, *Phys. Rev. Lett.* **61**, 310 (1988).

[110] P. S. Drell and E. D. Commins, *Phys. Rev. Lett.* **53**, 968 (1984).

[111] T. M. Wolfenden, P. E. G. Baird, and P. G. H. Sandars, *Europhys. Lett.* **15**, 731 (1991).

[112] T. P. Emmons, J. M. Reeves, and E. N. Fortson, *Phys. Rev. Lett.* **51**, 2089 (1983).

[113] M. J. D. Macpherson, K. P. Zetie, R. B. Warrington, D. N. Stacey, and J. P. Hoare, *Phys. Rev. Lett.* **67**, 2784 (1991).

[114] *Review of particle properties, Phys. Lett. B* **239**, 1 (1990).

[115] W. Marciano and J. Rosner, *Phys. Rev. Lett.* **65**, 2963 (1990).

[116] Ya. Zel'dovich, *Zh. Eksp. Teor. Fiz.* **33**, 1531 (1957) [*Sov. Phys. JETP* **6**, 1184 (1958)].

[117] W. R. Johnson, M. Idrees, and J. Sapirstein, *Phys. Rev. A* **35**, 3218 (1987).

[118] V. A. Dzuba, V. V. Flambaum, P. G. Silvestrov, and O. P. Sushkov, *Phys. Scr.* **35**, 69 (1987); *J. Phys. B* **20**, 3297 (1987).

[119] S. A. Blundell, D. S. Guo, W. R. Johnson, and J. Sapirstein, *At. Data Nucl. Data Tables* **37**, 103 (1987).

[120] S. A. Blundell, J. Sapirstein, and W. R. Johnson, *Phys. Rev. D* **45**, 1602 (1992).

[121] W. R. Johnson, S. A. Blundell, Z. W. Liu, and J. Sapirstein, *Phys. Rev. A* **37**, 1395 (1988).

[122] S. A. Blundell, W. R. Johnson, and J. Sapirstein, *Phys. Rev. A* **42**, 3751 (1990).

[123] A. B. Migdal, *Theory of Finite Fermi Systems and Applications to Atomic Nuclei* (Interscience, New York, 1967).

[124] V. A. Dzuba, V. V. Flambaum, and O. P. Sushkov, *Phys. Lett. A* **142**, 373 (1989).

[125] A. D. Fetter and J. D. Walecka, *Quantum Theory of Many-Particle Systems* (McGraw-Hill, New York, 1971).

[126] S. A. Blundell, W. R. Johnson and J. Sapirstein, *Phys. Rev. A* **43**, 3407 (1991).

[127] S. A. Blundell, W. R. Johnson, Z. W. Liu, and J. Sapirstein, *Phys. Rev. A* **40**, 2233 (1989).

[128] S. A. Blundell (unpublished).

[129] V. A. Dzuba, V. V. Flambaum, and O. P. Sushkov, *Phys. Lett. A* **141**, 147 (1989).

[130] S. A. Blundell, J. Sapirstein, and W. R. Johnson, *Phys. Rev. Lett.* **65**, 1411 (1990).

[131] S. A. Blundell, A. C. Hartley, Z. W. Liu, A.-M. Mårtensson-Pendrill, and J. Sapirstein, *Theor. Chim. Acta* **80**, 257 (1991).

[132] B. W. Lynn (private communication).

[133] A.-M. Mårtensson-Pendrill, *J. Phys. (Paris)* **46**, 1949 (1985).

[134] W. R. Johnson, D. S. Guo, M. Idrees, and J. Sapirstein, *Phys. Rev. A* **34**, 1043 (1986).

[135]   A. C. Hartley, E. Lindroth, and A.-M. Mårtensson-Pendrill, *J. Phys. B* **23**, 3417 (1990).

[136]   A. C. Hartley and P. G. H. Sandars, *J. Phys. B* **23**, 1961 (1990).

[137]   M. Brack, C. Guet, and H.-B. Håkansson, *Phys. Rep.* **123**, 275 (1985).

[138]   E. N. Fortson, S. Pollack, and L. Wilets, *Phys. Rev. C* **46**, 2587 (1992).

[139]   V. V. Flambaum, I. B. Khriplovich, and O. P. Sushkov, Phys. Lett. **146B**, 367 (1984).

[140]   P. A. Frantsuzov and I. Khriplovich, *Z. Phys. D* **7**, 297 (1988).

[141]   A. Ya. Kraftmakher, *Phys. Lett. A* **132**, 167 (1988).

[142]   V. V. Flambaum and I. B. Khriplovich, *Zh. Eksp. Teor. Fiz.* **89**, 1505 (1985) [*Sov. Phys. JETP* **62**, 872 (1985)].

[143]   M. G. Kozlov, *Phys. Lett. A* **130**, 426 (1988).

[144]   C. Bouchiat and C.-A. Piketty, *Europhys. Lett.* **2**, 511 (1986).

[145]   S. L. Gilbert and C. E. Wieman, *Phys. Rev. A* **34**, 792 (1986).

[146]   M. A. Bouchiat and J. Guena, *J. Phys. (Paris)* **49**, 2037 (1988).

[147]   M. E. Peskin and T. Takeuchi, *Phys. Rev. Lett.* **65**, 964 (1990).

[148]   P. Langacker and M. Luo, *Phys. Rev. D* **45**, 278 (1992).

[149]   C. Wieman (private communication).

[150]   S. Lameraux and E. N. Fortson, talk presented at the General Meeting of the American Physical Society, Washington DC, 12–15 April 1993.

[151]   J. A. R. Samson, in *Handbuch der Physik*, edited by W. Mehlhorn (Springer-Verlag, Berlin, 1982), Vol. 31, p. 123.

[152]   B. Sonntag and P. Zimmermann, *Rep. Prog. Phys.* **55**, 911 (1992).

[153]   V. Schmidt, *Rep. Prog. Phys.* **55**, 1483 (1992).

[154]   P. L. Altick and A. E. Glassgold, *Phys. Rev.* **133**, A632 (1964).

[155]   M. Ya. Amusia, V. K. Ivanov, N. A. Cherepkov, and L. V. Chernysheva, *Zh. Eksp. Teor. Fiz.* **66**, 1537 (1974) [*Sov. Phys. JETP* **39**, 752 (1974)]; M. Ya. Amusia, N. A. Cherepkov, Dj. Živanović, and V. Radojević, *Phys. Rev. A* **13**, 1466 (1976).

[156]   A. F. Starace and S. Shahabi, *Phys. Scr.* **21**, 368 (1980); *Phys. Rev. A* **25**, 2135 (1982).

[157]   W. R. Johnson and C. D. Lin, *Phys. Rev. A* **20**, 964 (1979).

[158]   W. R. Johnson, C. D. Lin, K. T. Cheng, and C. M. Lee, *Phys. Scr.* **21**, 409 (1980).

[159]   W. R. Johnson, K. T. Cheng, K.-N. Huang, and M. Le Dourneuf, *Phys. Rev. A* **22**, 989 (1980).

[160]   P. G. Burke and K. T. Taylor, *J. Phys. B* **8**, 2620 (1975).

[161]   D. L. Miller and A. F. Starace, *J. Phys. B* **13**, L525 (1980).

[162] C. H. Greene, *Phys. Rev. A* **23**, 661 (1981).

[163] H. P. Saha and C. D. Caldwell, *Phys. Rev. A* **40**, 7020 (1989).

[164] A. F. Starace, in *Handbuch der Physik*, edited by W. Mehlhorn (Springer-Verlag, Berlin, 1982), Vol. 31, p. 1.

[165] J. D. Jackson, *Classical Electrodynamics*, 2nd ed. (John Wiley & Sons, New York, 1975).

[166] P. A. M. Dirac, *Proc. Roy. Soc. London, Ser. A* **114**, 243 (1927).

[167] See, for example, J. J. Sakurai, *Modern Quantum Mechanics* edited by San Fu Tuan (Addison-Wesley, Redwood City, California, 1985).

[168] M. Ya. Amusia, *Atomic Photoeffect* (Plenum, New York, 1990).

[169] U. Fano and J. W. Cooper, *Rev. Mod. Phys.* **40**, 441 (1968).

[170] H. P. Kelly, *Phys. Rev.* **182**, 84 (1969).

[171] S. L. Carter, Ph.D. thesis, University of Virginia (1976).

[172] L. D. Landau and E. M. Lifshitz, *Quantum Mechanics (Non-Relativistic Theory)*, 3rd ed. (Pergamon Press, Oxford, 1977).

[173] See, for example, P. J. Davis and P. Rabinowitz, *Methods of Numerical Integration*, 2nd ed. (Academic Press, Orlando, 1984).

[174] C. Froese Fischer, *The Hartree-Fock Method for Atoms* (John Wiley & Sons, New York, 1977).

[175] L. M. Frantz, R. L. Mills, R. G. Newton, and A. M. Sessler, *Phys. Rev. Lett.* **1**, 340 (1958); B. A. Lippman, M. H. Mittleman, and K. M. Watson, *Phys. Rev.* **116**, 920 (1959).

[176] R. T. Pu and E. S. Chang, *Phys. Rev.* **151**, 31 (1966).

[177] H. J. Silverstone and M.-L. Yin, *J. Chem. Phys.* **49**, 2026 (1968).

[178] S. Huzinaga and C. Arnau, *Phys. Rev. A* **1**, 1285 (1970).

[179] T. Ishihara and R. T. Poe, *Phys. Rev. A* **6**, 111 (1972); **6**, 116 (1972).

[180] Z.-D. Qian, S. L. Carter, and H. P. Kelly, *Phys. Rev. A* **33**, 1751 (1986).

[181] J. J. Boyle, Ph.D. thesis, University of Virginia (1992).

[182] J. J. Boyle, *Phys. Rev. A* **48**, 2860 (1993).

[183] E. R. Brown, S. L. Carter, and H. P. Kelly, *Phys. Rev. A* **21**, 1237 (1980).

[184] L. J. Garvin, Ph.D. thesis, University of Virginia, 1983, pp. 128-50.

[185] Z. Altun, M. Kutzner, and H. P. Kelly, *Phys. Rev. A* **37**, 4671 (1988).

[186] M. Kutzner, Z. Altun, and H. P. Kelly, *Phys. Rev. A* **41**, 3612 (1990).

[187] See, for example, *Giant Resonances in Atoms, Molecules, and Solids*, edited by J. P. Connerade, J. M. Esteva, and R. C. Karnatak (Plenum, New York, 1987).

[188] P. Löwdin, *Phys. Rev.* **97**, 1474 (1955).

[189] U. Becker, D. Szostak, H. G. Kerkhoff, M. Kupsch, B. Langer, R. Wehlitz, A. Yagishita, and T. Hayaishi, *Phys. Rev. A* **39**, 3902 (1989).

[190] B. Kämmerling, H. Kossman, and V. Schmidt, *J. Phys. B* **22**, 841 (1989).

[191] U. Becker, in *Giant Resonances in Atoms, Molecules, and Solids*, edited by J. P. Connerade, J. M. Esteva, and R. C. Karnatak (Plenum, New York, 1987), p. 473.

[192] J. M. Bizau, D. Cubaynes, P. Gérard, and F. J. Wuilleumier, *Phys. Rev. A* **40**, 3002 (1989).

[193] G. Wendin, *J. Phys. B* **3**, 455 (1970); **3**, 466 (1970).

[194] J. J. Boyle, Z. Altun, and H. P. Kelly, *Phys. Rev. A* **47**, 4811 (1993).

[195] J. T. Costello, E. T. Kennedy, B. F. Sonntag, and C. L. Cromer, *J. Phys. B* **24**, 5063 (1991).

[196] H. P. Kelly, in *Advances in Theoretical Physics*, edited by K. A. Brueckner (Academic, New York, 1968), Vol. 2, pp. 75-169.

[197] H. P. Kelly, in *Photoionization and Other Probes of Many-Electron Interactions*, edited by F. J. Wuilleumier (Plenum, New York, 1976), pp. 83-109.

[198] H. P. Kelly, in *Fundamental Processes in Atomic Collision Physics I*, edited by H. Kleinpoppen, J. S. Briggs, and H. O. Lutz (Plenum, New York, 1985), pp. 239-268.

[199] M. Domke, G. Remmers, and G. Kaindl, *Phys. Rev. Lett.* **69**, 1171 (1992).

[200] R. P. Madden and K. Codling, *Astrophys. J.* **141**, 364 (1965).

[201] J. W. Cooper, U. Fano, and F. Prats, *Phys. Rev. Lett.* **10**, 518 (1963).

[202] U. Fano, *Phys. Rev.* **124**, 1866 (1961).

[203] E. J. Heller, W. P. Reinhardt, and H. A. Yamani, *J. Comp. Phys.* **13**, 536 (1973); J. T. Broad and W. P. Reinhardt, *Phys. Rev. A* **14**, 2159 (1976); E. J. Heller and H. A. Yamani, *Phys. Rev. A* **9**, 1201 (1974); *Phys. Rev. A* **9**, 1209 (1974); J. J. Wendoloski and W. P. Reinhardt, *Phys. Rev. A* **17**, 195 (1978).

[204] R. Moccia and P. Spizzo, *J. Phys. B* **18**, 3537 (1985); **20**, 1423 (1987); **21**, 1145 (1988); **23**, 3557 (1990); *Phys. Rev. A* **39**, 3855 (1989).

[205] F. Martin and A. Salin, *Chem. Phys. Lett.* **157**, 146 (1989); I. Sánchez and F. Martin, *J. Phys. B* **23**, 4263 (1990).

[206] C. Froese Fischer and H. P. Saha, *Can. J. Phys.* **65**, 772 (1987); C. Froese Fischer and M. Idrees, *J. Phys. B* **23**, 679 (1990).

[207] T. N. Chang and X. Tang, *Phys. Rev. A* **44**, 232 (1991).

[208] T. N. Chang, *Phys. Rev. A* **47**, 3441 (1993).

[209] T. N. Chang and X. Tang, *Phys. Rev. A* **46**, R2209 (1992).

[210] T. N. Chang and E. T. Bryan, *Phys. Rev. A* **38**, 645 (1988).

[211] T. N. Chang and Y. S. Kim, *Phys. Rev. A* **36**, 2609 (1986).

[212] L. Lipsky and A. Russek, *Phys. Rev.* **142**, 59 (1966); L. Lipsky and M. J. Conneely, *Phys. Rev. A* **14**, 2193 (1976).

[213] R. N. Zare, *J. Chem. Phys.* **45**, 1966 (1966); **47**, 3561 (1967).

[214] T. N. Chang and K. T. Chung, *Chinese J. Phys.* **27**, 425 (1989).

[215] C. Laughlin and G. A. Victor, *Atomic Physics 3*, eds. S. J. Smith and G. K. Walters (Plenum, New York, 1973), p. 247.

[216] C. H. Greene and L. Kim, *Phys. Rev. A* **36**, 2706 (1987).

[217] T. N. Chang, in *Relativistic Quantum Electrodynamics and Weak Interaction Effects in Atoms*, edited by W. Johnson, P. Mohr, J. Sucher, A.I.P. Conference Proceedings **189**, (American Institute of Physics, New York, 1989) p. 217.

[218] T. N. Chang, *Phys. Rev. A* **36**, 447 (1987).

[219] A. Burgess, *Proc. Phys. Soc. London* **81**, 442 (1963).

[220] P. G. Burke and D. D. McVicar, *Proc. Roy. Soc. London* **86**, 989 (1965).

[221] G. Bates and P. L. Altick, *J. Phys. B* **6**, 653 (1973).

[222] T. Scholz, P. Scott, and P. G. Burke, *J. Phys. B* **21**, L139 (1988).

[223] C. Schwartz, *Phys. Rev.* **124**, 1468 (1961); R. L. Armstead, *Phys. Rev.* **171**, 91 (1968).

[224] M. P. Ajmera and K. T. Chung, *Phys. Rev. A* **10**, 1013 (1974).

[225] D. W. Norcross, *J. Phys. B* **4**, 652 (1971).

[226] D. H. Oza, *Phys. Rev. A* **33**, 824 (1986).

[227] K. L. Bell and A. E. Kingston, *Proc. Phys. Soc. London, A* **90** , 895 (1967).

[228] M. P. Ajmera and K. T. Chung, *Phys. Rev. A* **12**, 475 (1976).

[229] A. L. Stewart, *J. Phys. B* **11**, 3851 (1978).

[230] C. H. Park, A. F. Starace, J. Tan, and C. D. Lin, *Phys. Rev. A* **33**, 1000 (1986).

[231] M. Crance and M. Aymar, *J. Phys. B* **18**, 3529 (1985).

[232] J. A. R. Samson, *Phys. Rep.* **28**, 303 (1976); J. A. R. Samson, Z. X. He, L. Yin, and G. N. Haddad, *J. Phys. B* **27**, 887 (1994).

[233] V. L. Jacobs, *Phys. Rev. A* **3**, 289 (1971); **4**, 939 (1971); **9**, 1938 (1974).

[234] A. Burgess and M. J. Seaton, *Mon. Not. Roy. Astron. Soc.* **120**, 121 (1960).

[235] A. Dalgarno, H. Doyle, and M. Oppenheimer, *Phys. Rev. Lett.* **29**, 1051 (1972); H. Doyle, M. Oppenheimer, and A. Dalgarno, *Phys. Rev. A* **11**, 909 (1975).

[236] R. F. Stebbings, F. B. Dunning, F. K. Tittel, and R. D. Rundel, *Phys. Rev. Lett.* **30**, 815 (1973).

[237] T. N. Chang and M. Zhen, *Phys. Rev. A* **47**, 4849 (1993).

[238] S. Salomonson, S. L. Carter, and H. P. Kelly, *Phys. Rev. A* **39**, 5111 (1989).

[239] R. Gersbacher and T. Broad, *J. Phys. B* **23**, 365 (1990).

[240] I. Sánchez and F. Martín, *J. Phys. B* **23**, 4263 (1990).

[241] P. Hamacher and J. Hinze, *J. Phys. B* **22**, 3397 (1989).

[242] C. Froese Fischer and M. Idrees, *J. Phys. B* **23**, 679 (1990).

[243] T. N. Chang, *Phys. Rev. A* **37**, 4090 (1988).

[244] T. N. Chang and L. Zhu, *Phys. Rev. A* **48**, R1725 (1993).

[245] G. Mehlman-Balloffet and J. M. Esteva, *Astrophys. J.* **157**, 945 (1969).

[246] J. M. Esteva, G. Mehlman-Balloffet, and J. Romand, *J. Quant. Spectrosc. Radiat. Transfer* **12**, 1291 (1972).

[247] H. C. Chi, K. N. Huang, and K. T. Cheng, *Phys. Rev. A***43**, 2542 (1991).

[248] J. Dubau and J. Wells, *J. Phys. B* **6**, 1452 (1973); **6**, L31 (1973).

[249] V. Radojevic and W. R. Johnson, *Phys. Rev. A***31**, 2991 (1985).

[250] P. F. O'Mahony and C. H. Greene, *Phys. Rev. A***31**, 250 (1985).

[251] D. W. Norcross and M. J. Seaton, *J Phys. B***9**, 2983 (1976).

[252] D. L. Moores, *Proc. Phys. Soc.* **91**, 830 (1967).

[253] R. M. Jopson, R. R. Freeman, W. E. Cooke, and J. Bokor, *Phys. Rev. Lett.* **51**, 1640 (1983).

[254] T. S. Yih, H. H. Wu, H. T. Chan, C. C. Chu, and B. J. Pong, *Chinese J. Phys.* **27**, 136 (1989).

[255] J. M. Preses, C. E. Burkhardt, W. P. Garver, and J. J. Leventhal, *Phys. Rev. A***29**, 985 (1984).

[256] R. E. Bonanno, C. W. Clark, and T. B. Lucatorto, *Phys. Rev. A***34**, 2082 (1986).

[257] D. M. P. Holland, K. Codling, J. B. West, and G. V. Marr, *J. Phys. B* **12**, 2465 (1979).

[258] V. Schmidt, in *X-Ray and Inner-Shell Processes, AIP Conf. Proc.* **215**, 559 (1990).

[259] F. C. Farnoux, G. Dujardin, L. Hellner, and M. Y. Adam, in *X-Ray and Inner-Shell Processes, AIP Conf. Proc.* **215**, 281 (1990).

[260] F. J. Wuilleumier, J. M. Bizau, D. Cubaynes, L. Journel, B. Rouvellou, M. Richer, K. H. Seibmann, P. Sladeczek, and P. Zimmermann, *Bull. Amer. Phys. Soc.* **37**, 1124 (1992).

[261] U. Becker and R. Wehlitz, *Phys. Scr. T* **41**, 127 (1992); U. Becker, O. Hemmers, B. Lanzer, I. Lee, A. Menzel, R. Wehlitz, and M. Ya Amusia, *Phys. Rev. A* **47**, R767 (1993).

[262] T. N. Chang, T. Ishihara, and R. T. Poe, *Phys. Rev. Lett.* **27**, 838 (1971); T. N. Chang and R. T. Poe, *Phys. Rev. A* **12**, 1432 (1975).

[263] S. L. Carter and H. P. Kelly, *J. Phys. B* **9**, L505 (1976).

[264] S. L. Carter and H. P. Kelly, *J. Phys. B* **9**, 1887 (1976).

[265] S. L. Carter and H. P. Kelly, *Phys. Rev. A* **16**, 1525 (1977).

[266] P. Winkler, *J. Phys. B* **10**, L693 (1977).

[267] S. L. Carter and H. P. Kelly, *Phys. Rev. A* **24**, 170 (1981).

[268] C. Pan and H. P. Kelly, *Phys. Rev. A* **39**, 6232 (1989).

[269] T. Ishihara, K. Hino, and J. H. McGuire, *Phys. Rev. A* **44**, R6980 (1991).

[270] Z. W. Liu and J. C. Liu, *Bull. Amer. Phys. Soc.* **38**, 1150 (1993).

[271] M. Ya Amusia, *Comments Atom. Mol. Phys.* **10**, 155 (1981).

[272] H. P. Kelly, in *X-Ray and Inner-Shell Processes, AIP Conf. Proc.* **215**, 292 (1990).

[273] J. C. Levin, D. W. Lindle, N. Keller, R. D. Miller, Y. Azuma, N. Berrah Mansour, H. G. Berry, and I. A. Sellin, *Phys. Rev. Lett.* **67**, 968 (1991).

[274] A. Dalgarno and H. R. Sadeghpour, *Phys. Rev. A* **46**, R3591 (1992).

[275] T. A. Carlson, *Phys. Rev.* **156**, 142 (1967).

[276] M. J. Van der Wiel and T. N. Chang, *J. Phys. B* **11**, L125 (1978).

[277] B. Brehm and K. Höfler, *Int. J. Mass Spectrom. Ion. Phys.* **17**, 371 (1975).

[278] D. M. P. Holland, K. Codling, and R. N. Chamberlain, *J. Phys. B* **14**, 839 (1981).

[279] H. P. Kelly, *Adv. Theor. Phys.* **8**, 305 (1983); *Phys. Scr.* **T17**, 109 (1987).

[280] R. M. Stermheimer, *Phys. Rev.* **115**, 1198 (1959).

[281] J. M. Bizau, F. Wuilleumier, D. Ederer, P. Koch, P. Dhez, S. Krummacher, and V. Schmidt, in *Proceedings of the 1st ECAP*, **2** (Heidelberg, 1981), p.1017.

[282] F. W. Byron, Jr. and C. J. Joachain, *Phys. Rev.* **164**, 1 (1967).

[283] R. L. Brown, *Phys. Rev. A* **1**, 586 (1970).

[284] V. Schmidt, N. Sandner, H. Kuntzemuller, P. Dhez, F. Wuilleumier, and E. Kallne, *Phys. Rev. A* **13**, 1748 (1976).

[285] G. R. Wight and M. J. Van der Wiel, *J. Phys. B* **9**, 1319 (1979).

[286] F. J. Wuilleumier (private communication, 1993).

[287] R. L. Simons and H. P. Kelly, *Phys. Rev. A* **22**, 625 (1980).

[288] A. Huetz, P. Selles, D. Waymel, and J. Mazeau, *J. Phys. B* **24**, 1917 (1991).

[289] H. Klar and M. Fehr, *Z. Phys. D* **23** 295 (1992).

[290] A. Wehlitz, F. Heiser, O. Hemmers, B. Langer, A. Menzel, and U. Becker, *Phys. Rev. Lett.* **67**, 3764 (1991).

[291] C. N. Yang, *Phys. Rev.* **74**, 764 (1948).

[292] J. Cooper and R. N. Zare, in *Lectures in Theoretical Physics*, edited by
       S. Geltman, K. T. Mahanthappa, and W. E. Britten (Gordon and Breach,
       New York, 1969), Vol. XI-C, pp. 317-337.

[293] M. Born and E. Wolf, *Principles of Optics* (Pergamon Press, New York,
       1959) Subsection 10.8.2.

[294] S. T. Manson and A. F. Starace, *Rev. Mod. Phys.* **54**, 389 (1982).

[295] U. Fano and D. Dill, *Phys. Rev. A* **6**, 185 (1972).

[296] D. Dill and U. Fano, *Phys. Rev. Lett.* **29**, 1203 (1972).

[297] D. Dill, *Phys. Rev. A* **7**, 1976 (1973).

[298] S. T. Manson and D. Dill, in *Electron Spectroscopy*, edited by
       C. R. Brundle and A. D. Baker (Academic Press, New York, 1978), Vol.
       2, pp. 157-195.

[299] M. Rotenberg, R. Bivens, N. Metropolis and J. K. Wooten Jr, *The 3-j
       and 6-j symbols* (M.I.T. Press, Cambridge, MA, 1959).

[300] J. Cooper and R. N. Zare, *J. Chem. Phys.* **48**, 942 (1968).

[301] W. Ong and S. T. Manson, *Phys. Rev. A* **20**, 2364 (1979).

[302] J. W. Cooper, *Phys. Rev.* **128**, 681 (1962).

[303] S. T. Manson and J. W. Cooper, *Phys. Rev.* **165**, 126 (1968).

[304] M. A. Chafee, *Phys. Rev.* **37**, 1233 (1931).

[305] Y.-Y. Yin and D. S. Elliot, *Phys. Rev. A* **45**, 281 (1992).

[306] A. F. Starace, R. H. Rast, and S. T. Manson, *Phys. Rev. Lett.* **38**, 1522
       (1977).

[307] E. S. Chang and K. T. Taylor, *J. Phys. B* **11**, L507 (1978).

[308] B. Langer, J. Viefhaus, O. Hemmers, A. Menzel, R. Wehlite, and
       U. Becker, in *ICAP XIII Book of Abstracts* (ICAP, Munich, 1992), p.
       A41.

[309] D. J. Kennedy and S. T. Manson, *Phys. Rev. A* **5**, 227 (1972).

[310] S. T. Manson, *J. Electron Spectrosc.* **37**, 37 (1985).

[311] D. W. Lindle, L. J. Medhurst, T. A. Ferrett, P. A. Heiman,
       M. N. Piancastelli, T. A. Carlson, P. C. Deshmukh, G. Nasreen, and
       S. T. Manson, *Phys. Rev. A* **38**, 2371 (1988).

[312] P. C. Deshmukh, V. Radojevic, and S. T. Manson, *Phys. Rev. A* **45**, 6339
       (1992).

[313] H. P. Kelly, *Phys. Rev.* **160**, 44 (1967).

[314] M. Ya. Amusia, N. A. Cherepkov, A. Tančič, S. G. Shapiro, and
       L. V. Chernysheva, *J. Exp. Theor. Phys.* **68**, 2023 (1975) (in Russian).

[315] M. Ya. Amusia, D. Davidovich, V. Radojevic, L. V. Chernysheva,
       N. A. Cherepkov, *Phys. Rev. A* **25**, 219 (1982).

[316] L. V. Chernysheva, G. F. Gribakin, V. K. Ivanov, and M. Yu. Kuchiev, *J. Phys. B* **21**, L419 (1988).

[317] M. Ya. Amusia, N. A. Cherepkov, L. V. Chernysheva, and S. G. Shapiro, *J. Phys. B* **9**, 531 (1976).

[318] N. H. March, W. H. Young, and S. Sampanthar, *The Many-Body Problem in Quantum Mechanics* (Cambridge University Press, Cambridge UK, 1967).

[319] H. P. Kelly, in *Atomic Physics III*, edited by P. G. H. Sandars (Plenum Press, New York, 1990).

[320] A. B. Migdal, *J. Exp. Theor. Phys.* **32**, 399 (1957) (in Russian).

[321] D. E. Golden and H. W. Bandel, *Phys. Rev.* **138**, A14 (1965).

[322] G. F. Gribakin, B. V. Gul'tsev, V. K. Ivanov, and M. Yu. Kuchiev, *J. Phys. B* **23**, 4505 (1990).

[323] D. J. Pegg, J. S. Tompson, R. N. Compton, and G. D. Alton, *Phys. Rev. Lett.* **59**, 2267 (1987).

[324] C. W. Walter and J. R. Peterson, *Phys. Rev. Lett.* **68**, 2281 (1992).

[325] M. Ya. Amusia, N. A. Cherepkov, V. A. Sosnivker, L. V. Chernysheva, and S. L. Sheftel, *J. Tech. Phys.* **97**, 745 (1990) (in Russian).

[326] M. Ya. Amusia, V. Dolmatov, *J. Exp. Theor. Phys.* **97**, 1129 (1990) (in Russian).

[327] M. Ya. Amusia and V. A. Kharchenko, *J. Phys. B* **14**, L219 (1981).

[328] M. Ya. Amusia, M. Yu. Kuchiev, and S. A. Sheinerman, *J. Exp. Theor. Phys.* **76**, 470 (1975) (in Russian).

[329] M. Yu. Kuchiev and S. A. Sheinerman, *Sov. Phys. Usp.* **32**, 569 (1989).

[330] M. Ya. Amusia, L. V. Chernysheva, G. F. Gribakin, K. L. Tsemekhman, *J. Phys. B* **23**, 393 (1990).

[331] *Polarizational Radiation of Atoms and Particles*, edited by V. N. Tzitovich and J. M. Oiringel (Plenum Press, New York, 1992).

[332] H. Ehrhardt, M. Schulz, T. Tekaat, K. Willmann, *Phys. Rev. Lett.* **48**, 1807 (1969).

[333] U. Amaldi, A. Egidi, R. Marconero, G. Pizzella, *Rev. Sci. Instrum.* **40**, 1001 (1969).

[334] I. E. McCarthy, E. Weigold, *Rep. Prog. Phys.* **51**, 299 (1988).

[335] R. K. Peterkop, *Theory of Ionization of Atoms by Electron Impact* (Colorado Associated University Press, Boulder, 1977).

[336] A. Lahmam-Bennani, H. F. Wellenstein, C. Dal Cappello, M. Rouault, A. Duguet, *J. Phys. B* **16**, 2219 (1983).

[337] A. Lahmam-Bennani, L. Avaldi, E. Fainelli, G. Stefani, *J. Phys. B* **21**, 2145 (1988).

390   *References*

[338]   A. Lahmam-Bennani, H. F. Wellenstein, C. Dal Cappello, A. Duguet, *J. Phys. B* **17**, 3159 (1984).

[339]   C. Dal Cappello, C. Tavard, A. Lahmam-Bennani, M. C. Dal Cappello, *J. Phys. B* **17**, 4557 (1984).

[340]   L. Avaldi, I. E. McCarthy, G. Stefani, *J. Phys. B* **22**, 3305 (1989).

[341]   F. W. Byron Jr, C. J. Joachain, B. Piraux, *J. Phys. B* **19**, 1201 (1986), and earlier papers quoted therein.

[342]   F. Mota Furtado, P. F. O'Mahony, *J. Phys. B* **21**, 137 (1988).

[343]   P. J. Redmond (unpublished); see L. Rosenberg, *Phys. Rev. D* **8**, 1833 (1973).

[344]   M. Brauner, J. Briggs, H. Klar, *J. Phys. B* **22**, 2265 (1989).

[345]   H. Klar, *Z. Phys. D* **16**, 231 (1990).

[346]   M. K. Baliyan, M. K. Srivastava, *Phys. Rev. A* **32**, 3098 (1985).

[347]   E. P. Curran, H. R. J. Walters, *J. Phys. B* **20**, 337 (1987).

[348]   E. P. Curran, C. T. Whelan, H. R. J. Walters, *J. Phys. B* **24**, L19 (1991).

[349]   H. Ehrhardt, G. Knoth, P. Schlemmer, K. Jung, *Phys. Lett. A* **110**, 92 (1985).

[350]   Sadhana Sharma, M. K. Srivastava, *Phys. Rev. A* **38**, 1083 (1988).

[351]   A. Franz and P. L. Altick, *J. Phys. B* **25**, 1577 (1992).

[352]   K. Jung, R. Müller-Fiedler, P. Schlemmer, H. Ehrhardt, H. Klar, *J. Phys. B* **18**, 2955 (1985).

[353]   P. Schlemmer, M. K. Srivastava, T. Rösel, H. Ehrhardt, *J. Phys. B* **24**, 2719 (1991).

[354]   Sadhana Sharma, M. K. Srivastava, *Phys. Rev. A* **35**, 1939 (1987).

[355]   T. Rösel, C. Dupre, J. Röder, A. Duguet, K. Jung, A. LahmamBennani, H. Ehrhardt, *J. Phys. B* **24**, 3059 (1991).

[356]   A. Pochat, R. J. Tweed, J. Peresse, C. Joachain, B. Piraux, F. W. Byron Jr., *J. Phys. B* **16**, L775 (1983).

[357]   F. W. Byron Jr, C. J. Joachain, B. Piraux, *J. Phys. B* **16**, L769 (1983).

[358]   F. Mota Furtado, P. F. O'Mahony, *J. Phys. B* **22**, 3925 (1989).

[359]   A. Duguet, A. Lahmam-Bennani, *Z. Phys. D* **23**, 383 (1992).

[360]   G. Stefani, L. Avaldi, R. Camilloni, *J. Phys. B* **23**, L227 (1990).

[361]   O. Robaux, R. J. Tweed, J. Langlois, *J. Phys. B* **24**, 4567 (1991).

[362]   A. Franz, P. L. Altick, *J. Phys. B* **25**, L257 (1992).

[363]   S. Nakazaki, *Advances in Atomic, Molecular, and Optical Physics* (Academic Press, New York, 1993) p. 1.

[364]   N. Anderson, J. W. Gallagher, and I. V. Hertel, *Phys. Rep.* **165**, 1 (1988).

[365]   Y. Itikawa, *Phys. Rep.* **143**, 69 (1986).

[366]   H. R. J. Walters, *Phys. Rep.* **116**, 1 (1984).

[367]   B. H. Bransden and M. R. C. McDowell, *Phys. Rep.* **30**, 207 (1977).

[368]   B. H. Bransden and M. R. C. McDowell, *Phys. Rep.* **46**, 249 (1978).

[369]   D. H. Madison and W. N. Shelton, *Phys. Rev. A* **7**, 499 (1973).

[370]   L. D. Thomas, G. Csanak, H. S. Taylor, and B. S. Yarlagadda, *J. Phys. B* **7**, 1719 (1974).

[371]   G. D. Meneses, F. J. daPaixao, and N. T. Padial, *Phys. Rev. A* **32**, 156 (1985).

[372]   K. Bartschat and D. H. Madison, *J. Phys. B* **20**, 1609 (1987).

[373]   R. Srivastava and W. Williamson, Jr, *Phys. Rev. A* **35**, 103 (1987).

[374]   R. Srivastava and A. K. Katiyar, *Phys. Rev. A* **35**, 1080 (1987).

[375]   K. Bartschat and D. H. Madison, *J. Phys. B* **20**, 5839 (1987).

[376]   A. W. Pangantiwar and R. Srivastava, *J. Phys. B* **20**, 5881 (1987).

[377]   K. Bartschat and D. H. Madison, *J. Phys. B* **21**, 153 (1988).

[378]   R. E. H. Clark, J. Abdallah Jr, G. Csanak, and S. P. Kramer, *Phys. Rev. A* **40**, 2935 (1989).

[379]   G. Csanak and D. C. Cartwright, *J. Phys. B* **22**, 2769 (1989).

[380]   G. D. Meneses, C. B. Pagan, and L. E. Machado, *Phys. Rev. A* **41**, 4740 (1990).

[381]   R. E. H. Clark, G. Csanak, and J. Abdallah Jr, *Phys. Rev. A* **44**, 2874 (1991).

[382]   D. H. Madison, K. Bartschat, and R. Srivastava, *J. Phys. B* **24**, 1839 (1991).

[383]   S. Trajmar, D. F. Register, D. C. Cartwright, and G. Csanak, *J. Phys. B* **25**, 4889 (1992).

[384]   R. E. H. Clark, G. Csanak, J. Abdallah Jr, and S. Trajmar, *J. Phys. B* **25**, 5233 (1992).

[385]   D. H. Madison, I. Bray, and I. E. McCarthy, *J. Phys. B* **24**, 3861 (1991).

[386]   D. H. Madison, K. Bartschat, and R. P. McEachran, *J. Phys. B* **25**, 5199 (1992).

[387]   M. R. C. McDowell, L. A. Morgan, and V. P. Myerscough, *J. Phys. B* **6**, 1435 (1974).

[388]   D. H. Madison, K. Bartschat, and J. L. Peacher, *Phys. Rev. A* **44**, 304 (1991).

[389]   D. H. Madison and K. H. Winters, *J. Phys. B* **16**, 4437 (1983).

[390]   J. E. Furst, W. M. K. P. Wijayaratna, D. H. Madison, and T. J. Gay, *Phys. Rev. A* **47**, 3775 (1993).

[391]   D. H. Madison and W. N. Shelton, *Phys. Rev. A* **7**, 514 (1973).

[392]   T. Zuo, R. P. McEachran and A. D. Stauffer, *J. Phys. B* **24**, 2853 (1991).

[393]   L. Fritsche, C. Kroner, and T. Reinert, *J. Phys. B* **25**, 4287 (1992).

[394]   H. L. Zhang and D. H. Sampson, *Phys. Rev. A* **47**, 208 (1993).

[395]   R. Szmytkowski, *J. Phys. B* **26**, 535 (1993).

[396]   R. Srivastava, T. Zuo, R. P. McEachran, and A. D. Stauffer, *J. Phys. B* **25**, 2409 (1992).

[397]   T. Zuo, R. P. McEachran, and A. D. Stauffer, *J. Phys. B* **25**, 3393 (1992).

[398]   R. Srivastava, T. Zuo, R. P. McEachran, and A. D. Stauffer, *J. Phys. B* **25**, 1073 (1992).

[399]   R. Srivastava, T. Zuo, R. P. McEachran, and A. D. Stauffer, *J. Phys. B* **25**, 3709 (1992).

[400]   R. Srivastava, R. P. McEachran, and A. D. Stauffer, *J. Phys. B* **25**, 4033 (1992).

[401]   B. D. Buckley and H. R. J. Walters, *J. Phys. B* **8**, 1693 (1975).

[402]   D. P. Dewangan and H. R. J. Walters, *J. Phys. B* **10**, 637 (1977).

[403]   K. H. Winters, *J. Phys. B* **11**, 149 (1978).

[404]   A. E. Kingston and H. R. J. Walters, *J. Phys. B* **13**, 4633 (1980).

[405]   F. W. Byron, C. J. Joachain, and R. M. Potoliege, *J. Phys. B* **18**, 1637 (1985).

[406]   D. H. Madison and K. H. Winters, *J. Phys. B* **20**, 4173 (1987).

[407]   A. M. Ermolaev and H. R. J. Walters, *J. Phys. B* **13**, L473 (1980).

[408]   D. H. Madison, *Phys. Rev. Lett.* **53**, 42 (1984).

[409]   D. H. Madison, J. A. Hughes, and D. S. McGinness, *J. Phys. B* **18**, 2737 (1985).

[410]   B. Marinkovic, V. Pejcev, D. Filipovic, and L. Vuskovic, *J. Phys. B* **24**, 1817 (1991).

[411]   B. Marinkovic, V. Pejcev, D. Filipovic, I. Cadez, and L. Vuskovic, *J. Phys. B* **25**, 5179 (1992).

[412]   S. K. Srivastava and L. Vuskovic, *J. Phys. B* **13**, 2633 (1980).

[413]   L. Vuskovic, M. Zuo, G. F. Shen, B. Stumpf, and B. Bederson, *Phys. Rev. A* **40**, 133 (1989).

[414]   J. F. Williams, *J. Phys. B* **8**, 1683 (1975); **8**, 2191 (1975).

[415]   T. W. Shyn and S. Y. Cho, *Phys. Rev. A* **40**, 1315 (1989).

[416]   J. F. Williams, *J. Phys. B* **9**, 1519 (1976).

[417]   J. F. Williams and B. A. Willis, *J. Phys. B* **8**, 1641 (1975).

[418]   J. Callaway, *Phys. Rep.* **45**, 89 (1978).

[419] D. H. Madison and J. Callaway, *J. Phys. B* **20**, 4197 (1987).

[420] H. R. J. Walters, *J. Phys. B* **14**, 3499 (1981).

[421] H. R. J. Walters, *J. Phys. B* **21**, 1277 (1988).

[422] I. Bray and A. T. Stelbovics, *Phys. Rev. A* **46**, 6995 (1992).

[423] I. E. McCarthy and A. T. Stelbovics, *Phys. Rev. A* **28**, 2693 (1983).

[424] I. E. McCarthy and E. Weigold, *Advances in Atomic, Molecular, and Optical Physics*, Vol. 27 (Academic Press, New York, 1991), p. 165.

[425] I. Bray, D. A. Konovalov, and I. E. McCarthy, *Phys. Rev. A* **43**, 5878 (1991).

[426] I. Bray, D. A. Konovalov, and I. E. McCarthy, *Phys. Rev. A* **44**, 7179 (1991).

[427] T. T. Scholz, H. R. J. Walters, P. G. Burke, and M. P. Scott, *J. Phys. B* **24**, 2097 (1991).

[428] A. Lahmam-Bennani, *J. Phys. B* **24**, 2401 (1991), and references therein.

[429] H. Ehrhardt, K. Jung, G. Knoth, and P. Schlemmer, *Z. Phys. D* **1**, 3 (1986).

[430] P. Selles, A. Huetz, and J. Mazeau, *J. Phys. B* **20**, 5195 (1987).

[431] P. Schlemmer, T. Rösel, K. Jung, and H. Ehrhardt, *Phys. Rev. Lett.* **63**, 252 (1989).

[432] T. Rösel, R. Bär, K. Jung, and H. Ehrhardt, in *Invited Papers and Progress Reports, European Conference on (e, 2e) Collisions and Related Problems*, ed. H. Ehrhardt (Universität Kaiserslautern, Kaiserslautern, Germany, 1989), pp. 69-81 (unpublished).

[433] F. Gélébart and R. J. Tweed, *J. Phys. B* **23**, L641 (1990); F. Gélébart, P. Defrance, and J. Peresse, *ibid.* **23**, 1337 (1990); M. Cherid, F. Gélébart, A. Pochat, R. J. Tweed, X. Zhang, C. T. Whelan, H. R. J. Walters, *Z. Phys. D* **23**, 347 (1992); A. Pochat, X. Zhang, C. T. Whelan, H. R. J. Walters, R. J. Tweed, F. Gélébart, M. Cherid, and R. J. Allan, *Phys. Rev. A* **47**, R3483 (1993).

[434] L. Frost, P. Freienstein, and M. Wagner, *J. Phys. B* **23**, L715 (1990); T. Rösel, C. Dupré, J. Röder, A. Duguet, K. Jung, A. Lahmam-Bennani, and H. Ehrhardt, *ibid.* **24**, 3059 (1991).

[435] P. Selles, J. Mazeau, and A. Huetz, *J. Phys. B.* **23**, 2613 (1990).

[436] M. Brauner, J. S. Briggs, H. Klar, J. T. Broad, T. Rösel, K. Jung, and H. Ehrhardt, *J. Phys. B.* **24**, 657 (1991).

[437] T. J. Hawley-Jones, F. H. Read, S. Cvejanovic, P. Hammond, and G. C. King, *J. Phys. B* **25**, 2393 (1992); A. J. Murray, M. B. Woolf, and F. H. Read, *ibid.* **25**, 3021 (1992); A. J. Murray and F. H. Read, *Phys. Rev. Lett.* **69**, 2912 (1992); *Phys. Rev. A* **47**, 3724 (1993).

[438] T. Rösel, J. Röder, L. Frost, K. Jung, and H. Ehrhardt, *J. Phys. B* **25**, 3859 (1992); T. Rösel, P. Schlemmer, J. Röder, L. Frost, K. Jung, and H. Ehrhardt, *Z. Phys. D* **23**, 359 (1992).

[439] T. Rösel, J. Röder, L. Frost, K. Jung, H. Ehrhardt, S. Jones, and D. Madison, *Phys. Rev. A* **46**, 2539 (1992).

[440] G. W. Wannier, *Phys. Rev.* **90**, 817 (1953); **100**, 1180 (1955).

[441] E. P. Wigner, *Phys. Rev.* **73**, 1002 (1948).

[442] A. Temkin, *Phys. Rev. Lett.* **49**, 365 (1982); M. K. Srivastava and A. Temkin, *Phys. Rev. A* **43**, 3570 (1991), and references therein.

[443] I. Vinkalns and M. Gailitis, in *Proc. 5th Int. Conf. on the Physics of Electronic and Atomic Collisions* (Leningrad, Nauka, 1967), Abstracts pp. 648-50.

[444] A. R. P. Rau, (a) *Phys. Rev. A* **4**, 207 (1971); (b) *Phys. Rep.* **110**, 369 (1984).

[445] (a) R. Peterkop, *J. Phys. B.***4**, 513 (1971); (b) R. Peterkop, *J. Phys. B.***16**, L587 (1983); (c) R. Peterkop and P. B. Tsukerman, *Zh. Eksp. Teor. Fiz.* **31**, 374 (1970) [*Sov. Phys. JETP* **31**, 374 (1970)].

[446] C. H. Greene and A. R. P. Rau, *Phys. Rev. Lett.* **48**, 533 (1982); *J. Phys. B* **16**, 99 (1983); H. Klar and W. Schlecht, *J. Phys. B* **9**, 1699 (1976); T. A. Roth, *Phys. Rev. A* **5**, 476 (1972).

[447] U. Fano, *J. Phys. B* **14**, L401 (1974).

[448] H. Klar, *Z. Phys. A* **307**, 75 (1982); J. M. Feagin, *J. Phys. B* **17**, 2433 (1984); N. Simonović and P. Grujić, *ibid.* **20**, 3427 (1987).

[449] D. S. F. Crothers, *J. Phys. B* **19**, 463 (1986).

[450] M. Gailitis and R. Peterkop, *J. Phys. B.* **22**, 1231 (1989), and references therein.

[451] S. Watanabe, *J. Phys. B.* **24**, L39 (1991); A. K. Kazansky and V. N. Ostrovsky, *ibid.* **25**, 2121 (1992).

[452] F. H. Read, in *Electron Impact Ionization,* edited by T. D. Mark and G. H. Dunn (Berlin, Springer, 1985), and references therein; *J. Phys. B.* **17**, 3965 (1984).

[453] M. S. Lubell, *Phys. Rev. A* **47**, R2450 (1993), and references therein; X. Q. Guo, D. M. Crowe, M. S. Lubell, F. C. Tang, A. Vasilakis, J. Slevin, and M. Eminyan, *Phys. Rev. Lett.* **65**, 1857 (1990).

[454] M. R. H. Rudge, *Rev. Mod. Phys.* **40**, 564 (1968).

[455] C. Bottcher, *Adv. At. Mol. Phys.* **20**, 241 (1985); **25**, 303 (1988).

[456] R. K. Peterkop, *Opt. Spektrosk.* **13**, 153 (1962) [*Opt. Spectrosc.* (USSR) **13**, 87 (1962)].

[457] M. R. H. Rudge and M. J. Seaton, *Proc. Roy. Soc. London, Ser. A* **283**, 262 (1965).

[458] S. Jetzke, J. Zaremba, and F. H. M. Faisal, *Z. Phys. D* **11**, 63 (1989); S. Jetzke and F. H. M. Faisal, *J. Phys. B.* **25**, 1543 (1992).

[459] H. Ehrhardt, M. Schulz, T. Tekaat, and K. Willmann, *Phys. Rev. Lett.* **22**, 89 (1969).

[460] H. Ehrhardt, K. H. Hesselbacher, K. Jung, and K. Willmann, *J. Phys. B.* **5**, 1559 (1972); E. Schubert, K. Jung, and H. Ehrhardt, *ibid.* **14**, 3267 (1981).

[461] P. Fournier-Lagarde, J. Mazeau, and A. Huetz, *J. Phys. B.* **17**, L591 (1984).

[462] (a) P. L. Altick, *J. Phys. B.* **18**, 1841 (1985); (b) K. D. Shaw and P. L. Altick, *ibid.* **19**, 3161 (1986); (c) P. L. Altick and T. Rösel, *ibid.* **21**, 2635 (1988).

[463] P. Selles, J. Mazeau, and A. Huetz, *J. Phys. B.* **20**, 5183 (1987).

[464] C. Pan and A. F. Starace, (a) *Phys. Rev. Lett.* **67**, 185 (1991); (b) *Phys. Rev. A* **45**, 4588 (1992); (c) **47**, 2389 (1993).

[465] S. Jones, D. H. Madison, and M. K. Srivastava, *J. Phys. B.* **25**, 1899 (1992).

[466] J. Botero and J. Macek, *Phys. Rev. Lett.* **68**, 576 (1992).

[467] F. W. Byron Jr, and C. J. Joachain, *Phys. Rep.* **179**, 211 (1989).

[468] A. Messiah, *Quantum Mechanics* (Wiley, New York, 1966), Vol. II; P. Roman, *Advanced Quantum Theory* (Addison-Wesley, Reading, MA, 1965).

[469] A. F. Fetter and M. K. Watson, *Adv. Theor. Phys.* **1**, 115 (1965).

[470] J. S. Bell and E. J. Squires, *Phys. Rev. Lett.* **3**, 96 (1959); M. Namiki, *Prog. Theor. Phys. (Kyoto)* **22**, 843 (1960).

[471] D. J. Thouless, *The Quantum Mechanics of Many-Body Systems* (Academic Press, New York, 1961).

[472] R. T. Pu and E. S. Chang, *Phys. Rev.* **151**, 31 (1966); H. P. Kelly, *ibid.* **160**, 44 (1967); **171**, 54 (1968).

[473] M. Knowles and M. R. C. McDowell, *J. Phys. B.* **6**, 300 (1973); B. S. Yarlagadda, C. Csanak, H. S. Taylor, B. Schneider, and R. Yaris, *Phys. Rev. A* **7**, 146 (1973); M. S. Pindzola and H. P. Kelly, *ibid.* **9**, 323 (1974); **11**, 221 (1975).

[474] N. T. Padial, G. D. Meneses, F. J. da Paixao, Gy. Csanak, and D. C. Cartwright, *Phys. Rev. A* **23**, 2194 (1981); M. Ya. Amusia, N. A. Cherepkov, L. V. Chernysheva, D. M. Davidovic, and V. Radojevic, *ibid.* **25**, 219 (1982).

[475] C. Pan and H. P. Kelly, *Phys. Rev. A* **41**, 3624 (1990).

[476] H. P. Kelly, *Adv. Theor. Phys.* **2**, 75 (1968); *Adv. Chem. Phys.* **14**, 129 (1969).

[477]   K. Smith, *The Calculation of Atomic Collision Processes* (Wiley-Interscience, New York, 1971).

[478]   L. H. Thomas, *Proc. Roy. Soc. London, Ser. A* **114**, 561 (1927).

[479]   R. Shakeshaft and L. Spruch, *Rev. Mod. Phys.* **51**, 369 (1979).

[480]   N. Bohr, *Danske. K. Vidensk. Selsk. Mat. Fys. Meddr.* **18** No.8.

[481]   R. M. Drisko, Ph.D. thesis, Carnegie Institute of Technology, unpublished (1955).

[482]   M. Lieber (private communication, 1987).

[483]   K. Detmann and G. Liebfried, *Z. Physik* **218**, 1 (1968).

[484]   G. Doolen, *Phys. Rev* **166** No.5 (25 February 1968).

[485]   J. H. McGuire, J. C. Straton, W. C. Axmann, T. Ishihara, and E. Horsdal, *Phys. Rev.* Lett. **62**, 2933 (1989). An earlier calculation was done by J. S. Briggs and K. Taulbjerg, *J. Phys. B* **12**, 2565 (1979).

[486]   J. Palinkas, R. Schuch, H. Cederquist and O. Gustafsson, *Phys. Rev. Lett.* **22**, 2464 (1989).

[487]   P. R. Simony, Ph.D. thesis, Kansas State University, unpublished (1981).

[488]   J. H. McGuire, *Adv. At. Mol. and Optical Physics* **29**, 217 (1991).

[489]   E. Horsdal-Pedersen, C. L. Cocke, and M. Stockli, *Phys. Rev. Lett.* **50**, 1910 (1983). Also, H. Vogt, W. Schwab, R. Schuch, M. Schluz and E. Justininano, *Phys. Rev. Lett.* **57**, 2256 (1986).

[490]   J. H. McGuire and O. L. Weaver, *J. Phys.* **17**, L583, (1984).

[491]   O. L. Weaver and J. H. McGuire, *Phys. Rev. A* **32**, 1435 (1985).

[492]   J. H. McGuire, *Indian J. Physics* **62B**, 261 (1988).

[493]   E. Horsdal-Pedersen, P. Loftager and J. L. Rasmussen, *J. Phys. B* **15**, 7461 (1982).

[494]   R. Shakeshaft and J. Wadehra, *Phys. Rev. A* **22**, 968 (1980).

[495]   A. Igarashi and N. Toshima, *Phys. Rev. A* **46**, R1159 (1992).

[496]   J. H. McGuire, N. C. Sil and N. C. Deb, *Phys. Rev. A* **34**, 685 (1986).

[497]   E. P. Wigner, *Phys. Rev.* **70**, 15 (1946); **70**, 606 (1946).

[498]   E. P. Wigner and L. Eisenbud, *Phys. Rev.* **72**, 29 (1947).

[499]   A. M. Lane and R. G. Thomas, *Rev. Mod. Phys.* **30**, 252 (1958).

[500]   G. Breit, *Encyclopedia of Physics* Vol. 41/1, ed. S. Flugge (Springer-Verlag, Berlin, 1959), p. 107.

[501]   R. F. Barrett, L. C. Biedenharn, M. Danos, R. P. Delsanto, W. Greiner, and H. G. Wahsweiler, *Rev. Mod. Phys.* **45**, 44 (1973).

[502]   P. G. Burke, A. Hibbert, and W. D. Robb, *J. Phys. B* **4**, 153 (1971).

[503]   K. Bartschat and P. G. Burke, *J. Phys. B* **20**, 3191 (1987).

[504] B. I. Schneider, *Chem. Phys. Lett.* **31**, 237 (1975).

[505] P. G. Burke, I. Mackey, and I. Shimamura, *J. Phys. B* **10**, 2497 (1977).

[506] C. J. Gillan, O. Nagy, P. G. Burke, L. A Morgan, and C. J. Noble, *J. Phys. B* **20**, 4585 (1987).

[507] J. Tennyson, *J. Phys. B* **19**, 4255 (1986).

[508] K. Higgins, P. G. Burke, and H. R. J. Walters, *J. Phys. B* **23**, 1345 (1990).

[509] J. Tennyson, C. J. Noble, and P. G. Burke, *Int. J. Quantum Chem.* **29**, 1033 (1986).

[510] K. L. Bell, P. G. Burke, and A. E. Kingston, *J. Phys. B.* **10**, 3117 (1977).

[511] M. J. Seaton, *J. Phys. B.* **18**, 2111 (1985).

[512] B. K. Sarpal, S. E. Branchett, J. Tennyson, and L. A. Morgan, *J. Phys. B.* **24**, 3685 (1991).

[513] D. S. C. Allison, P. G. Burke, and W. D. Robb, *J. Phys. B.* **5**, 55 (1972).

[514] J. C. Light and R. B. Walker, *J. Chem. Phys.* **65**, 4222 (1976).

[515] B. I. Schneider, M. LeDourneuf, and P. G. Burke, *J. Phys. B.* **12**, L365 (1979).

[516] P. G. Burke, P. Francken, and C. J. Joachain, *J. Phys. B.* **24**, 761 (1991).

[517] M. Dörr, M. Terao-Dunseath, J. Purvis, C. J. Noble, P. G. Burke, and C. J. Joachain, *J. Phys. B.* **25**, 2809 (1992).

[518] P. G. Burke and W. D. Robb, *Adv. Atom. Molec. Phys.* **11**, 143 (1975).

[519] P. G. Burke and K. A Berrington, *Atomic and Molecular Processes: an R-Matrix approach* (Institute of Physics Publishing, 1993).

[520] C. Bloch, *Nucl. Phys.* **4**, 503 (1957).

[521] J. Tennyson, P. G. Burke, and K. A. Berrington, *Comput. Phys. Commun.* **47**, 207 (1987).

[522] P. J. A. Buttle, *Phys. Rev.* **160**, 719 (1967).

[523] W. Kohn, *Phys. Rev.* **74**, 1763 (1948).

[524] J. L. Jackson, *Phys. Rev.* **83**, 301 (1951).

[525] R. S. Oberoi and R. K. Nesbet, *Phys. Rev. A* **8**, 215 (1973); **9**, 2804 (1974).

[526] C. H. Greene, *Phys. Rev. A* **28**, 2209 (1983); **32**, 1880 (1985).

[527] K. L. Baluja, P. G. Burke, and L. A. Morgan, *Comput. Phys. Commun.* **27**, 299 (1982).

[528] C. J. Noble and V. M. Burke, *Comput. Phys. Commun.*, to be submitted (1994).

[529] U. Fano, *Phys. Rev.* **140**, A67 (1965).

[530] P. G. Burke, *Comput. Phys. Commun.* **1**, 241 (1970).

[531] A. Hibbert, *Comput. Phys. Commun.* **1**, 359 (1970); **2**, 180 (1971).

[532] V. M. Burke, P. G. Burke, and N. S. Scott, *Comput. Phys. Commun.* **69**, 76 (1992).

[533] P. G. Burke, V. M. Burke, and K. Dunseath, *J. Phys. B*, to be submitted (1994).

[534] V. M. Burke, K. Dunseath, and P. G. Burke, *Comput. Phys. Commun.* to be submitted (1994).

[535] H. Nussbaumer and P. J. Storey, *Astron. Astrophys.* **64**, 139 (1978); **89**, 308 (1980).

[536] H. Nussbaumer, M. Pettini, and P. J. Storey, *Astron. Astrophys.* **102**, 351 (1981).

[537] K. L. Baluja, A. Hibbert, and M. Mohan, *J. Phys. B* **19**, 3613 (1986).

[538] K. A. Berrington, P. G. Burke, A. Hibbert, M. Mohan, and K. L. Baluja, *J. Phys. B.* **21**, 339 (1988).

[539] N. S. Scott and P. G. Burke, *J. Phys. B.* **13**, 4299 (1980).

[540] N. S. Scott and K. T. Taylor, *Comput. Phys. Commun.* **25**, 347 (1982).

[541] A. K. Pradhan and K. A. Berrington, *J. Phys. B.* **26**, 157 (1993).

[542] P. G. Burke and T. G. Webb, *J. Phys. B.* **3**, L131 (1971).

[543] J. Callaway and J. W. Wooten, *Phys. Rev. A* **9**, 1924 (1974); **11**, 1118 (1975).

[544] P. G. Burke, C. J. Noble, and M. P. Scott, *Proc. Roy. Soc. London, Ser. A* **410**, 289 (1987).

[545] M. P. Scott, T. T. Scholz, H. R. J. Walters, and P. G. Burke, *J. Phys. B.* **22**, 3055 (1989).

[546] P. G. Burke, K. A. Berrington, C. V. Sukumar, *J. Phys. B.* **14**, 289 (1981).

[547] J. Callaway, K. Unnikrishnan, and D. H. Oza, *Phys. Rev. A* **36**, 2576 (1987).

[548] A. E. Kingston, W. C. Fon, and P. G. Burke, *J. Phys. B.* **9**, 605 (1976).

[549] W. E. Kauippila, W. R. Ott, and W. L. Fite, *Phys. Rev. A* **1**, 1099 (1970).

[550] R. L. Long Jr., D. M. Cox, S. J. Smith, *J. Res. NBST A* **72**, 521 (1968).

[551] J. F. Williams, *J. Phys. B.* **14**, 1197 (1981).

[552] W. L. van Wyngaarden and H. R. J. Walters, *J. Phys. B.* **19**, L53 (1986); **19**, 929 (1986); **19**, 1827 (1986).

[553] M. P. Scott and P. G. Burke, *J. Phys. B.* **26**, L191 (1993).

[554] B. Odgers, M. P. Scott, P. G. Burke, *J. Phys. B,* to be published (1994).

[555] M. LeDourneuf, J.-M. Launey, and P. G. Burke, *J. Phys. B.* **23**, L559 (1990).

[556] M. Dörr, P. G. Burke, C. J. Joachain, C. J. Noble, J. Purvis, and M. Terao-Dunseath, *J. Phys. B.* **26**, L275 (1993).

[557] J. Purvis, M. Dörr, M. Terao-Dunseath, C. J. Joachain, P. G. Burke, and C. J. Noble, *Phys. Rev. Lett.* **71**, 3943 (1993).

[558] N. S. Scott, A. McMinn, P. G. Burke, V. M. Burke, and C. J. Noble, *Comput. Phys. Commun.* **78**, 67 (1993).

[559] J. J. Chang, *J. Phys. B.* **8**, 2327 (1975).

[560] J. J. Chang, *J. Phys. B.* **10**, 3335 (1971).

[561] P. H. Norrington and I. P. Grant, *J. Phys. B.* **14**, L261 (1981).

[562] P. H. Norrington and I. P. Grant, *J. Phys. B.* **20**, 4869 (1987).

[563] I. P. Grant and H. M. Quiney, *Adv. At. Mol. Phys.* **23**, 37 (1988).

[564] A. I. Akhiezer and V. B. Berestetsky, *Quantum Electrodynamics* (Interscience Press, New York, 1965).

[565] F. A. Parpia, I. P. Grant, and C. F. Fischer, in preparation (1991).

[566] K. G. Dyall, I. P. Grant, C. T. Johnson, F. A. Parpia, and E. P. Plummer, *Comput. Phys. Commun.* **55**, 425 (1989).

[567] I. P. Grant and B. J. McKenzie, *J. Phys. B.* **13**, 2671 (1980).

[568] K. J. Reed and R. J. W. Henry, *Phys. Rev. A* **8**, 2348 (1989).

[569] M. J. Seaton, private communication (1990).

[570] W. P. Wijesundera, F. A. Parpia, I. P. Grant, and P. H. Norrington, *J. Phys. B.* **24**, 1803 (1991).

[571] A. K. Bhatia and U. Feldman, *J. Appl. Phys.* **53**, 4711 (1982).

[572] C. E. Moore, in *Atomic Energy Levels as Derived From the Analyses of Optical Spectra*, Natl. Bur. Stand. Ref. Data Ser., (US) Circ. No. 35 (US GPO, Washington, DC, 1971), Vol. III.

[573] J. P. Croskery, N. S. Scott, K. L. Bell, and K. A. Berrington, *Comput. Phys. Commun.* **27**, 385 (1982).

[574] F. Calogero, *Variable Phase Approach to Potential Scattering* (Academic Press, New York, 1967).

[575] N. S. Scott, P. G. Burke, and K. Bartschat, *J. Phys. B.* **16**, L361 (1983).

[576] D. W. O. Heddle, *J. Phys. B.* **8**, L33 (1975).

[577] D. W. O. Heddle, *J. Phys. B.* **11**, L711 (1978).

[578] K. Bartschat and P. G. Burke, *J. Phys. B.* **19**, 1231 (1986).

[579] W. P. Wijesundera, I. P. Grant, and P. H. Norrington, *J. Phys. B.* **25**, 2143 (1992).

[580] C. E. Kuyatt, J. A. Simpson, and S. Mielczarek, *Phys. Rev. A.* **138**, 385 (1965).

[581] I. P. Zapesochnyi and O. B. Shpenik, *Zh. Eksp. Teor. Fiz.* **50**, 890 (1966) [*Sov. Phys. JETP* **23**, 592 (1966)].

[582] S. D. Rockwood, *Phys. Rev. A* **40**, 1823 (1973).

[583] M. Düweke, N. Kirchner, E. Reichert, and E. Staudt, *J. Phys. B.* **6**, L208 (1973).

[584] T. W. Ottley and H. Kleinpoppen, *J. Phys. B.* **8**, 621 (1975).

[585] P. D. Burrow and J. A. Michejda, in *Electron and Photon Interactions with Atoms* edited by H. Kleinpoppen and M. R. C. McDowell (Plenum, New York, 1976).

[586] K. Albert, C. Christian, T. Heindorff, E. Reichert, and S. Schön, *J. Phys. B.* **10**, 3733 (1977).

[587] K. Jost and B. Ohnemus, *Phys. Rev. A.* **19**, 641 (1979).

[588] D. S. Newman, M. Zubek, and G. C. King, *J. Phys. B.* **18**, 985 (1985).

[589] D. W. Walker, *J. Phys. B.* **2**, 356 (1969).

[590] D. W. Walker, *J. Phys. B.* **3**, 788 (1970).

[591] D. W. Walker, *J. Phys. B.* **3**, L123 (1970).

[592] D. W. Walker, *J. Phys. B.* **8**, L161 (1975).

[593] A. Temkin, *Phys. Rev.* **107**, 1004 (1957).

[594] C. W. McCutchen, *Phys. Rev. A.* **112**, 1848 (1958).

[595] K. Bartschat, *J. Phys. B.* **18**, 2519 (1985).

[596] W. P. Wijesundera, I. P. Grant, and P. H. Norrington, *J. Phys. B.* **25**, 2165 (1992).

[597] N. F. Mott and H. S. W. Massey, *The Theory of Atomic Collisions*, (Clarendon Press, Oxford, 1965).

[598] P. G. Burke and M. J. Seaton, in *Methods in Computational Physics* (Academic Press, New York, 1971), Vol. 10, p. 1.

[599] M. A. Crees, M. J. Seaton, and P. M. H. Wilson, *Comput. Phys. Commun.* **15**, 23 (1978).

[600] K. A. Berrington, P. G. Burke, K. Butler, M. J. Seaton, P. J. Storey, K. T. Taylor, and Yu Yan, *J. Phys. B.* **20**, 6379 (1987).

[601] W. Eissner and M. J. Seaton, *J. Phys. B.* **5**, 2187 (1972).

[602] M. S. Pindzola, D. C. Griffin and C. Bottcher, *Phys. Rev. A* **39**, 2385 (1989).

[603] J. Mitroy, D. C Griffin, D. W. Norcross, and M. S. Pindzola, *Phys. Rev. A* **38**, 3339 (1988).

[604] D. C. Griffin and M. S. Pindzola, *Phys. Rev. A* **42**, 248 (1990).

[605] B. A. Huber, Ch. Ristori, P. A. Hervieux, M. Maurel, C. Guet, and J. J. Andrä, *Phys. Rev. Lett.* **67**, 1407 (1991).

[606] M. S. Pindzola, D. C. Griffin, and N. R. Badnell, in 17th ICPEAC Abstracts of Contributed Papers for Brisbane Australia (1991), p. 285.

[607] D. C. Griffin, M. S. Pindzola, and N. R. Badnell, *Phys. Rev. A* **47**, (1993).

[608] N. R. Badnell, M. S. Pindzola, D. C. Griffin, *Phys. Rev. A* **43**, 2250 (1991).

[609] E. K. Wahlin, J. S. Thompson, G. H. Dunn, R. A. Phaneuf, D. C. Gregory, and A. C. H. Smith, *Phys. Rev. Lett.* **66**, 157 (1991).

[610] X. Q. Guo, E. W. Bell, J. S. Thompson, G. H. Dunn, M. E. Bannister, R. A. Phaneuf, A. C. H. Smith, *Phys. Rev.* **47**, R9 (1993).

[611] S. M. Younger, *Phys. Rev. A* **22**, 111 (1980).

[612] H. Jakubowicz and D. L. Moores, *J. Phys. B.* **14**, 3733 (1981).

[613] D. L. Moores and K. J. Reed, *Phys. Rev. A* **39**, 1747 (1989).

[614] D. W. Mueller, T. J. Morgan, G. H. Dunn, D. C. Gregory, and D. H. Crandall, *Phys. Rev. A* **31**, 2905 (1985).

[615] M. S. Pindzola and D. C. Griffin, *Phys. Rev. A* **46**, 2486 (1992).

[616] R. G. Montague and M. F. A. Harrison, *J. Phys. B.* **17**, 2707 (1984).

[617] M. S. Pindzola, D. C. Griffin, and C. Bottcher, *J. Phys. B.* **16**, L355 (1983).

[618] S. M. Younger, *Phys. Rev. A* **35**, 2841 (1987).

[619] M. S. Pindzola, D. C. Griffin, and C. Bottcher, *Phys. Rev. A* **27**, 2331 (1983).

[620] D. C. Gregory and D. H. Crandall, *Phys. Rev. A* **27**, 2338 (1983).

[621] D. C. Griffin, M. S. Pindzola, and C. Bottchcr, *Phys. Rev. A* **36**, 3642 (1987).

[622] D. C. Gregory, L. J. Wang, F. W. Meyer, and K. Rinn, *Phys. Rev. A* **35**, 3256 (1987).

[623] K. J. LaGattuta and Y. Hahn, *Phys. Rev. A* **24**, 2273 (1981).

[624] S. S. Tayal and R. J. W. Henry, *Phys. Rev. A* **39**, 3890 (1989).

[625] M. H. Chen, K. J. Reed, and D. L. Moores, *Phys. Rev. Lett.* **64**, 1350 (1990).

[626] A. Müller, G. Hofmann, K. Tinschert, and E. Salzborn, *Phys. Rev. Lett.* **61**, 1352 (1988).

[627] R. J. W. Henry, *J. Phys. B.* **12**, L309 (1979).

[628] S. S. Tayal and R. J. W. Henry, *Phys. Rev. A* **42**, 1831 (1990).

[629] K. J. Reed and M. H. Chen, *Phys. Rev. A* **45**, 4519 (1992).

[630] M. S. Pindzola and D. C. Griffin, *Phys. Rev. A* **36**, 2628 (1987).

[631] See, for example, J. D. Bjorken and S. D. Drell *Relativistic Quantum Mechanics* (McGraw-Hill, New York, 1964), p. 54.

[632] See, for example, Albert Messiah *Quantum Mechanics,* Vol. II (North-Holland, Amsterdam, 1965) p. 927.

[633] E. R. Cohen and B. N. Taylor, *CODATA Bull.* **63** (1986).

Index

absorption effects, 246
all-order methods, 71, 73, 75, 76
anapole moment, 68, 86
angular momentum transfer
    analysis, 169
anticommutation relations, 45
argon
    double photoionization cross
        section, 158, 159, 165
    electron scattering, 281
        elastic cross section, 197
        phase shifts, 196
    photoelectron angular
        distribution, 180
    triply differential cross section,
        221
atomic polarization, 245
atomic structure, vii, 1

B-splines, 55, 127, 131
barium, photoionization cross
    section, 120
beryllium
    correlation energy, 7
    photoionization cross section, 144
beryllium-like C^{2+}, photoionization
    cross section, 147
Bethe ridge, 217
Bloch equation, 8, 19
Bloch operator, R-matrix theory,
    307
Breit interaction, 13, 48, 84
Bremsstrahlung, 210

Brillouin-Wigner perturbation
    theory, 14
Brillouin-Wigner resolvent, 16
Brueckner orbitals, 72, 73, 82

cadmium
    electron scattering, 246
calcium
    electron scattering elastic cross
        section, 198
carbon, C^{2+}, beryllium-like,
    photoionization cross section,
    367
carbon, C^{3+}, lithium-like,
    dielectronic capture, ionization,
    367
cesium
    dipole matrix elements, 73
    electron scattering, 346
    energies, 70, 75
    hyperfine constants, 76
    hyperfine splitting, 73
    photoelectron angular
        distribution, 177
chlorine, photoelectron angular
    distribution, 176, 178
close-coupling calculations
    electron scattering, 253
    photoionization, 114
collision strength, 311
Cooper-Zare formula, 173
copper-like ions, energies, 58, 63
coupled-cluster approach, 75
coupled-cluster theory, 10, 37

dielectric constant, 295
dipole matrix elements
  length, 134, 153
  velocity, 134, 153
dipole moment, 70
  length gauge, 97
  parity nonconserving, 68
  velocity gauge, 97
Dirac equation, 40
Dirac Hamiltonian, 13
Dirac matrices, 41
distorted waves, 238
distorting potentials, 240
Dyson equation, 191

effective charge, 269
electron impact ionization
  argon (+3), 278
  asymptotic boundary conditions, 263
  doubly differential cross section, 274
  helium atom (+2), 280
  helium atom (+3), 276
  hydrogen atom (+2), 280
  hydrogen atom (+5), 275
  krypton (+3), 278
  lithium atom (+2), 280
  many-body perturbation expansion, 268
  near threshold (+26), 261
  neon (+3), 278
  partial wave expansion (+3), 265
  target dependence (+26), 261
  threshold behavior, 285
  threshold law (+1), 261
  triply differential cross section (+26), 261
  Wannier-Peterkop-Rau (WPR) theory, 277, 279
  Wannier-Peterkop-Rau (WPR) theory (+1), 284
  xenon (+3), 278
electron scattering, 213, 237, 261, 305, 325, 350, 359
  autoionization, 349

close-coupling calculations, 253
  cross section, 310, 311
  elastic collisions, 193, 247, 248, 250, 252, 256, 258
  excitation, 349
  fine structure effects, 311
  inelastic, 208
  inelastic collisions, 246, 247, 249, 251, 253, 254, 257, 259
  inelastic scattering by Fe^+, 311
  intermediate energies, 314
  ionization, 349
  psuedostate expansion, 314
  resonance, 349
  scattering amplitude, 193
  triply differential cross section, target dependence, 261
electron-atom and electron-ion scattering, 183
electron-impact ionization, 359
exchange effects, 242, 246
excitation-autoionization, 360

final-state correlations, photoionization, 104
finite basis sets, 54, 126
fluorine, F^{2+}, doubly-ionized, electron impact ionization, 363

graphical representation, 21
Green's function, 244
ground-state correlations, photoionization, 104

Hartree-Fock
  mixed-parity, 77
  potential, 42, 109, 129, 155, 188, 192
  scattering amplitudes, 172
helium
  double photoionization cross section, 163
  electron scattering, 217, 277, 279, 280, 283–285
  photoionization cross section, 126, 141, 144, 145

helium (*cont.*)
  (positive ion)-electron scattering,
    phase shifts, 140
  triply differential cross section,
    219, 222–224, 231–233
hydrogen
  electron scattering, 250, 251, 256,
    257, 276, 279, 283, 284, 316–318,
    320
  multiphoton processes, 323
  negative ion, photoionization
    cross section, 141
  photoelectron angular
    distribution, 175
  triply differential cross section,
    226, 227, 229
hydrogen-electron scattering, phase
  shifts, 138
hyperfine constants, 76
hyperfine splittings, 72

impulse approximation, 217
intermediate normalization, 15, 50
ionization energies, 75
iron, Fe^{15+}, sodium-like, ionization,
  364

Kramers-Kronig relation, 295
krypton, electron scattering, 282
krypton, oxygen-like, electron
  scattering, 332

lead, electron scattering, 343
linked-cluster expansion, 97
lithium, electron scattering, 283, 284

magnesium, photoionization cross
  section, 147, 148
mercury, electron scattering, 335
mixed-parity Hartree-Fock, 77
mixed-parity MBPT, 77, 80
multiphoton ionization, 320, 323
  above threshold ionization, 324
  high frequency stabilization, 324
multiphoton processes, 149, 320
muonic hydrogen
  electron scattering, 206
  potential, 207

negative ion formation, 201
neon
  electron scattering, 281
  triply differential cross section,
    219
nitrogen, photoelectron angular
  distribution, 177
normalization
  intermediate, 100
  momentum scale, 96
  perturbative correction, 101
nuclear MBPT, 4

optical limit, 215

parity nonconservation, 65
parity-favored transitions, 170
parity-unfavored transitions, 170
partial wave expansion, 265
perturbation theory
  close-coupling method, 349
  convergence, 248
  distorted wave approximation,
    220, 238, 261, 349
  first order, 244
  Rayleigh-Schrödinger, 50
  second order, 220, 243, 245
photoionization, viii, 91
  angular distribution, 167
  configuration interaction, 135
  coupled-equations, 114
  cross section, 94
  double, 150
  double-excitation, 125
  generalized resonance technique,
    120
  polarization effects, 117
  relaxation effects, 117
  single, 93
platinum, photoionization cross
  section, 113
polarizability, 102, 131, 153
positron–atom elastic scattering,
  198
post-collision interaction, 210
potassium, photoelectron angular
  distribution, 177

potassium-like ions
  electron scattering, 355
  excitation, 355
potential
  $V^{(N-1)}$, 109, 155, 160
  $V^{(N-2)}$, 155, 160
  Hartree-Fock, 109, 129, 155, 188,
    192
projection operators, 15, 50

radiation damping, 350
radon, photoelectron angular
  distribution, 182
random phase approximation, 60,
  71
  with exchange, 204
Rayleigh-Schrödinger perturbation
  theory, 18, 50
Redmond phase, 223
relativistic effects, 13, 39, 69, 242,
  325
renormalization constant, 193
rubidium, photoelectron angular
  distribution, 177

screening potential, 269
second Born amplitude, 294
second quantization, 21, 45
self-energy operator, 70, 72–74, 82,
  190
sodium
  electron scattering, 247–249,
    252–254, 258, 259
  photoelectron angular
    distribution, 177

sodium-like ions
  electron scattering, 357
  excitation, 357
spherical spinors, 41
structural radiation, 72

Thomas process for mass transfer,
  287
Thomas ridge, 293
threshold law
  Wannier-Peterkop-Rau (WPR)
    theory, 261, 277
time-evolution operator, 98
triply differential cross section, 215
tungsten
  photoionization cross section, 123
  singly-ionized, electron impact
    ionization, 363

valence electrons, 51
vector transition polarizability, 87

weak nuclear charge, 67, 69, 88
Wick's theorem, 23, 100

xenon
  electron scattering, 282
  photoelectron angular
    distribution, 180
  photoionization cross section, 119
  Xe^{6+}, cadmium-like, ionization,
    364

zinc, copper-like, energies, 45, 57, 59